Jaime Gil Aluja and Ana Maria Gil Lafuente

Towards an Advanced Modelling of Complex Economic Phenomena

Studies in Fuzziness and Soft Computing, Volume 276

Editor-in-Chief

Prof. Janusz Kacprzyk
Systems Research Institute
Polish Academy of Sciences
ul. Newelska 6
01-447 Warsaw
Poland
E-mail: kacprzyk@ibspan.waw.pl

Further volumes of this series can be found on our homepage: springer.com

Vol. 262. Zsófia Lendek, Thierry Marie Guerra, Robert Babuška, and Bart De Schutter
Stability Analysis and Nonlinear Observer Design Using Takagi-Sugeno Fuzzy Models, 2010
ISBN 978-3-642-16775-1

Vol. 263. Jiuping Xu and Xiaoyang Zhou
Fuzzy-Like Multiple Objective Decision Making, 2010
ISBN 978-3-642-16894-9

Vol. 264. Hak-Keung Lam and Frank Hung-Fat Leung
Stability Analysis of Fuzzy-Model-Based Control Systems, 2011
ISBN 978-3-642-17843-6

Vol. 265. Ronald R. Yager, Janusz Kacprzyk, and Prof. Gleb Beliakov (eds.)
Recent Developments in the Ordered Weighted Averaging Operators: Theory and Practice, 2011
ISBN 978-3-642-17909-9

Vol. 266. Edwin Lughofer
Evolving Fuzzy Systems – Methodologies, Advanced Concepts and Applications, 2011
ISBN 978-3-642-18086-6

Vol. 267. Enrique Herrera-Viedma, José Luis García-Lapresta, Janusz Kacprzyk, Mario Fedrizzi, Hannu Nurmi, and Sławomir Zadrożny
Consensual Processes, 2011
ISBN 978-3-642-20532-3

Vol. 268. Olga Poleshchuk and Evgeniy Komarov
Expert Fuzzy Information Processing, 2011
ISBN 978-3-642-20124-0

Vol. 269. Kasthurirangan Gopalakrishnan, Siddhartha Kumar Khaitan, and Soteris Kalogirou (Eds.)
Soft Computing in Green and Renewable Energy Systems, 2011
ISBN 978-3-642-22175-0

Vol. 270. Christer Carlsson and Robert Fullér
Possibility for Decision, 2011
ISBN 978-3-642-22641-0

Vol. 271. Enric Trillas, P. Bonissone, Luis Magdalena and Janusz Kacprzyk
Combining Experimentation and Theory, 2011
ISBN 978-3-642-24665-4

Vol. 272. Oscar Castillo
Type-2 Fuzzy Logic in Intelligent Control Applications, 2012
ISBN 978-3-642-24662-3

Vol. 273. Rudolf Seising and Veronica Sanz (Eds.)
Soft Computing in Humanities and Social Sciences, 2012
ISBN 978-3-642-24671-5

Vol. 274. Irina Georgescu
Possibility Theory and the Risk, 2012
ISBN 978-3-642-24739-2

Vol. 275. Alexander Rotshtein and Hanna Rakytyanska
Fuzzy Evidence in Identification, Forecasting and Diagnosis, 2012
ISBN 978-3-642-25785-8

Vol. 276. Jaime Gil Aluja and Ana Maria Gil Lafuente
Towards an Advanced Modelling of Complex Economic Phenomena, 2012
ISBN 978-3-642-24811-5

Jaime Gil Aluja and Ana Maria Gil Lafuente

Towards an Advanced Modelling of Complex Economic Phenomena

Pretopological and Topological Uncertainty Research Tools

Springer

Authors

Prof. Jaime Gil Aluja
Royal Academy of Economic and
Financial Sciences (RACEF)
Via Laietana, 32, 4°, despacho n° 91
08003 Barcelona
Spain
E-mail: secretaria@racef.es

Prof. Ana Maria Gil Lafuente
University of Barcelona
Department of Economics and
Business Organization
Faculty of Economics and Business
Diagonal 690
08034 Barcelona
Spain
E-mail: amgil@ub.edu

ISBN 978-3-642-43123-4 ISBN 978-3-642-24812-2 (eBook)

DOI 10.1007/978-3-642-24812-2

Studies in Fuzziness and Soft Computing ISSN 1434-9922

Typeset by Scientific Publishing Services Pvt. Ltd., Chennai, India.

Printed on acid-free paper

9 8 7 6 5 4 3 2 1

springer.com

Contents

Part II: Towards an Idea of Uncertain Topological Space

Part III: Operative Techniques for Economy and Management

Introduction

The Paths towards Knowledge

The search for order and stability has traditionally been one of the fundamental objectives of **economic science**. We do not believe that there is any economist, whatever the school in which they have based their knowledge, who does not establish their decisions by thinking of obtaining an **equilibrium** or by thinking of breaking an existing equilibrium. But always with the idea of finding another, which may be more favourable for the interests which they wish to defend.

If we focus on the field of the **formalizing** of the phenomena in the sphere of knowledge a marked confusion can be observed in researchers, when a reality full of disarray which makes life **unstable** can be seen to be treated in the same way as in situations of equilibrium, enveloped in stability. It is this way because it results as difficult to accept that society, the economy and the activity of businesses such as those we have known up to now have no possibilities of survival in the immediate future in which many deep changes will be inevitable. When faced with this panorama there are no lack of studies which aim to find solutions by undertaking new paths in their economic studies, in which **fluctuations** and **instability** are taking an ever more fundamental position.

In effect, in a context of changes such as those which we are currently witnessing, who is capable of predicting the **evolution of events** with the necessary precision of a prophet? Maybe we should be satisfied with **less** and **better** employ that which is available to us. For this, a brief reflexion based on the possibilities which the answers offered by the laboratories where the new scientific discoveries are taught may result as useful. This confirms to us that research activity has reached a crossroads at which the future of humanity is at stake. On the one hand the **geometric conception** of knowledge and on the other the **Darwinian conception**. On the one side the sublime, monotonous and well-known reiterative songs, renewed only in their forms. The imposition of some pre-established beliefs from the blinding light of the Newtonian dawn, in which the reduction of the operation of the world to the predictability of a mechanism is dreamed of. And on the other side the emptiness of the unknown. The varying and at times dissonant whispers of notes which sound disjointed and incoherent. The attraction of adventure. The invitation to jump from the edge of a cliff where the distance to fall cannot be seen, only guided by the hope of opening new horizons. The response to the calling of **Ludwig Boltzmann**, of **Bertran Russell**, of **Lukasiewicz**,of **Zadeh**, of **Lorenz**, of **Prigogine** and of **Kaufmann**. The rejection of the yoke and burden of predestination and the proclamation of freedom of choice which time and again clashes against the wall of doubt.

Karl Popper[1] stated "that all events are caused by an event, in such a way that all events may be predicted or explained.... On the other hand, common sense attributes to sane adults the capacity to freely choose between various paths....". This kind of interior contradiction a greater problem which William James[2] designated "dilemma of determinism". When we move this dilemma to physics or economics then what is known as "the time paradox" appears, in which we gamble neither more nor less than **our relationship with the world** or **with society**. In effect, Is society already written or is it found to be in permanent construction? Over time economic science has fed on the knowledge supplied by physics and as from the Newtonian Dynamic to quantum physics temporal symmetry is accepted, without distinction between past and future, economic science has been seen impregnated by a nature of no time references. On the other hand, our **perception** of economic phenomena leads us to think that past and future play a different part.

The time paradox "was identified by the Viennese physicist **Ludwig Boltzmann**, who believed it possible to follow the example of **Charles Darwin** in biology and give an evolutionist description of physical phenomena. His attempt had the effect of demonstrating the contradictions between the laws of Newtonian Physics -based on the equivalence between past and future- and all attempts of evolutionist formulation which asserted an essential distinction between future and past"[3]. At present, however, this perception of reality and of time has changed, above all since the birth and development of **non-equilibrium physics** with concepts such as self-organisation and **dissipative structures**. We are constantly more conscious that the deepest roots of the new ways of looking with which researchers scrutinise and examine social systems, economics and management are to be found in the 19th century when the first essences of **evolutionism** were born.

In effect, in his fundamental work "On the Origin of Species" published in 1859, Darwin considered that the **fluctuations** in biological species, thanks to natural selection, give rise to an **irreversible** biological evolution. That a self-organisation of systems with a growing complexity takes place between the association of **fluctuations** (which is similar to the idea of chance, we would say uncertainty) and **irreversibility**.

The evolutionary description is found associated with the concept of **entropy**, which, in thermodynamics, allows us to distinguish between **reversible** and **irreversible** processes. In 1865, Clausius[4] associated entropy with the second principle of thermodynamics. His formulation of the two principles of thermodynamics is the following: "The energy of the universe is constant". "The entropy of the universe tends to a maximum". Faced with the energy which it continuously retains, entropy allows the establishment of a distinction between

[1] Popper, K.: L'univers irrésolu. Plaidoyer pour l'indéterminisme. Ed. Hermann. París, 1984, page. XV.

[2] James, W.: The dilemma of determinism, in The Will to Believe. Ed. Dover. New York, 1956.

[3] Prigogine, I.: *La fin des certitudes. Spanish translation with the title* «El fin de las certidumbres». Ed. Taurus. Buenos Aires, 1997, page 8.

[4] Clausius, R.: *Ann. Phys.* CXXV, 1865, page 353.

reversible processes (constant entropy) and irreversible processes (entropy created). Therefore, in an isolated system the **entropy** increases when irreversible processes exist and remains constant in the presence of reversible processes. So, entropy reaches a maximum value when the system is nearing equilibrium and the irreversible process ends. It was **Ludwig Boltzmann** (1844-1906) who established a relationship between entropy and probability through the famous formula $H = K$. In P.

In 1872 Boltzmann published his famous "H Theorem". This theorem exposed how in the heart of a population of particles, the collisions between these modifies the distribution of the value of this function H at each moment until a minimum which corresponds with that which has come to be called the **Maxwell-Boltzmann equilibrium distribution** is reached. In this state the collisions no longer modify the distribution of velocities in the population and the magnitude H remains constant. In this way the collisions between particles lead to **equilibrium**.

As much in the case of **Darwin** as in that of **Boltzmann**, **chance** (or, if prefered, uncertainty) and **evolution** are intimately related, but the results of their respective investigations lead to contrasting conclusions. For **Boltzmann**, the probability reaches its maximum when **uniformity** is being reached, while for **Darwin** evolution leads to new **self-organised structures**.

In comparison with these approaches, and **following** the **traditional physics** prototype in which complete and certain knowledge are linked and in which from certain initial conditions the predictability of the future and the possibility of returning to the past are guaranteed, marginal economic theory is supported by the **mechanics of movement**, which describe processes of a **reversible** and **determinist** character, where the direction of time plays no part whatsoever and in which there is no place for either uncertainty or irreversibility. To conclude, in classic studies, economic and management systems constitute **great automata**. But the incorporation of **instability** is causing a substantial change in which the concept of "economic law" acquires a new **meaning**.

It is true that some phenomena arising from the life of states, institutions and companies can be described through the use of **determinist equations**. But, on the other hand, others entail uncertain processes or, in any case, **stochastic processes**. Not only do they possess **laws** but also **facts** which do not result as a consequence of the laws and instead redefine their **possibilities**. It may occur that our own existence, with all its complexity, is also engraved in the primordial event baptised with the name of the Big-Bang. Ilya Prigogine[5] asked if time made its first appearance with the Big-Bang of if time previously existed in our universe. In this way the frontier of our knowledge, reasoning and speculation are difficult to define. The Big-Bang may be conceived as an event associated with instability, which implies that it is the starting point of our universe, but not of time. Therefore, time does not have a beginning and it is possible that it does not have an end. The reality is that economic science, which has searched so much for that which is **permanent**, **symmetry** and **laws**, has instead found that which is **mutable, irreversible** and **complex**.

---

[5] Prigogine, I.: La fin des certitudes. Spanish version, Ed. Taurus. Buenos Aires, 1997, pages. 11-12.

In this search, the scholars of economics and management are discovering processes in which the **transition of chaos to order** takes place, that is to say sequences which lead towards **self-organisation**. The question posed is **how** this creation of new structures takes place, which is to say this **self- organisation**. So, given the entropy of a system, if it is disturbed in such a way that a state remains sufficiently **near to equilibrium** the system itself responds by reestablishing the initial situation. This is a **stable system**. But, if a state is taken **far enough from equilibrium**, it enters a situation of **instability** in relation with the disturbance. This point is habitually named the bifurcation point[6]. From this, new phenomena take place which may correspond to behaviour far from the original. In this context determinist processes have no usefulness for predicting **which path** will be chosen between those existing and bifurcation. In many of these bifurcations **symmetry rupture** is produced.

The notions of time and of determinism have been present in western thought since pre-Socratic times, causing deeply felt tension when attempting to give an impulse to **objective knowledge** at the same time as promoting the humanist ideal of **freedom**. Science would fall into a contradiction if it were to opt for a determinist concept when we find ourselves involved in the task of developing a free society. Neither science and certitude nor ignorance and possibility can be identified. A scientific activity is coming to light in which investigations are not limited to the study of **simplified** and **ideal** phenomena but which are determined to unravel the **secrets** of societies written in a **real, essentially complex** world.

Irruption of Mathematical Spaces on Management Studies

Over many decades, economists have lived facing away from these ideas, closing the door to renewal and blocking the arrival of this breath of fresh air. Because of this, economic thought remained deeply-rooted as it had initially been, between 1880 and 1914, in **mechanistic mathematics**. The classic mechanics of Lagrange was used, which gave an impression of severity, compared with that which Perroux called the "laxity of the economic discourse". But on the other hand, the thoughts of researchers remain trapped by certain **economic laws**, parallel to the **laws of nature**, which prevent them from exercising one of their most prized treasures: **imagination**. The inevitable consequence is that the automatisms of the mathematics of determinism have exerted great prestige and still prevail today in many spheres of scientific activity in economics and business management.

But the search for new formal structures has not disappeared from the restless spirits of many researchers. This is the case of Lotfi A. Zadeh. The developments of physics and chaos and instability mathematics have provided, we believe, the important finding of Zadeh[7], who, with his fuzzy sets has created a fundamental change in the panorama of research within the spheres of social sciences. And in the evolutionary onrush of new proposals, introduced with thanks to him, concepts

6 In a boolean environment the term bifurcation makes sense, but in multivalent logic trifurcation, pentafurcation..., or endecafurcation, can be produced, amongst others.

7 Zadeh, L.: «Fuzzy Sets». Information and Control, 8th June 1965, pages 338-353.

as deeply-rooted as **profitability**, **economy**, **productivity**... expressed by cardinal functions, are losing their particular attraction in favour of other notions such as **relation**, **grouping**, **assignment** and **ordering**, which now acquire a new sense. This displacement is fundamental, because it means the transfer of **non-arithmetical** elements, considered contemporary in traditional studies, to the privileged position they currently occupy.

Little by little we are being provided with an arsenal of operative instruments of a non-numerical nature, in the shape of models and algorithms, capable of providing answers to the "aggressions" which our economics and management systems must withstand, coming from an environment full of turmoil. Despite this, it seems that the agglutinating element that constitutes the basic support on which the findings accomplished and those in the immediate future may settle has not yet been found. Can **the notions of pretopology and topology** help to achieve this objective?

In the work which we are presenting, we dare to propose a set of elements from which we hope arise focuses capable of renewing those structures of **economic thought** which are upheld by the **geometrical idea**, so deeply-rooted in the worshipping circles of the orthodox which, monopolising the means of power, assign privileges and deny beauty.

The concepts of **pretopology** and **topology**, habitually marginalized in economics and management studies, have centred our interest in recent times. We consider that it is not possible to conceive formal structures capable of representing **the Darwinism concept of economic behaviour** today without recurring to this fundamental generalisation of metric spaces in one way or another.

In our attempts to find a solid base to the structures proposed for the **treatment of economic phenomena**, we have frequently resorted to the **theory of clans** and the theory of affinities with results which we believe to be satisfactory. We would like to go further, establishing, if possible, the connection between their axiomatics at the same time as developing some **uncertain pretopologies** and **topologies** capable of linking previously unconnected theories, at the same time easing the creation of other new theories. Our aspirations are as ambitious as our enthusiasm is unflagging. We think that even though the flight does not reach so high, we will be capable of reaching levels sufficient enough to capture the attention of those searching for new paths towards a knowledge closer to the complex realities of modern times.

Now seems the right time to remember that in a "crude way", **topology** can be conceived as a branch of science which studies **space**. It therefore analyzes the idea of space and searches for properties **common to all spaces**. For this, it begins from the most general concept of space and studies the properties which belong as much to three-dimensional euclidean space R^3 and the n- dimensionals R^n as to the infinite-dimensional space H of Hilbert, to non-euclidean space and to the geometrics of Riemann, to only quote the most well-known. Therefore, topology does not act directly in either the linking operations between real numbers or in their generalisations.

Brief Summary of This Work

We believe that to arrive at such a high generalisation demands a certain trajectory to, in this way, imbue ourselves with the meaning and possibilities of these structures full of so much abstraction. The immersion in pretopological studies could be a good entry point towards the objectives which we pursue. For this reason, the axioms of the most general of pretopologies has been the starting point for our work.

In the first part of this study we have made a brief reference to ordinary pretopology to later move on to fuzzification. To better assume the meaning of pretopologies, we have begun with a principle example based on very singular financial products which have been passing all "tests" to the extent that new axioms were added to those of the most general pretopology until Moore's uncertain pretopology was reached. The proposed pretopologies do not usually always pass the obstacles which new axioms provide. We have wanted to develop a scheme of this nature, in an attempt to span the range of phenomena that economic and management reality may pose wider. Most of the time, fortunately, it will not be necessary to make use of Moore's uncertain pretopology. In many cases isotone uncertain pretopology or distributive uncertain pretopology will be enough.

Being able to arrive at Moore's uncertain pretopology has been a basic element for our algorithms as it has allowed us, with all of the necessary adaptations, to reach Moore closings and thanks to this we have been able to use, in the sphere which concerns us, the theory of affinities.

The task undertaken has obliged the following of a certain trajectory. Therefore, after a brief description of the fundamental concepts of uncertain ordinary pretopology, the conditions necessary for the existence of isotone uncertain pretopologies have been established. From here the existing relationships in a system are expressed by using a fuzzy graph. We have considered that this was susceptible to treatment until the "closes" were to be found on one side and the "opens" on another.

The traditional study of the fuzzy graph through its $\propto$-cuts has provided a **set of boolean graphs** with their range of Moore closing and their corresponding closes. In this first approximation uncertainty appears as a consequence of the possibility of accepting distinct levels. Simultaneously it is noted that **another kind of uncertainty** is born from the subjectivity in the assignment of valuations for each of the elements of the fuzzy subset. Moving into this field may lead, we hope, towards revealing proposals. Based on this reasoning we have proposed to deal with this approach in its **integrity**.

At the beginning of this task a crossroads appeared at which an important decision for the later development of our work needed to be taken. We refer to the **sense which should be given** to the referential from which the set of base elements among which the functional application will be performed should arise. There may be two interpretations. **The first** considers a "crisp" referential while the second is developed from **a referential of fuzzy subsets**.

During this work, as much in reference to pretopologies as topologies dealt with in the second part, we have been able to show that both focuses result as useful for **decision making** in the economics and management sphere. The choice of one or the other depends on the information available and the objectives to achieve.

We have started the second part of our work expounding the axioms which allow the definition of a topology in the most general way, which later allows an interesting game when passing to uncertainty. The notions of **filter base** and **base** on the one hand and the concept of **neighbourhood** on the other have helped us in this task. Immediately afterwards we have assumed the task of showing certain relationships of a singular importance which may exist between two or more topological spaces. In this way we will develop the concept of topological continuity and, leaning on the notion of **inverse of an application**, that of **homeomorphism**. Upon ending this block of knowledge in this way we hope to have brought the basic elements to achieve one of our main objectives to the surface: **uncertain topology**.

We will dedicate the second block of this part to uncertain topology. The **two focuses** noted in the study of pretopologies in uncertainty now become evident under the same axioms. The transformation of deterministic structures into uncertain structures has captured all of our attention. But uncertainty does not always appear for the same reason and choosing one or another, consigning the others to oblivion, would be showing a lack of scientific sensitivity.

We are able to isolate three ways in which to incorporate uncertainty. We have tested two of them in the pretopology of uncertainty, using a fuzzy subset as a descriptor of a physical or mental object and starting from a **referential set formed by the referentials of the fuzzy subset** in the first form and constructing the **referential set with fuzzy subsets** in the second form. The third and last form of incorporating uncertainty consists in converting the fuzzy elements into booleans by means of a breaking down in levels by $\propto$-cuts. We wish to express our conviction that we cannot expect that the paths of access to uncertainty are finished with that which we expound. On the contrary, we consider that these open doors should still provide interesting possibilities for future development.

The conglomeration of elements put forward in the first and second part of this work constitutes a formal body which, we believe, possesses a high level of homogeneity. The **axioms** established for the **pretopologies** and **topologies**, from which it has been able to establish a great number of properties, should lead us to a **third and final part** closer to the formal objectives and material of business economics and management. In this part we have proposed the connection of some operative instruments already used by ourselves with elaborated formal structures. In this way, we have dedicated each one of the blocks of which this part consists to two interesting but incomplete theories: the **theory of clans** and the **theory of affinities**. We consider that the connection of the previously existing with that achieved can provide unquestionable advantages at the time of their use for economic and management problem solving.

With this intention we connect the **axioms** of the **theory of clans** with those established for the distinct topologies. We have been able to verify that certain concepts, basic in the sphere of **topology**, result as valid when they are transferred to the study of **clans**. For our own tranquility we perform a "test" using some problems form the area of economics management, which validate in practise as much as from a technical perspective.

We end this third part with a block destined to recall the most significant elements of the theory of affinities to, once adequately restructured, affect pretopological spaces. In this sense the comparison between the isotone pretopology and Moore closing has constituted a good starting point. From here it results as essential, to our understanding, to find the sources from which the closes may be found and subsequently Moore closings. The connection of these elements with pretopologies and topologies therefore results as immediate. With this the definition of affinity acquires its greatest meaning.

However, the practical needs, in the sense of capacity of use of this host of knowledge to deal with economics and management problems, demands the availability of **operative methods**. These would arrive in the form of algorithms of an alternative use: that of the **maximum inverse correspondence** and that of the **maximum complete sub-matrices.** We consider that with the presentation of these algorithms and with their use in a revealing supposition we can close our work, albeit in a provisional way.

With this work we aim to begin the task of finding the **bases** on which to **place the structure of economics and management systems**, taking into account that the deep mutations produced in these go beyond, on many occasions, the limits within which the own strengths of the system allows a return to positions of equilibrium. Going beyond these limits means the birth of a new context, **not connected to lineal processes**. Our proposal is aimed at demonstrating the wide possibilities which, in this sense, **uncertain pretopological and topological spaces** acquire.

A Return to the Origin?

It seems that Epicuro was the first to expose the problem of the inseparability of the **determinist** world of the atoms and human **freedom**. It is true that the formulation of the **laws of nature** contributes an important element in not denying evolution in the name of the truth of being, but on the contrary trying to describe the movements characterised by a speed which varies with the passing of time. Despite their formulation, these laws entail the **supremacy** of the **being** over **evolution**, as evident in Newton's law which linked force and acceleration: it is deterministic and reversible in time. But, in spite of the fact that **Newtonian physics** was relegated by the two great discoveries of the 20th century, **quantum mechanics** and **relativity**, his **determinism** and **temporal symmetry** have survived. As it is known, quantum mechanics does not describe trajectories but wave functions, but its fundamental equation, the **Schrödinger equation**, is determinist and of reversible time.

But, if for a great quantity of physicists, amongst whom one can find **Einstein**, the problem of determinism and of time has been resolved ("time is just an illusion"), for philosophers is continues to be a question mark on which the sense of human existence depends. In this way **Henri Bergson**[8] asserts that "time postpones or, better said, is a postponement. Therefore it must be elaboration. Will it not be then the vehicle for creation and election? Does the existence of time not then prove that there is indetermination in things?" In this way, for **Bergson** realism and indeterminism walk hand in hand. **Karl Popper** also considers that "the determinism of Laplace -confirmed as it appears to be by the determinism of physical theories and his brilliant success- is the most solid and serious obstacle in the way of an explanation and an apology of human freedom, creativity and responsibility".[9]

With these precedents we have undertaken the task of developing this book which aims towards the objective of widening the perspective of economics and management studies, from a description of certain **spaces** capable of granting a place for the geometrical concepts of certainty and reversibility and also the innovation which uncertainty or irreversibility may mean. If in the most well-known treatise of economics and management idealised systems of economics and management are described which are stable and reversible, we have aimed to move closer to the world in which we live characterised by instability and evolution, which brings a certain complexity. From here the transition to **uncertain pretopologies** and **topologies**. We believe that even within chaotic systems an **order** may exist. This order is that which we are searching for, reaching far from the past into the future.

---

8 Bergson, H.: «Le possible et le réel», in: Oeuvres. Presses Universitaires de France. París, 1970, page 1333.

9 Popper, K. : L'univers irrésolu. Plaidoyer pour l'indéterminisme. Ed. Hermann. París, 1984, page 2.

Precedents: Intuitive and Axiomatic Aspects of Topology

Brief Historical Overview of Topology

Since several decades ago economic science has dived into many distinct fields of knowledge to find the necessary elements with the objective of a better understanding of the complexity inherent to the systems in which stable equilibriums overflow. Maybe the moment has arrived to return to a path already timidly undertaken on many occasions but then abandoned, we believe prematurely, without fully achieving the desired objectives. We are refering to **topology**.

The word **topology** comes from the Greek terms *τοποσ* (place) and *λογοσ* (study). It first appeared in 1847 when **Johann Benedikt Listing** used it in one of his works.

In its origins topology included work relative to the properties which physical or mental objects support/maintain when they are submitted to continuous transformations such as deformations, doubling, lengthening, etc... but not breakage (for example,a triangle is topologically equivalent to a sphere). In short, it could be said that it deals with the study of soft and gradual changes, of the analysis of the **non-broken**.

But its existence is previous to the word with which it is known today. Indeed, it is considered that the first to use it was **Gottfried Wilhelm Leibniz** (1646-1716). Among his discoveries is found the mathematical principle of continuity which he called "análysis situs" (position analysis) in the year 1679. It is also known as "rubber band geometry". When Leibniz refered to análysis situs he wanted to demonstrate a genuine geometrical analysis in the sense of expressing the "place", the "position", against the algebraic analysis which deals with magnitudes. In this way, while the cartesian coordinates were refered to as specific **quantities**, Leibniz aimed to deal with the geometry of sets independent to the quantities which the elements of these sets could define. **Topology** constitutes a kind of geometry in which longitudes, angles, faces and forms are infinitely changeable. All of the geometrical forms studied in our youth are the same for a topologist. **Topology** studies those properties of **forms which do not change** when **reversible continuous transformations** take place. In what concerns us, we emphasize his interest for the problems of groupings, limits, neighbourhood and morphology of **economics and management objects**.

There is no doubt, however, that the person who established the solid bases of what would be later known as topology was **Bernhard Riemann** (1826-1866), disciple of **Carl Friedrich Gauss** (1777-1855). In his doctoral thesis Riemann established fundamental topological concepts among which that of "extended magnitude several times" stands out, the origin of what would later become known as **topological space**. In this way the idea of **functions space** arose (all the possible forms of a function in a given dominion) and also that of **space of positions of a geometrical figure**. In relation to the study of surfaces, it associated whole numbers to spaces which had previously been defined. These are the **Betti numbers**, which have been the seed of **algebraic topology**. The development of Riemann's ideas was limited principally to the lack of an element which later arose with force. We are refering to the **set theory**. An important contribution appeared with the work of **Georg Cantor** (1845-1918) thanks to the systematization of the set theory and the notions of "accumulation point", "closed set", "open set" and "dimensions", amongst others. Later, once into the 20th century, significant contributions were made by **David Hilbert** (1862-1943), **Maurice Frechet** (1878-1973) and **Frigyes Riest** (1880-1956), amongst others. Without doubt the key figure of the topology that we know today was **Felix Hausdorff** (1868-1942). His definition of topology, formulated in 1914, would serve as a base for that accepted at present times, formulated by **Alexandroff** in 1928. He made it evident that when mathematicians demonstrate theorems of analysis they always use similar methods and that the important thing was not metric properties but the relationships of proximity between the points and subsets. He dispensed with the notion of distance, although he accepted those **minimum properties** which allowed the consideration of **proximity between points and subsets.**

Some Classical Approaches of Topology

Over time some curious problems which, in some way, illustrate a part of the content of topology have been transmitted. Among those most well-known we may mention "the Möbius band", "the Seven Bridges of Königsberg", "the Four Colour Problem" and "the Three Bodies in Space".

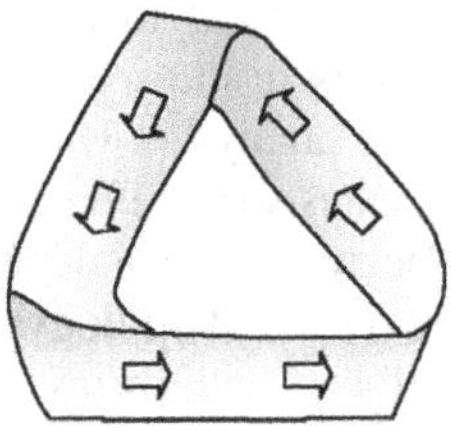

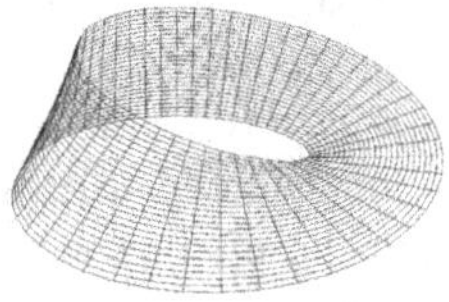

1. In 1858, the German mathematician A. F. Möbius demonstrated that if you take a strip of paper long enough, you twist one of its extremes 180 degrees and then stick the two extremes together to create a ring, you obtain an object with a single face which, apparently, is to perform the impossible. From every perspective and no matter how many times you turn the surface you always find one continuous face.

If we cut the strip in half longitudinally until we reach the starting point of the cut then we are not left with two closed bands but only with one. This is what is known as "cutting into ribbons".

Topologists are used to taking simple figures as a base to create more complicated surfaces, generalising the results in three-dimensions, with a view to include figures of four, five..., *n* dimensions. In this way a flat surface such as a sheet of paper is considered as if it were a perforated sphere which has been stretched and squashed until the aforementioned sheet of paper is created.

It is difficult to recognise such figures when they have been stretched in a fanciful way. To be able to identify them it is normal to characterise each topological type by means of **simple invariant properties**. One of these, relative to a surface, is the **number of edges**. A second is the **number of faces**. A Möbius band is an example of **a face**. The third characteristic invariant is the aforementioned **Betti number**, defined by the **maximum number of transversal cuts** which may be made in a surface without dividing it, taking into account that a transversal cut begins and ends at the edge. The ribbon cuts also allow the finding of the Betti number of a surface and are performed in a way that starts and ends at one point on the surface, always avoiding the edge. The maximum number of ribbon cuts possible without dividing a surface with an edge leads to the Betti number. The Betti number of a Möbius band is one. Each ribbon cut intercepts only one transversal which shows that named the **fundamental relationship of duality**. **S. Lefschetz** generalised this relationship to n dimensions and in 1927 formulated, the "duality theorem" which has been considered one of the milestones of topological development of the 20th century. As a final point, the numbers corresponding to the three topological invariants are always the same for an object, although its form may be changed with stretching but always without breakage or rejoining.

If a band formed after performing three 180 degree twists in one of its extremes before fastening the two extremes is considered, it can be observed that this band also has only one edge and a single face, a transversal cut is the only possibility. The Betti number is therefore one. Essentially, we find ourselves before an ordinary Möbius band (a single twist of 180 degrees), although there is a difference in the way it is situated in relation with three-dimensional space. Indeed, while here the edge of the band shows a knot, in the ordinary Möbius band there is a simple curve without knots.

2. The problem of the **Bridges of Königsberg** is found immersed in net theory and graph theory. It is one of the oldest of topology and it is normally presented by means of a graph formed by four vertices and seven non-orientated arcs.

In the 17th century the Prussian city of Königsberg had been built around an island named Kneiphof, joined to each bank of the river Pregel by two bridges (and therefore, four bridges in total) and by another bridge to a neighbouring island which, in turn, was connected to the riverbanks by another two bridges.

The problem was finding the way of crossing each of the seven bridges without crossing any of them more than once. The local people saw this problem as a riddle without an answer. It was in 1736 when the Swiss mathematician **Leonhard Euler**, interested in the problem, found an ingenious answer to the impossible by using a mathematical demonstration of the impossibility of such a route, which then created a number n of bridges. His explanation consisted of a revealing example of the deceptive simplicity of topological approaches. The problem of Königsberg is linked with the well-known exercise of creating a specific figure on a sheet of paper without lifting the pencil nor passing more than once over the same line.

The reasoning of Euler is very simple. He began with the representation of the problem by means of a **graph** in which the **vertices** substitute the parts of solid ground and the **arcs or edges** substitute the bridges. He named those vertices which result in an even number of trajectories **even vertices** and those which result in an odd number **odd vertices**. The number of journeys necessary to traverse a connecting graph is equal to half the number of odd vertices. As it is not possible to construct a graph with an odd number of odd vertices (each arc has two vertices) the problem of the bridges of Königsberg does not have a solution. It would be necessary to add another arc, which is to say, another bridge.

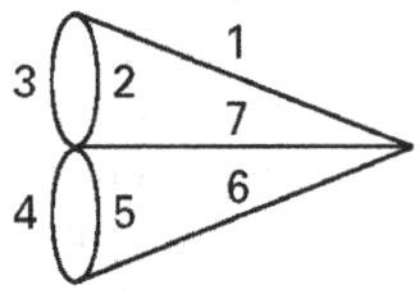

Impossible
(Königsberg Bridges)

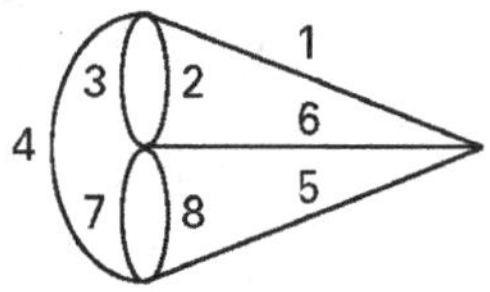

Possible

The idea of Euler consisted of creating a **connected graph in which circuits do not exist**. In a graph of this nature the number of vertices is equal to the number of arcs plus one. For this one begins with a graph in which those arcs which produce **circuits** are eliminated without deleting any vertex. The maximum number of arcs destroyed in a way in which all the vertices remain connected is the Betti number of the graph.

We will look at this analytically. Let us assume a graph with initial V vertices and A arcs of which B is eliminated. Therefore, we have:

$$V = 1 + (A - B)$$

$$B = 1 + (A - V)$$

It can be graphically proven that the Betti number of the graph of the bridges of Königsberg is four, as it is necessary to eliminate 4 arcs for circuits not to exist in the graph. Let us take a look at this:

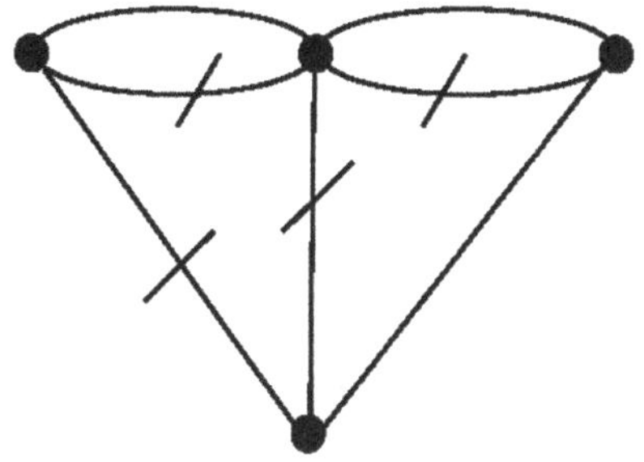

A = seven arcs (seven bridges)
V = four vertices (four areas of separated land)
$B = 1 + 7 - 4 = 4$.

Although the idea which supports the naming of the "Betti number" had already been used prior to that by **G. R. Kirchhoff in 1847** and by **James Clerk Maxwell in 1873**, it was **Enri Poincaré** who settled on the name in 1895 in honour of the mathematical physicist **Enrico Betti** (1823-1892) who, in 1871, had created the Riemann connection numbers.

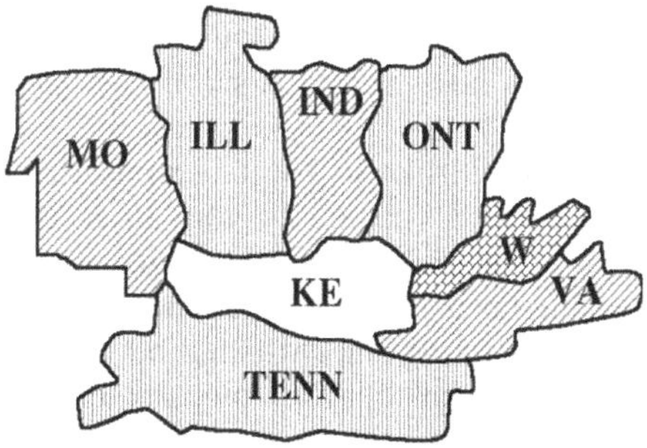

3. The well-known four- colour problem is also placed in the field of graphs. It is as simple as it is unsolvable and as unsolvable as it is difficult to demonstrate its unsolvability. In short it is stated that, given a geographical map each country should be coloured in a different colour from those adjacent to it, using the least number of colours. It is necessary to take into account that a single point of contact does not presuppose the existence of a border. So, it must be demonstrated that four colours are enough to colour any map of a single plan.

For many countries only three colours were necessary, and for some well-known cases such as the American State of Kentucky and its surrounding states four colours are necessary.

In 1946 the Belgian mathematician **S. M. de Backer** tested that any map with a number of countries equal or lower to 35 could be coloured, in the stated circumstances, with no more than four colours. Later other researchers managed to increase the number of countries significantly, although a general demonstration for any quantity of countries is still not known.

It is evident that the **four colour theorem** was enunciated only for the cases of flat or spherical surfaces, as other distinct problems arise for other geometrical figures.

4. In 1887 King Oscar of Sweden offered a prize of 2,500 Krona to whoever gave the answer to the question of "if the solar system is stable". In celestial mechanics the interaction of two bodies does not pose a problem, but that of three presents great difficulties. **Poincaré** won the prize in 1890 with the memoir "on the Three-body Problem and the Equations of dynamics".

The movement of two bodies (the Sun and the Earth, for example) is periodical. The period is of one year. This shows that these bodies cannot crash into one another, nor can they infinitely move away from one another. They have not previously done this and therefore they can never do this. **Periodicity** constitutes a very useful element for controlling stability.

In the third chapter of his memoir, **Poincaré** strived to explain **the existence** of **periodical solutions** for differential equations. Assuming that at a determined moment the system is found in a specific state and that at a later time it once again returns to the same state. All positions and velocities are the same as before. Therefore, the **movement** which has driven a state to itself once again should be repeated, once and again: it is a periodical movement. Given a point in a many dimensional space, as time passes the point will move giving rise to a curve. When will the curve form a closed loop? As the question does not affect either the **form**, the **size** or the **position** of the loop, the answer corresponds to topology. In this way the existence of periodical solutions depends on the topological properties of the **relationship** between the position of a point now and its position in a later period.

A simple example can be revealing. If we would like to know if an artificial satellite has a periodical orbit, instead of following all of its trajectory all around the Earth with a telescope we focus on it in a way that "sweeps" a plane which goes from north to south, from one point of the horizon to the other and which is in line with the centre of our planet. We take note of the place which it has passed the first time, its speed and its direction. We must then wait only **focusing on the plane**. Periodicity demands that it must once again pass by the same point, at the same speed and in the same direction. Therefore, instead of observing all states, just a few are enough. This surface is known as the **Poincaré section**

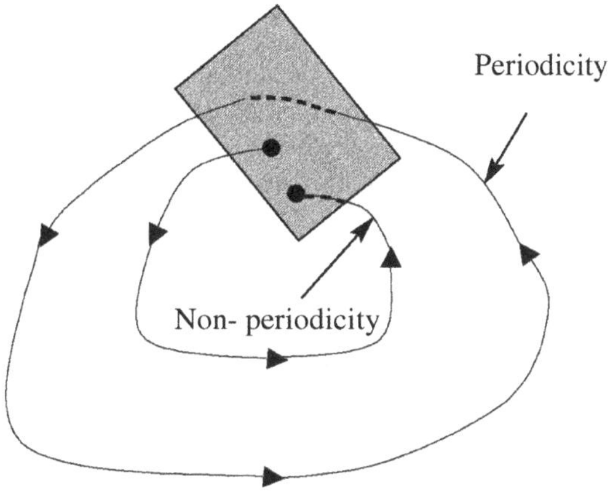

From this idea, **Poincaré** went deep into another approach, known as **Hill's reduced model**. Three bodies are considered, one of which possesses a mass so small that it does not affect the other two, of large mass. In exchange these do affect the first (for example, an interstellar particle and two planets). The two large bodies move forming elliptical paths around their mutual centre of gravity, but the tiny body moves oscillating from one side to another with nothing to change its direction. **Poincaré** used his **superficial section method** to try to find periodical movements of the tiny body, but what he found was a very complicated and counterintuitive behaviour: the system began activity in one state, followed a curve which took another state when it returned to the Poincaré section, then another and another successively. The system, in summary, passed through the Poincaré section with an uncertain sequence of points. Poincaré had found a chaotic panorama.

The main reason for that which we have just stated has been to illustrate the idea of topology from an anecdotal historical perspective, at the same time as allowing the introduction of some basic ideas which have survived over time.

In the last fifty years there has been an increase of interest in the use of topology to deal with a large number of problems by a great number of mathematicians and also of physicists and engineers. Due to this, important applications in the study of **flow nets** and **closed circuits** (Kirchhoff), in **magnetic fields** (Maxwell) and also in studies of the disposition of colours and design of printed electronic circuits, amongst others have taken place. We believe that it is time that **economists** and **management specialists** are also interested by this important branch of mathematics, as maybe it would be possible to represent **economics and management systems** through **topological spaces**, finding the "laws", if they exist, of the mutations which take place when the limits capable of producing a return to equilibrium are passed, but the new situations do not give rise to a traumatic rupture with the past. The work which we are currently presenting follows this path.

We know of **the use of topology** to represent problems expressed through **"non-lineal" differential equations**, which is to say equations with derivatives **without** effects proportional to the causes. The simplification which linearity brings has historically come imposed on many occasions by the difficulty, when not the impossibility, of its calculation and solution. **Economics and management phenomena** rarely present an adopted lineal form in reality which is why there is **an urgent need** for the recourse to complex non-lineal schemes.

That which we have just stated should not lead us to think that topology is an insurmountably difficult part of mathematics. The opposite is in fact the case, above all in the basic ideas which they report. In effect, many topological concepts are habitually employed. From here we arrive at the notions of **interior** and **exterior**, **connection** and **non-connection**, known and fostered in schools, even from the youngest infants.

In the texts which are dedicated to explaining topology there exists the habit of presenting some **theorems** which are naturally susceptible to draw the attention of readers and arouse their curiosity. Here we have five of these:

a) **The wind cannot simultaneously blow in all areas of the Earth**. There must exist an area without wind at all times. This place may be, for example, the South Pole.
b) If the wind is blowing in all areas of the northern hemisphere at a determined moment, then at this moment it must blow in all directions at the equator. For example, a place must exist where the wind blows north-easterly.
c) Under the same circumstances at least **two diametrically opposite points** must exist at the equator where the wind blows in radically opposite directions.
d) At a determined moment there exists at least one point **on the Earth and its antipode** which have the same temperature and the same humidity.
e) If the Earth were divided into three single great powers then at least **one of them would never witness a sunset**, given that in at least one of the three cases it would be a given fact that of two of their points one should be an antipode of the other.

Each one of these theorems is a consequence of mathematical discoveries of a most general nature. In this way the first two are particular cases of the **general "fixed point" theorem**, which was formulated in 1904 by the Dutch mathematician **L. E. J. Brouwer**[1]. The theorems c) and d) are a direct result of the general "**antipodal point**" theorem formulated for n dimensions in 1933 by the Polish researchers **K. Borsuk** and **S. M. Ulam**. The last of these theorems was deduced from another **more general for n dimensions**, elaborated in 1930 by the Soviet mathematicians **L. Lusternick and L. Schnirelmann**.

The content of chapter I of the great work published with the name of the author **Nicolas Bourbaki**[2] results of great interest. **Eléments de Mathématique**,

[1] Shinbrot, M.: «Teoremas del punto fijo», in *Matemáticas en el mundo moderno* (various authors). Ed. Blume. Madrid, 1974, pages. 165-171.

[2] Bourbaki, N.: *Eléments de Mathématique. Livre III Topologie Générale.* Hermann & Cie. Editeurs. París, 1940. As it is known, N. Bourbaki has been used as the pseudonym of an important group of mainly French mathematicians (between 10 and 20 depending on the moment in time).

whose third book is integrally dedicated to topological structures. Its clarity has given us a a lot in our attempt to develop some algorithms based on uncertain topology.

Considerations on the Subject of the Idea of Topology

The aim of everything that we have just shown is to "create" the necessary conditions in order to stimulate the study of topology. If we have achieved to open scientific curiosity then it is possible that the need appears for an explanation of why **interest exists** for **topological spaces**. We will do this through some brief ideas.

We remind ourselves that in sets theory it is only possible to study the relationship between a point and a set. Therefore, this relationship remains limited to the possession of the set or not.

However, upon considering R^2 in the set S (which is a subset of E) formed by an open ball, the three points p_1, p_2, p_3 in the first figure on this page.

P_1 can be "seen" **inside S**, p_2 **on the border of S** and p_3 **outside S**. This initial interpretation has its foundation in the notion of the proximity of p_1 to S and also to its complement.

In this way the **adherence point** of set S is born, considered as that which **is not outside**. In other words, it is that which is as close to S as we would like.

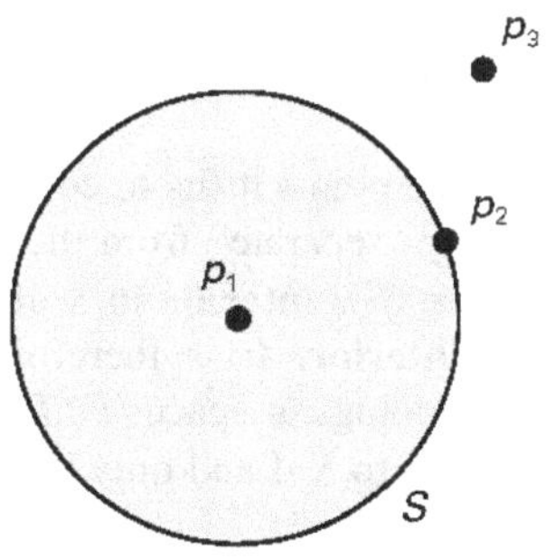

From now on, we will designate a topological space by using $(E, T(E))$. So, if S is a subset of E, it would be said that a point $p \in E$ is adherent to S when any **range** U of p crosses S. This is equivalent to saying that **no** range of p **exists** which is totally included in $E - S$. Therefore, any point of the set S is an adherence point of S. The set of adherence points of S is its **adherent or adherence application**.

We will now move on to the concept of **accumulation**. If we consider in R^2 the set S formed by the union of the circle and point p_2, the points p_1, p_2, p_3 of the following figure are all **adherence points**.

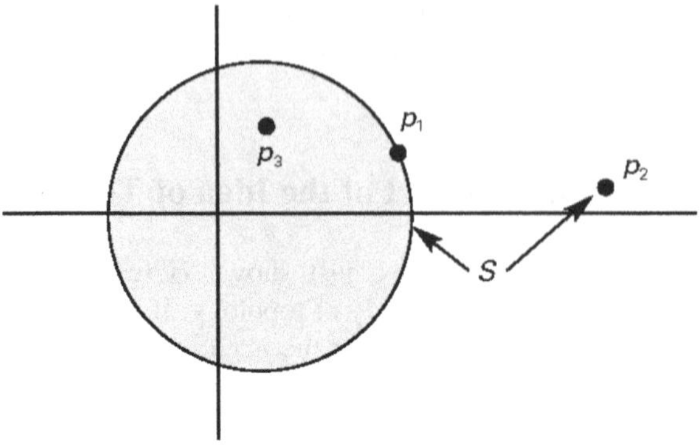

However, the points p_1, p_3 have a distinct sense to p_2. In effect, "near" to p_1, p_3 other points of S can be found, which does not occur with p_2, which is **isolated**. From here the notions of **accumulation point** and **isolation point**.

Given a topological space $(E, T(E))$ and also given a subset of E, S, we will say that $p \in E$ is an **accumulation point of S**, if any range U of p contains a point of S distinct to p. The set of all of the accumulation points of S are called **accumulation** or **set derived from S**. The points which do not fulfil this condition are **isolated points**.

We will now separate the adherence points according to proximity as much of the set S as the complement of this set. Those points which are "near" to both the set S and its complement are named **border points**. In other words, when all the range U of p crosses S and its complement $E - S$. The **set of border points** is called **border**.

Finally, among the adherence points it fits to consider those found "within" the set S and, therefore which may "separate" from the complementary set $E - S$. In this way, therefore, a point $p \in S$ is **interior to** S, if S is a range of p. The set of points inside S are named **interior**. It is therefore possible to formulate the following theorem: Given a topological space $(E, T(E))$ and a subset S of E , one would say that $p \in S$ is **interior to** S if and only if it is not an adherence point of its complement $E - S$.

It does not result difficult to accept that the **interior** has the **dual properties** of **adherence**.

We have made reference to a concept that of range of a point, for which some remarks are convenient. We will do this through its most elemental properties:

1 The point p belongs to all range of p.
2 If U is a range of p, any set $T \supset U$ is a range of p.
3 If U_1 and U_2 are ranges of p, the same occurs with the intersection $U_1 \cap U_2$. This is generalizable to a finite number of ranges of p.
4 A range U of p is also a range of all points x of a suitable range F of p.

The axioms 1) and 4) have constituted decisive axioms for a **generalisation** of topological spaces. In this way it has been able to define a topological space $T(E)$, from a set E when a system $U(p)$ of subsets of E corresponds to each element p of E, named ranges of U of p, which verify the following axioms:

1 $p \in U$, for all of the range $U \in U(p)$.
2 If $U \in U(p)$ and $F \supset U$, then $F \in U(p)$.
3 If $U_1, U_2 \in U(p)$, then $U_1 \cap U_2 \in U(p)$; $E \in U(p)$
4 For each $U \in U(p)$, there exists an $F \in U(p)$, so that $U \in U(q)$ for all $q \in F$.

In this way, a set E together with a topology $T(E)$ is named topological space $(E, T(E))$. The elements of E take the name of **points** of topological space.

The axioms 1) and 4) have their correspondence with the axioms of the **Hausdorff range**, introduced by F. Hausdorff in his fundamental work of 1914 "Grundzüge der Mengenlehre". In effect, a topological space $(E, T(E))$ is said to be of Hausdorff when it fulfils one of the following axioms, equivalent to each other.

a) If $p \neq q$, with p, q being two points of E, the ranges $U \in U(p)$ and $F \in U(q)$ exist, with $U \cap F = \emptyset$ (Hausdorff separation axiom).
b) The intersection of all the closed ranges of a point p only contains p.

The first of these axioms is named of separation as the points p and q remain separated by the ranges U and F.

Elemental Notions in Topological Spaces

Everything that we have just stated has the purpose of bringing to the surface sufficient elements in order to establish some definitions which we will summarise.

If S is a subset of the referential E:

1 A point $p \in E$ is named **interior** of S when a range $U \in U(p)$ exists which belongs entirely to S. In classic nomenclature, the set of all the interiors is known with the denomination of **opening** of S.
2 A point $p \in E$ is named **exterior** of S when a range $U \in U(p)$ exists which belongs entirely to the **compliment** of S. The set of points exterior to S are known as **exterior** of S.
3 A point $p \in E$ is a **border** point of S, or with respect to S, when in each range of p there exists points which belong to S and to the compliment of S. The set of all these points p of S is named the **border** of S.

Of these definitions it can be deduced that all points $p \in E$ unmistakeably belong to one of the three classes previously described. At the same time it can be said that the **exterior of S** coincides with the **interior of the compliment of S**.

With these three definitions presented, we find ourselves able to move on to the important concept of **adherence**. A point is $p \in E$ named adherence point of S if in each range of p there are points of S. The set of all the adherence points of S is designated **closure** or **close of S**. The **close** of S comes given by the union of the opening of S with the border of S.

In the same way, we will say that a set S is named **open** when it fulfils one of the following conditions, equivalent to each other:

a) All points of S are interior.
b) The opening of S coincides with S. It is enough that the opening contains S.
c) The border of S is contained within S.

A set S is named **closed** when it fulfils one of the following conditions, also equivalent to each other:

a) The set S contains all its adherence points.
b) The set S coincides with the close of S. It is enough to demand that the close of S is contained within S.
c) The border of S is found contained in S.

It is possible that a set S is open and closed at the same time.

A simple figure can help us to visualise the equivalence of the three conditions, as much in relation with the **open** as with the **closed**.

Let us look at this through the following formulae:

$E = S$ + complement S.
= open of S + close of the complement of S.
= close of S + open of the complement of S.

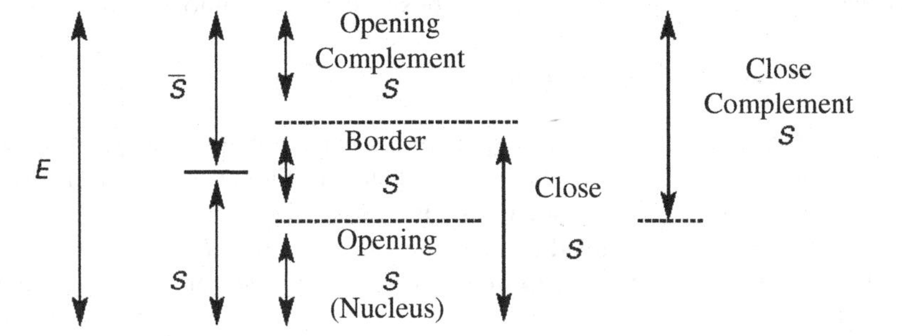

We can conclude by making the following theorems:

1 A set S is open if, and only if, the complement of S is closed. A set S is closed if, and only if, the complement of S is open.
2 A set S is open if, and only if, it is a range of all its points.

Everything which we have just stated brings us to consider a "**topological space** as a set in which we have selected a series of subsets, which we call open, and which fulfil the same properties as the open subsets of any **metric space**"[3].

Brief Reference to Metric Spaces

Remember that a **metric space** is a pair *(E, d)* formed by the set *E* and an application
$d(E \cdot E \rightarrow R)$ which fulfils the following axioms:

a) $d(p, q) \geq 0$, $d(p, q) = 0$ if and only if $p = q$.
The distance between the two points is never negative and is only zero if and only if the two points are the same.

b) $d(p, q) = d(q, p)$, for any $p, q \in E$.
The distance is symmetric.

c) $d(p, q) + d(q, r) \geq d(p, r)$, for any $p, q, r \in E$.
The distance fulfils triangular inequality.

The elements of the set *E* are named **points of space** and *d* **metric** on *E*.

For indicative purposes only we will remind ourselves of two of the spaces which appear with the greatest frequency in the works consulted or studied. These are **euclidean space** and **Hilbert's H space**.

R^n Euclidean space is perhaps the most widespread metric space. Its points are given under the form e = $(p_1, p_2, \ldots..p_n)$, in which $p_1, p_2, \ldots..p_n$ are arbitrary real numbers. The distance in these spaces comes from the well-known formula:

$$d(e, n) = \sqrt{\sum_{i=1}^{n}(q_i - p_i)^2}$$

It is unquestionable that the first two axioms are fulfilled. Let us see if the same occurs with the third, which is to say:

$$\sqrt{\sum_{i=1}^{n}(r_i - p_i)^2} \leq \sqrt{\sum_{i=1}^{n}(q_i - p_i)^2} + \sqrt{\sum_{i=1}^{n}(r_i - p_i)^2}$$

Let us make $(q_i - p_i) = a_i$, $(r_i - q_i) = b_i$. The previous expression can be deduced from:

3 Mascaró, F.; Monterde, J.; Nuño, J. J., and Sivera, R.: *Introducció a la topología*. Ed. Universitat de Valencia, 1997, page. 51.

$$\sqrt{\sum_{i=1}^{n}(a_i + b_i)^2} \leq \left(\sqrt{\sum_{i=1}^{n} a_i^2} + \sqrt{\sum_{i=1}^{n} b_i^2}\right)^2 = \sum_{i=1}^{n} a_i^2 + \sum_{i=1}^{n} b_i^2 + 2\sqrt{\sum_{i=1}^{n} a_i^2 \cdot \sum_{i=1}^{n} b_i^2}$$

$$2\sum_{i=1}^{n} a_i \cdot b_i \leq 2\sqrt{\sum_{i=1}^{n} a_i^2 \cdot \sum_{i=1}^{n} b_i^2}$$

$$\sum_{i=1}^{n}(a_i \cdot b_i)^2 \leq \sum_{i=1}^{n} a_i^2 \cdot \sum_{i=1}^{n} b_i^2$$

This final inequality is that known as the Cauchy-Schwarz inequality.

We will now, finally, refer to **Hilbert's H space**. This space is formed by all the successions of real numbers $h = (p_1, p_2 ....)$, so that the sum of its squares $\sum p_i^2 (i = 1, 2, ...)$ is convergent. In this case the distance is the same as the euclidean:

$$d(h, \eta) = \sqrt{\sum (q_i - p_i)^2}$$

Here it is warranted to state that the convergence of the infinite series has moved inside the root.

In this way one has:

$$\sum_{i=1}^{N}(q_i - p_i)^2 = \sum_{i=1}^{N} q_i^2 + \sum_{i=1}^{N} p_i^2 - 2\sum_{i=1}^{N} p_i \cdot q_i$$

Upon creating N, following the established hypothesis, the first two addends remain enclosed, whereas the third is in this way as a consequence of the Cauchy-Schwarz inequality:

$$\left(\sum_{i=1}^{N} p_i \cdot q_i\right)^2 \leq \sum_{i=1}^{N} p_i^2 \cdot \sum_{i=1}^{N} q_i^2$$

Therefore, in this way, the series placed within the root is convergent and the distance is found to be adequately defined, as the axioms 1) and 2) fulfil this need with simple observation and 3) is easily tested by the steps to the limit in the euclidean space formula.

From this brief reminder we should indicate that **metric spaces** are not general enough to describe whatever "form" of space, as others exist in which it is not possible to make each pair of elements of the real numbers such as distance between each coincide. To this respect it is enough to observe the axiomatic of the metric spaces to confirm the intervention of real numbers in them. In a crude way we can say that **all metric space** is a **topological space**. In other words, a defined metric space on a set in R induces a topological space. On the other hand the reciprocal proposal that every topological space proceeds from an adequate metric space is not true.

In this brief summary we have tried, with a selection of some historical and elemental aspects as close as possible to intuition, to drive **topological reasoning** to that point in which a strict formal rigor is imposed, capable of describing one of the outlines with a greater level of abstraction than modern mathematics provides: **topological spaces** apt for the treatment of economics and management systems.

To achieve this it has been considered as opportune to begin the basic content of this work with the description of pretopological axiomatics, first from a perspective which, abusing language, could be called **orthodox**, to later move on to its transformation attempting to generalise to deal with situations immerse in a context of **uncertainty**: both aspects will constitute the content of the first part of our work.

Part I

From Pretopology to Uncertain Pretopological Axiomatics

Chapter 1
Pretopology

1 The Notion of Combinatorial Pretopology

In our journey through the study of **spaces** apt to formalise economics and management problems, we centre our attention on pretopological study, before dealing with topology, putting forward the distinct axioms relative to the different levels of pretopology. We will begin with the most general[1].

The axiomatic of ordinary pretopology is simple. Let us remind ourselves.

Given a finite referential E and the set of its parts $P(E)$ ("power set"), Γ of $P(E)$ in $P(E)$ is considered a functional application. Therefore, it is said that Γ is a pretopology in E if, and only if, the following axioms are fulfilled:

1. $\mathbf{\Gamma} \varnothing = \varnothing$
2. $\forall A_j \in P(E)$:

$$A_j \subset \mathbf{\Gamma} A_j$$

In which A_j is an element of the "power set".

The second condition entails that $\mathbf{\Gamma} E = E$.

We will move on to develop the first example. With the set E being of financial products, each of which have some certain characteristics or not:

$$E = \{a, b, c\}$$

and the **set of its parts**, which is to say all of the possible groupings of these three products taken one by one, two by two and the group of three. One has:

$$P(E) = \{ \varnothing, a, b, c, ab, ac, bc, E\}$$

The following figure shows a pretopology of $P(E)$ in $P(E)$, in which certain relationships can be observed between the possible groups of financial products which fulfil the established axioms.

---

[1] To this respect, the work of Gil Aluja, J.: «La pretopología en la gestión de la incertidumbre». Speech for the purpose of entering as Doctor «Honoris Causa» of the Universidad de León Ed. Secretaría de Publicaciones y Medios Audiovisuales. León, 2002, pages. 58-73.may be consulted.

J. Gil Aluja & A.M. Gil Lafuente: Towards an Advanced Modelling, STUDFUZZ 276, pp. 3–27.
springerlink.com

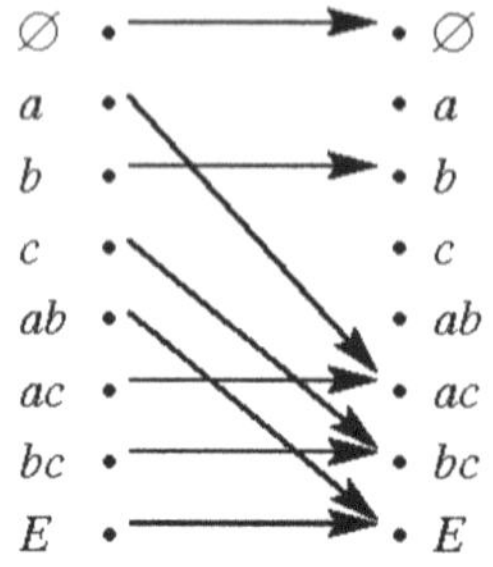

It is easy to check that the previous axioms are verified:

$\mathbf{\Gamma}\varnothing = \varnothing$ $\mathbf{\Gamma}\{a\} = \{a,c\} \supset \{a\}$ $\mathbf{\Gamma}\{b\} = \{b\} \supset \{b\}$
$\mathbf{\Gamma}\{c\} = \{b,c\} \supset \{c\}$ $\mathbf{\Gamma}\{a,b\} = E \supset \{a,b\}$ $\mathbf{\Gamma}\{a,c\} = \{a, c\} \supset \{a,c\}$
$\mathbf{\Gamma}\{b,c\} = \{b,c\} \supset \{b,c\}$ $\mathbf{\Gamma}\ E = E \supset E$

The demand of this condition entails that in the functional application each financial product or group of financial products is found related with another product or group of products with which it comprises.

In continuation we will show a second example in which we begin with a reflexive graph in matrix form as in the following:

Γ	a	b	c	d	e
a	1				
b		1	1	1	1
c		1	1	1	
d		1		1	
e	1	1	1	1	1

One has:

$\mathbf{\Gamma}\varnothing = \varnothing$ $\mathbf{\Gamma}\{a\} = \{a\}$ $\mathbf{\Gamma}\{b\} = \{b,c,d,e$ $\mathbf{\Gamma}\{c\} = \{b,c,d\}$
$\mathbf{\Gamma}\{d\} = \{b,d\}$ $\mathbf{\Gamma}\{e\} = E$ $\mathbf{\Gamma}\{a,b\} = E$ $\mathbf{\Gamma}\{a,c\} = \{a,b,c,d\}$
$\mathbf{\Gamma}\{a,d\} = \{a,b,d\}$ ...., $\mathbf{\Gamma}\{b,c,d,e\} = E$ $\mathbf{\Gamma}\ E = E \supset E$

Once these examples have been demonstrated, we will move on to expound some of the already presented notions linked to the concept of pretopology, now formulated from a focus and nomenclature which we will follow throughout this work.

We will begin by saying that the pair $(E, \mathbf{\Gamma})$ will normally be named **pretopological space** and the functional application $\mathbf{\Gamma}\ A_j$ for all the elements of $P(E)$ is habitually known as an **adherent** or **adherence application**.

In this way, it can be confirmed in the example the adherence of **c** is **bc**, the adherence of **ac** is **ac**.

Another concept, that of **interior application** or **interior**, which is designated as δA_j, merits our special attention.

Let $(E, \mathbf{\Gamma})$ be a pretopological space and $A_j \in P(E)$. It is defined as **interior** of A_j:

$$\delta A_j = \overline{\Gamma \overline{A_j}}$$

in which – indicates the negation.

From the first of our examples we can calculate $\delta\,\{a, b\}$:

$$\delta\{a, b\} = \overline{\Gamma \overline{\{a, b\}}} = \overline{\Gamma\{c\}} = \overline{\{b, c\}} = \{a\}$$

therefore, $\{a\}$ is the interior of $\{a, b\}$.

In the same way the interiors of all of the elements of this example are obtained.

$$\delta\varnothing = \left(\overline{\Gamma\overline{\varnothing}}\right) = (\overline{\Gamma E}) = \overline{E} = \varnothing$$

$$\delta\{a\} = \overline{(\Gamma\ \{\overline{a}\})} = \overline{\Gamma\ \{b, c\}} = \overline{\{b, c\}} = \{a\}$$

$$\delta\{b\} = \overline{(\Gamma\ \{\overline{b}\})} = \overline{\Gamma\ \{a, c\}} = \overline{\{a, c\}} = \{b\}$$

$$\delta\{c\} = \overline{(\Gamma\ \{\overline{c}\})} = \overline{\Gamma\ \{a, b\}} = \overline{E} = \varnothing$$

$$\delta\{a, b\} = \overline{(\Gamma\ \{\overline{a, b}\})} = \overline{\Gamma\ \{c\}} = \overline{\{b, c\}} = \{a\}$$

$$\delta\{a, c\} = \overline{(\Gamma\ \{\overline{a, c}\})} = \overline{\Gamma\ \{b\}} = \overline{\{b\}} = \{a, c\}$$

$$\delta\{b, c\} = \overline{(\Gamma\ \{\overline{b, c}\})} = \overline{\Gamma\ \{a\}} = \overline{\{a, c\}} = \{b\}$$

$$\delta E = \overline{(\Gamma\ \overline{E})} = \overline{\Gamma\varnothing} = \overline{\varnothing} = E$$

which may be expressed by means of the following graph:

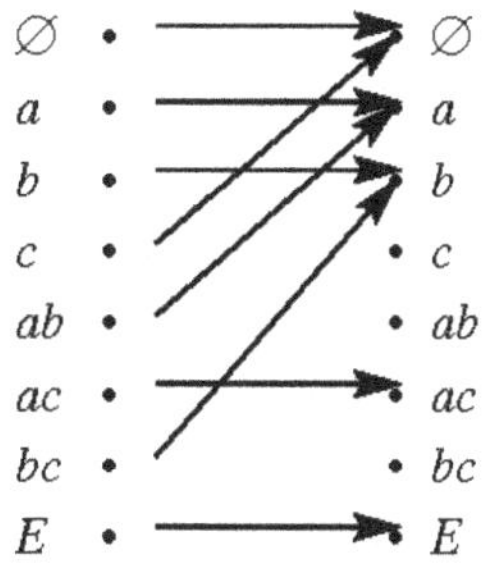

In this functional application each product or group of financial products is found related with a part of the group or with its own product or group`.

Evidently one has:

$$\Gamma A_j = \overline{\delta \overline{A_j}}$$

As:

$$(\overline{\Gamma \overline{A_j}} = \delta A_j) \Rightarrow (\overline{\Gamma A_j} = \delta \overline{A_j}) \Rightarrow (\Gamma A_j = \overline{\delta \overline{A_j}})$$

Therefore:

$$\Gamma \varnothing = \varnothing \quad \text{and} \quad \delta E = E$$

From an opposite perspective one has:

$$\overline{A_j} = \Gamma \overline{A_j}$$

We now find ourselves in disposition to define an **open subset** and a **closed subset**.

It is said that A_j is an open subset if:

$$A_j = \delta A_j$$

and it is said that A_j is a closed subset if:

$$A_j = \mathbf{\Gamma} A_j$$

We will return to the first of the pretopologies to see which are open and which are closed subsets.

The closed are:

$\mathbf{\Gamma}\varnothing = \varnothing$ $\quad$ $\mathbf{\Gamma}\{b\}=\{b\}$ $\quad$ $\mathbf{\Gamma}\{a, c\}=\{a, c\}$ $\quad$ $\mathbf{\Gamma}\{b, c\}=\{b, c\}$ $\quad$ $\mathbf{\Gamma} E = E$

and the open are:

$\delta \varnothing = \varnothing$ $\quad$ $\delta\{a\}=\{a\}$ $\quad$ $\delta\{b\}=\{b\}$ $\quad$ $\delta\{a, c\}=\{a, c\}$ $\quad$ $\delta E = E$

In our example it can be observed that the the **open** and **closed subsets** are formed by products or groups of financial products which are related to one another.

In a pretopology, a subset of $P(E)$ can be open and closed at the same time. From this, let a necessary and sufficient condition for a subset to be both open and closed at the same time be:

$\mathbf{\Gamma} A_j = \delta A_j$ which is to say $\Gamma A_j = \overline{\Gamma \overline{A_j}}$ and $\overline{\Gamma A_j} = \Gamma \overline{A_j}$

We see the following easily demonstrated properties:

$$(A_j = \delta A_j) \Rightarrow (A_j = \boldsymbol{\Gamma} A_j)$$

$$(A_j = \boldsymbol{\Gamma} A_j) \Rightarrow (A_j = \delta A_j)$$

In effect, for the first:

$$A_j = \delta A_j = \overline{\Gamma \overline{A_j}}$$

in which:

$$\overline{A_j} = \Gamma \overline{A_j}$$

The same would be done for the second.

We will finish, for now, expounding the following property.

$\forall A_j, A_k \in P(E)$:

$$((A_j \subset A_k) \Rightarrow (\Gamma A_j \subset \Gamma A_k)) \Leftrightarrow ((A_j \subset A_k) \Rightarrow (\delta A_j \subset \delta A_k))$$

We perform the verification:

$$(A_j \subset A_k) \Rightarrow (\overline{A_k} \subset \overline{A_j}) \Rightarrow (\Gamma \overline{A_k} \subset \Gamma \overline{A_j}) \Rightarrow (\overline{\Gamma \overline{A_j}} \subset \overline{\Gamma \overline{A_k}}) \Rightarrow (\delta A_j \subset \delta A_k)$$

$$(A_j \subset A_k) \Rightarrow (\overline{A_k} \subset \overline{A_j}) \Rightarrow (\delta \overline{A_k} \subset \delta \overline{A_j}) \Rightarrow (\overline{\delta \overline{A_j}} \subset \overline{\delta \overline{A_k}}) \Rightarrow (\Gamma A_j \subset \Gamma A_k)$$

All that we have just shown allows us to deal with **certain pretopological spaces** of special interest. We will begin with isotone pretopological space.

2 Isotone Pretopological Space

A pretopological space $(E,\boldsymbol{\Gamma})$ such as:

$\forall A_j, A_k \in P(E)$:

$$(A_j \subset A_k) \Rightarrow (\Gamma A_j \subset \Gamma A_k)$$

is named isotone in inclusion and also monotone in inclusion, although the term **isotone** alone is habitually used.

Also, as we have seen, when given the previous property the following is also fulfilled:

$$(A_j \subset A_k) \Rightarrow (\delta A_j \subset \delta A_k)$$

Therefore, in this type of pretopological space, the isotone inclusion is given for both closes and opens.

Let us move on to expound a supposition of isotone pretopological space[2]. For this we will continue with the example of a set of financial products such as :

$$E = \{a, b, c, d\}$$

From the set of its parts $P(E)$ we have constructed the following functional application which, as we will confirm, is an isotone pretopology.

$$\Gamma\varnothing = \varnothing \quad \Gamma\{a\} = \{a, d\} \quad \Gamma\{b\} = \{a, b\} \quad \Gamma\{c\} = \{c, d\} \quad \Gamma\{d\} = \{c, d\}$$
$$\Gamma\{a, b\} = \{a, b, d\} \quad \Gamma\{a, c\} = \{a, c, d\} \quad \Gamma\{a, d\} = \{a, c, d\} \quad \Gamma\{b, c\} = E$$
$$\Gamma\{b, d\} = E \quad \Gamma\{c, d\} = \{c, d\} \quad \Gamma\{a, b, c\} = E \quad \Gamma\{a, b, d\} = E$$
$$\Gamma\{a, c, d\} = \{a, c, d\} \quad \Gamma\{b, c, d\} = E \quad \Gamma E = E$$

If the corresponding calculations are performed δ is also obtained.

$$\delta\varnothing = \left(\overline{\Gamma\overline{\varnothing}}\right) = (\overline{\Gamma E}) = \overline{E} = \varnothing$$
$$\delta\{a\} = \left(\overline{\Gamma\ \overline{\{a\}}}\right) = \overline{\Gamma\{b, c, d\}} = \overline{E} = \varnothing$$
$$\delta\{b\} = \left(\overline{\Gamma\ \overline{\{b\}}}\right) = \overline{\Gamma\{a, c, d\}} = \overline{\{a, c, d\}} = \{b\}$$
$$\delta\{c\} = \left(\overline{\Gamma\ \overline{\{c\}}}\right) = \overline{\Gamma\{a, b, d\}} = \overline{E} = \varnothing$$

....

$$\delta\{a, c, d\} = \left(\overline{\Gamma\overline{\{a, c, d\}}}\right) = \overline{\Gamma\{b\}} = \overline{\{a, b\}} = \{c, d\}$$
$$\delta\{b, c, d\} = \left(\overline{\Gamma\overline{\{b, c, d\}}}\right) = \overline{\Gamma\{a\}} = \overline{\{a, d\}} = \{b, c\}$$
$$\delta E = \left(\overline{\Gamma\overline{E}}\right) = \overline{\Gamma\varnothing} = \overline{\varnothing} = E$$

The figures on the following page represent $\mathbf{\Gamma} A_j$ and δA_j.

In the field of finance in which we develop our example, the **relationships** of products and groups of financial products which form the isotone pretopology are of such a nature that when a group is made up of some products which form a part of a second group, the group of products related to the first also forms a part of the group of financial products related to the second.

Let us confirm if the isotonia is verified:

[2] Kaufmann, A.: «Prétopologie ordinaire et prétopologie floue». *Note de Travail 115*. La Tronche, 1983, page. 10.

$\{a\} \subset \{a, b\}$	$\Gamma\{a\} = \{a, d\} \subset \Gamma\{a, b\} = \{a, b, d\}$
$\{a\} \subset \{a, c\}$	$\Gamma\{a\} = \{a, d\} \subset \Gamma\{a, c\} = \{a, c, d\}$
$\{a\} \subset \{a, d\}$	$\Gamma\{a\} = \{a, d\} \subset \Gamma\{a, d\} = \{a, c, d\}$
$\{a\} \subset \{a, b, c\}$	$\Gamma\{a\} = \{a, d\} \subset \Gamma\{a, b, c\} = E$
$\{a\} \subset \{a, b, d\}$	$\Gamma\{a\} = \{a, d\} \subset \Gamma\{a, b, d\} = E$
$\{a\} \subset \{a, c, d\}$	$\Gamma\{a\} = \{a, d\} \subset \Gamma\{a, c, d\} = \{a, c, d\}$
$\{b\} \subset \{a, b\}$	$\Gamma\{b\} = \{a, b\} \subset \Gamma\{a, b\} = \{a, b, d\}$
$\{b\} \subset \{b, c\}$	$\Gamma\{b\} = \{a, b\} \subset \Gamma\{b, c\} = E$
$\{b\} \subset \{b, d\}$	$\Gamma\{b\} = \{a, b\} \subset \Gamma\{b, d\} = E$
$\{b\} \subset \{a, b, c\}$,	$\Gamma\{b\} = \{a, b\} \subset \Gamma\{a, b, c\} = E$
$\{b\} \subset \{a, b, d\}$,	$\Gamma\{b\} = \{a, b\} \subset \Gamma\{a, b, d\} = E$
$\{b\} \subset \{b, c, d\}$,	$\Gamma\{b\} = \{a, b\} \subset \Gamma\{b, c, d\} = E$
$\{c\} \subset \{a, c\}$	$\Gamma\{c\} = \{c, d\} \subset \Gamma\{a, c\} = \{a, c, d\}$
$\{c\} \subset \{b, c\}$	$\Gamma\{c\} = \{c, d\} \subset \Gamma\{b, c\} = E$
$\{c\} \subset \{c, d\}$	$\Gamma\{c\} = \{c, d\} \subset \Gamma\{c, d\} = \{c, d\}$
$\{c\} \subset \{a, b, c\}$,	$\Gamma\{c\} = \{c, d\} \subset \Gamma\{a, b, c\} = E$
$\{c\} \subset \{a, c, d\}$,	$\Gamma\{c\} = \{c, d\} \subset \Gamma\{a, c, d\} = \{a, c, d\}$
$\{c\} \subset \{b, c, d\}$,	$\Gamma\{c\} = \{c, d\} \subset \Gamma\{b, c, d\} = E$
$\{d\} \subset \{a, d\}$	$\Gamma\{d\} = \{c, d\} \subset \Gamma\{a, d\} = \{a, c, d\}$
$\{d\} \subset \{b, d\}$	$\Gamma\{d\} = \{c, d\} \subset \Gamma\{b, d\} = E$
$\{d\} \subset \{c, d\}$	$\Gamma\{d\} = \{c, d\} \subset \Gamma\{c, d\} = \{c, d\}$
$\{d\} \subset \{a, b, d\}$,	$\Gamma\{d\} = \{c, d\} \subset \Gamma\{a, b, d\} = E$
$\{d\} \subset \{a, c, d\}$,	$\Gamma\{d\} = \{c, d\} \subset \Gamma\{a, c, d\} = \{a, c, d\}$
$\{d\} \subset \{b, c, d\}$,	$\Gamma\{d\} = \{c, d\} \subset \Gamma\{b, c, d\} = E$
$\{a, b\} \subset \{a, b, c\}$	$\Gamma\{a, b\} = \{a, b, d\} \subset \Gamma\{a, b, c\} = E$
$\{a, b\} \subset \{a, b, d\}$	$\Gamma\{a, b\} = \{a, b, d\} \subset \Gamma\{a, b, d\} = E$
$\{a, c\} \subset \{a, b, c\}$	$\Gamma\{a, c\} = \{a, c, d\} \subset \Gamma\{a, b, c\} = E$
$\{a, c\} \subset \{a, c, d\}$	$\Gamma\{a, c\} = \{a, c, d\} \subset \Gamma\{a, c, d\} = \{a, c, d\}$
$\{a, d\} \subset \{a, b, d\}$	$\Gamma\{a, d\} = \{a, c, d\} \subset \Gamma\{a, b, d\} = E$
$\{a, d\} \subset \{a, c, d\}$	$\Gamma\{a, d\} = \{a, c, d\} \subset \Gamma\{a, c, d\} = \{a, c, d\}$
$\{b, c\} \subset \{a, b, c\}$	$\Gamma\{b, c\} = E \subset \Gamma\{a, b, c\} = E$
$\{b, c\} \subset \{b, c, d\}$	$\Gamma\{b, c\} = E \subset \Gamma\{b, c, d\} = E$
$\{b, d\} \subset \{a, b, d\}$	$\Gamma\{b, d\} = E \subset \Gamma\{a, b, d\} = E$
$\{b, d\} \subset \{b, c, d\}$	$\Gamma\{b, d\} = E \subset \Gamma\{b, c, d\} = E$
$\{c, d\} \subset \{a, c, d\}$	$\Gamma\{c, d\} = \{c, d\} \subset \Gamma\{a, c, d\} = \{a, c, d\}$
$\{c, d\} \subset \{b, c, d\}$	$\Gamma\{c, d\} = \{c, d\} \subset \Gamma\{b, c, d\} = E$

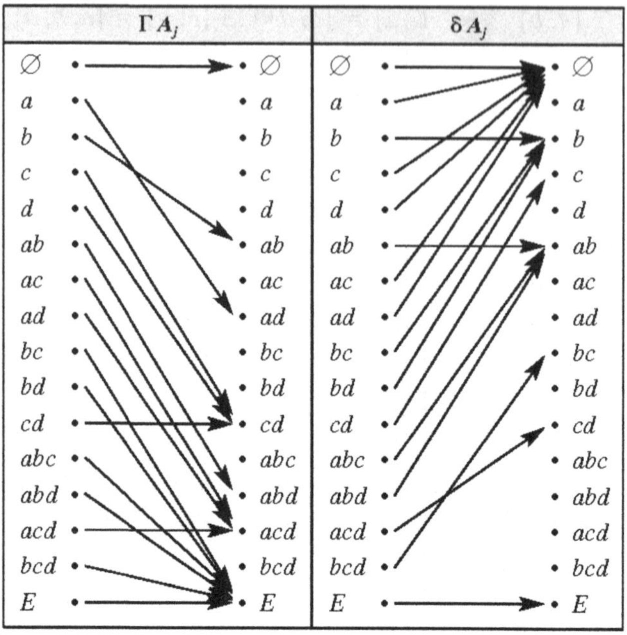

There is no doubt whatsoever that we find ourselves before an isotone pretopology. A simple glance at the previous figures relative to **Γ** and δ illustrates this isotonia. In continuation we will move on to expound another type of pretopological space, distributive pretopology.

3 Distributive Pretopology

A pretopological space $(E,\mathbf{\Gamma})$ such as:

$$\forall\, A_j, A_k \in P(E):$$

$$\Gamma(A_j \cup A_k) \;=\; \Gamma A_j \cup \Gamma A_k$$

is called distributive. This definition is equivalent to the following, taking into account the well-known De Morgan's theorems:

$$\delta(A_j \cap A_k) \;=\; \delta A_j \cap \delta A_k$$

Let us look at an example[3] in a referential of financial products:

$$E = \{a, b, c, d\}$$

[3] The duly adapted examples in this epigraph have been taken from the theorems contained in the work of Kaufmann, A.: «Prétopologie ordinaire et prétopologie floue». Note de travail 115. La Tronche, 1983, pages. 27-32.

The following application is considered functional:

$\Gamma\varnothing = \varnothing \quad \Gamma\{a\} = \{a\} \quad \Gamma\{b\} = \{a, b\} \quad \Gamma\{c\} = \{c, d\} \quad \Gamma\{d\} = \{d\}$
$\Gamma\{a, b\} = \{a, b\} \quad \Gamma\{a, c\} = \{a, c, d\} \quad \Gamma\{a, d\} = \{a, d\}$
$\Gamma\{b, c\} = E \quad \Gamma\{b, d\} = \{a, b, d\} \quad \Gamma\{c, d\} = \{c, d\} \quad \Gamma\{a, b, c\} = E$
$\Gamma\{a, b, d\} = \{a, b, d\} \quad \Gamma\{a, c, d\} = \{a, c, d\} \quad \Gamma\{b, c, d\} = E \quad \Gamma E = E$

And therefore, also:

$\delta\varnothing = \varnothing \quad \delta\{a\} = \varnothing \quad \delta\{b\} = \{b\} \quad \delta\{c\} = \{c\} \quad \delta\{d\} = \varnothing$
$\delta\{a, b\} = \{a, b\} \quad \delta\{a, c\} = \{c\} \quad \delta\{a, d\} = \varnothing \quad \delta\{b, c\} = \{b, c\}$
$\delta\{b, d\} = \{b\} \quad \delta\{c, d\} = \{c, d\} \quad \delta\{a, b, c\} = \{a, b, c\} \quad \delta\{a, b, d\} = \{a, b\}$
$\delta\{a, c, d\} = \{c, d\} \quad \delta\{b, c, d\} = \{b, c, d\} \quad \delta E = E$

The following figures represent, respectively, the adherence **Γ** and the interior δ of the proposed distributive pretopology.

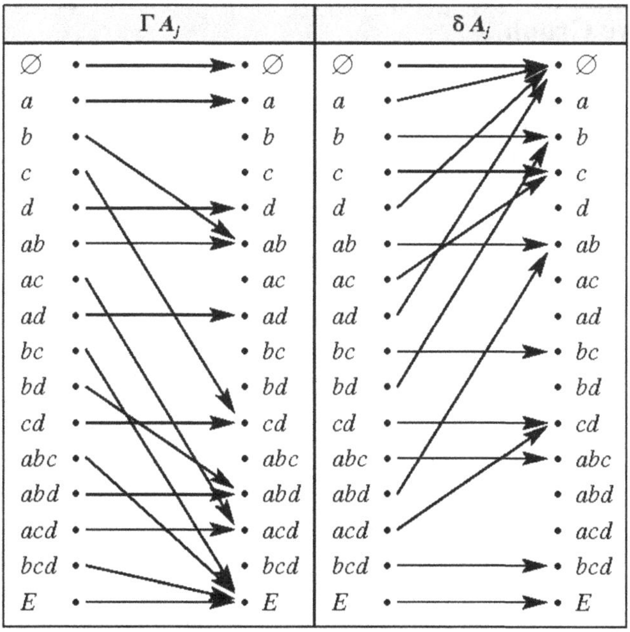

It can be confirmed that the established axiom is always fulfilled. In effect:

$\Gamma\{a, b\} = \{a, b\} \quad \Gamma\{a\} \cup \Gamma\{b\} = \{a\} \cup \{a, b\} = \{a, b\}$
$\Gamma\{a, c\} = \{a, c, d\} \quad \Gamma\{a\} \cup \Gamma\{c\} = \{a\} \cup \{c, d\} = \{a, c, d\}$

....

$\Gamma\{b, c, d\} = E \quad \Gamma\{b\} \cup \Gamma\{c\} \cup \Gamma\{d\} = \{a, b\} \cup \{c, d\} \cup \{d\} = E$
$\Gamma E = E \quad \Gamma\{a\} \cup \Gamma\{b\} \cup \Gamma\{c\} \cup \Gamma\{d\} = \{a\} \cup \{a, b\} \cup \{c, d\} \cup \{d\} = E$

These relationships between products and groupings of financial products assume the property of joining any two groups, the **grouped products** are found related with the same products with which each product of the **individually considered** groupings is found.

Let us look at an important property: A distributive pretopological space (E,Γ) is isotone (the reciprocal is not certain). That is to say:

$$\forall A_j, A_k \in P(E):$$

$$(\Gamma(A_j \cup A_k) = \Gamma A_j \cup \Gamma A_k) \Rightarrow ((A_j \subset A_k) \Rightarrow (\Gamma A_j \subset \Gamma A_k))$$

The demonstration is simple:

Let $A_j \cup \Gamma A_k$. If we assume $A_j \subset A_k$ then $A_j \subset A_k = A_j \cup A_k = A_k$ and $\Gamma(A_j \cup A_k) = \Gamma A_k$. It can also be written that $\Gamma A_j \cup \Gamma A_k = \Gamma A_k$, therefore, $\Gamma A_j \subset \Gamma A_k$.

4 Correspondence between Distributive Pretopology and Reflexive Graph

Since many years ago we have become accustomed to representing the relationships which are produced in economics and management systems through the **concept of graphs**, be this expressed in matrix or sagittate form. The possibilities which properties of **pretopological spaces** offer allow the widening of the field of treatment of complex problems with the help of **graph theory**. This is why it is interesting to explore the relationship between these elements of non-numerical mathematics.

When we have presented some of the topological approaches which have come to us throughout the years, it has been possible to observe, for example, the use of graph theory for the suitable focus for the problem of the bridges of Königsberg, amongst others.

It will therefore not be a surprise that we dedicate special attention to those elements which may serve to lighten the load of the inherent abstraction of pretopological spaces.

As it is known, a **finite reflexive graph** (E,Γ) complies with the previous property, $\Gamma(A_j \cup A_k) = \Gamma A_j \cup \Gamma A_k$. For this reason it may be represented by a distributive pretopology and reciprocally. Therefore, a biunivocal correspondence[4] exists between one and the other. Here we have an aspect with a significant value for a greater comprehension and explanation of economics and management phenomena.

Perhaps an example may help to see the light in this area which is so interesting. For this we will begin with a distributive pretopology of the referential:

$$E = \{a, b, c, d, e\}$$

[4] Notice that, in general, the same notation (E,Γ) is used for graphs in the Berge sense as for pretopological spaces.

given by the following **adherence**:

$\Gamma\varnothing=\varnothing$	$\Gamma\{a\}=\{a\}$	$\Gamma\{b\}=\{b,c,d,e\}$	$\Gamma\{c\}=\{b,c,d\}$
$\Gamma\{d\}=\{b,d\}$	$\Gamma\{e\}=E$	$\Gamma\{a,b\}=E$	$\Gamma\{a,c\}=\{a,b,c,d,\}$
$\Gamma\{a,d\}=\{a,b,d\}$	$\Gamma\{a,e\}=E$	$\Gamma\{b,c\}=\{b,c,d,e\}$	$\Gamma\{b,d\}=\{b,c,d,e\}$
$\Gamma\{b,e\}=E$	$\Gamma\{c,d\}=\{b,c,d\}$	$\Gamma\{c,e\}=E$	$\Gamma\{d,e\}=E$
$\Gamma\{a,b,c\}=E$	$\Gamma\{a,b,d\}=E$	$\Gamma\{a,b,e\}=E$	$\Gamma\{a,c,d\}=\{a,b,c,d\}$
$\Gamma\{a,c,e\}=E$	$\Gamma\{a,d,e\}=E$	$\Gamma\{b,c,d\}=\{b,c,d,e\}$	$\Gamma\{b,c,e\}=E$
$\Gamma\{b,d,e\}=E$	$\Gamma\{c,d,e\}=E$	$\Gamma\{a,b,c,d\}=E$	$\Gamma\{a,b,c,e\}=E$
$\Gamma\{a,b,d,e\}=E$	$\Gamma\{a,c,d,e\}=E$	$\Gamma\{b,c,d,e\}=E$	$\Gamma E=E$

We will also provide the **interior** application:

$\delta\varnothing=\varnothing$ $\delta\{a\}=\varnothing$ $\delta\{b\}=\varnothing$ $\delta\{c\}=\varnothing$ $\delta\{d\}=\varnothing$ $\delta\{e\}=\varnothing$
$\delta\{a,b\}=\varnothing$ $\delta\{a,c\}=\varnothing$ $\delta\{a,d\}=\varnothing$ $\delta\{a,e\}=\{a\}$ $\delta\{b,c\}=\varnothing$
$\delta\{b,d\}=\varnothing$ $\delta\{b,e\}=\{e\}$ $\delta\{c,d\}=\varnothing$ $\delta\{c,e\}=\varnothing$ $\delta\{d,e\}=\varnothing$
$\delta\{a,b,c\}=\varnothing$ $\delta\{a,b,d\}=\varnothing$ $\delta\{a,b,e\}=\{a,e\}$ $\delta\{a,c,d\}=\varnothing$
$\delta\{a,c,e\}=\{a\}$ $\delta\{a,d,e\}=\{a\}$ $\delta\{b,c,d\}=\varnothing$ $\delta\{b,c,e\}=\{c,e\}$
$\delta\{b,d,e\}=\{e\}$ $\delta\{c,d,e\}=\varnothing$ $\delta\{a,b,c,d\}=\varnothing$ $\delta\{a,b,c,e\}=\{a,c,e\}$
$\delta\{a,b,d,e\}=\{a,e\}$ $\delta\{a,c,d,e\}=\{a\}$ $\delta\{b,c,d,e\}=\{b,c,d,e\}$ $\delta E=E$

The only opens in this distributive pretopology are $\varnothing$, $\{b, c, d, e\}$ and E and the only closes are $\varnothing$, $\{a\}$ and E.

We believe it to be interesting to state the cited existing connection between distributive pretopology and reflexive graph[5]. If the adherent application $\boldsymbol{\Gamma}$ is considered as a **distributive pretopology**, by hypothesis one has $\boldsymbol{\Gamma}(A_j \cup A_k)= \boldsymbol{\Gamma} A_j \cup \boldsymbol{\Gamma} A_k$ and also $\delta(A_j \cap A_k)= \delta A_j \cap \boldsymbol{\Gamma} A_k$ which is also a property that is given in the **reflexive graphs**.

It is easy to find the adherence from a graph. For this we will begin from the **sagittate reflexive graph** which we have produced following the previous **adherence**:

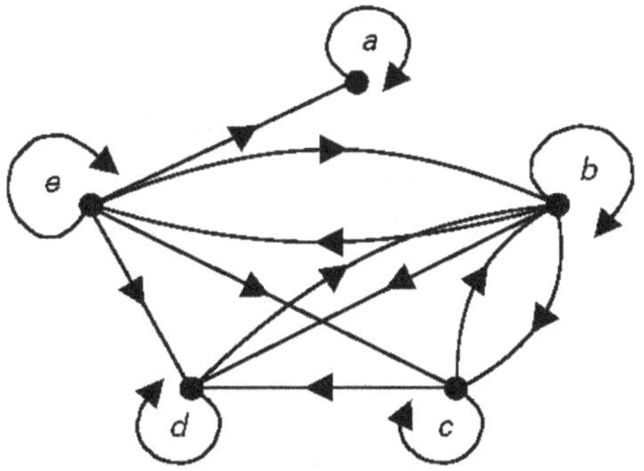

[5] We refer to the graph in the Berge sense, which is to say a functional application of E in E. Certain mathematicians consider graphs with application (functional or not) of E1 over E2, in which E1 and E2 may be different.

If we place the set of the parts $P(E)$ in rows and the elements of the referential E as columns the following matrix is found, which shows the corresponding distributive pretopology:

	a	*b*	*c*	*d*	*e*
a	1				
b		1	1	1	1
c		1	1	1	
d		1		1	
e	1	1	1	1	1
ab	1	1	1	1	1
ac	1	1	1	1	
ad	1	1		1	
ae	1	1	1	1	1
bc		1	1	1	1
bd		1	1	1	1
be	1	1	1	1	1
cd		1	1	1	
ce	1	1	1	1	1
de	1	1	1	1	1
abc	1	1	1	1	1
abd	1	1	1	1	1
abe	1	1	1	1	1
acd	1	1	1	1	
ace	1	1	1	1	1
ade	1	1	1	1	1
bcd		1	1	1	1
bce	1	1	1	1	1
bde	1	1	1	1	1
cde	1	1	1	1	1
abcd	1	1	1	1	1
abce	1	1	1	1	1
abde	1	1	1	1	1
acde	1	1	1	1	1
bcde	1	1	1	1	1
E	1	1	1	1	1

We know the importance of reflexive graphs for the solution of diverse economics and management problems only too well, and also their decisive importance for the obtaining of maximum sub-relationships of similarity[6] and forgotten effects[7], amongst others. All of this has been possible thanks to **distributive pretopologies**.

Let us take a look at yet more properties.

In a **distributive pretopological space** each finite intersection of opens is an open.

$\forall A_j, A_k \in P(E)$:

$$(A_j = \delta A_j \text{ y } A_k = \delta A_k) \Rightarrow (A_j \cap A_k = \delta (A_j \cap A_k))$$

Which results as immediate, given that $A_j \cap A_k = \boldsymbol{\delta} A_j \cap \boldsymbol{\delta} A_k = \boldsymbol{\delta} (A_j \cap A_k)$ with the pretopology being distributive.

We evidently have the dual property:

$\forall A_j, A_k \in P(E)$:

$$(A_j = \Gamma A_j \text{ y } A_k = \Gamma A_k) \Rightarrow (A_j \cup A_k = \Gamma (A_j \cup A_k))$$

We will now move on to perform an important observation.

Two distributive pretopologies in E may have the same opens (respectively closes) **without having to be identical**. In effect, if we begin from the referential:

$$E = \{a, b, c\}$$

and we have the following distributive pretopological space $(E,\boldsymbol{\Gamma}_1)$:

$\Gamma_1 \varnothing = \varnothing$	$\Gamma_1 \{a\} = \{a, b\}$	$\Gamma_1 \{b\} = \{b, c\}$	$\Gamma_1 \{c\} = \{c\}$
$\Gamma_1 \{a, b\} = E$	$\Gamma_1 \{a, c\} = E$	$\Gamma_1 \{b, c\} = \{b, c\}$	$\Gamma_1 E = E$

and another distributive pretopological space $(E,\boldsymbol{\Gamma}_2)$:

$\Gamma_2 \varnothing = \varnothing$	$\Gamma_2 \{a\} = \{a, c\}$	$\Gamma_2 \{b\} = \{b, c\}$	$\Gamma_2 \{c\} = \{c\}$
$\Gamma_2 \{a, b\} = E$	$\Gamma_2 \{a, c\} = E$	$\Gamma_2 \{b, c\} = \{b, c\}$	$\Gamma_2 E = E$

where $(\boldsymbol{\Gamma}_1 \{a\} = \{a, b\}) \neq (\boldsymbol{\Gamma}_2 \{a\} = \{a, c\})$.

If this is crossed with δ, we obtain:

$\delta_1 \varnothing = \varnothing$	$\delta_1 \{a\} = \{a\}$	$\delta_1 \{b\} = \varnothing$	$\delta_1 \{c\} = \varnothing$
$\delta_1 \{a, b\} = \{a, b\}$	$\delta_1 \{a, c\} = \{a\}$	$\delta_1 \{b, c\} = \{c\}$	$\delta_1 E = E$

where $(\delta_1 \{b, c\} = \{c\}) \neq (\delta_2 \{b, c\} = \{b\})$.

---

[6] See, for example, the supposed differentiation of financial products in Gil Lafuente, A. M.: *Nuevas Estrategias para el análisis financiero en la empresa*. Ed. Ariel, 2001, pages. 339-344.

[7] Gil Lafuente, A. M.: *Nuevas Estrategias para el análisis financiero en la empresa*. Ed. Ariel, 2001, pages. 452-469.

These two distributive pretopologies are different, $(E,\mathbf{\Gamma}_1) \neq (E,\mathbf{\Gamma}_2)$, but on the other hand they have the same closes:

$$\{\varnothing, \{c\}, \{b, c\}, E\}$$

But they are also different with δ, even though they have the same opens:

$$\{\varnothing, \{a\}, \{a, b\}, E\}$$

In the most general way it would be said that the set of the opens (respectively of the closes) **does not specify** a single pretopology.

This property can have a special importance if it is taken into account that, continuing with our example, even though the circumstance that the same groups are found related with each other is produced in two sets or groups of financial products, this does not mean that the relationships between all the groups of one will be the same as those of the other.

5 Comparison of Pretopologies

On certain occasions it can be useful or necessary to perform a certain **comparison** between two or more pretopological spaces, above all taking into account the growing interest to establish **precedences** when there are different groups of relationships between financial products or distinct structural relationships correspondent to economics or management systems with plural elements. The following considerations may result as being of special interest in this way.

We will consider pretopologies, be they isotone or not. Let us assume two pretopologies $\mathbf{\Gamma}_1$ and $\mathbf{\Gamma}_2$ in E. It is said that $\mathbf{\Gamma}_1$ is **finer** than $\mathbf{\Gamma}_2$ if and only if:

$\forall A_j \in P(E)$:

$$\Gamma_1 A_j \subset \Gamma_2 A_j$$

or that which results as equivalent:

$\forall A_j \in P(E)$:

$$\delta_1 A_j \supset \delta_2 A_j$$

The relationship " $\mathbf{\Gamma}_1$ finer than $\mathbf{\Gamma}_2$ " is a relationship of partial order in the set of pretopologies in E. As is custom, ≤ is written:

$$(\Gamma_1 \preccurlyeq \Gamma_2) \Rightarrow (\Gamma_1 A_j \subset \Gamma_2 A_j) \Rightarrow (\delta_1 A_j \supset \delta_2 A_j)$$

If $\Gamma_1 \leq \Gamma_2$ it is also said that " Γ_2 is thicker than Γ_1 ". Let us look at an example[8].

A	$\varnothing$	$\{a\}$	$\{b\}$	$\{c\}$	$\{a, b\}$	$\{a, c\}$	$\{b, c\}$	E
$\Gamma_1 A$	$\varnothing$	$\{a, b\}$	E	$\{c\}$	$\{a, b\}$	E	E	E
$\Gamma_2 A$	$\varnothing$	E	E	$\{b, c\}$	$\{a, b\}$	E	E	E
$\delta_1 A$	$\varnothing$	$\varnothing$	$\varnothing$	$\{c\}$	$\{a, b\}$	$\varnothing$	$\{c\}$	E
$\delta_2 A$	$\varnothing$	$\varnothing$	$\varnothing$	$\{c\}$	$\{a\}$	$\varnothing$	$\varnothing$	E

It is easily verified that " Γ_2 is thicker than Γ_1 ". It will be written that:

$$(\Gamma_1 \preccurlyeq \Gamma_2) \Leftrightarrow (\delta_1 \succcurlyeq \delta_2)$$

The "finer" relationship of partial order (respectively, "thicker") can be strict if:

$$\forall A_j \in P(E) \qquad A_j \neq \varnothing \qquad A_j \neq E:$$

$$\Gamma_1 A_j \subset\subset \Gamma_2 A_j$$

or:

$$\delta_1 A_j \supset\supset \delta_2 A_j$$

It is said that " Γ_1 is strictly finer than Γ_2 ".

In our supposition, to which we have repeatedly turned, it is said that in $\Gamma_2 A_j$ a same group of financial products is found related to another group with a greater number of units than in $\Gamma_1 A_j$.

Let us move on to another comparative property.

Beginning with Γ_1 and Γ_2 , two pretopologies in E. Let $A_b^{(1)}$ be the open set of Γ_1 and $A_b^{(2)}$ the set of the opens of Γ_2 . Therefore:

$$(\forall A_j \in P(E): \quad \Gamma_1 A_j \subset \Gamma_2 A_j) \Rightarrow (A_b^{(2)} \subset A_b^{(1)})$$

To demonstrate this we move to the interior applications δ_1 and δ_2. Let us assume $A_j \subset A_b^{(2)}$. Therefore $\delta_2 A_j = A_j \subset \delta_1 A_j$. From where $\delta_1 A_j = A_j$. Therefore $A_j = A_b^{(1)}$.

Let us look at an example.

$$E = \{a, b, c\}$$

[8] Kaufmann, A.: «Prétopologie ordinaire et prétopologie floue». *Note de Travail 115*. La Tronche, 1983, pages. 33-35.

$\Gamma_1 \varnothing = \varnothing$ $\Gamma_1\{a\} = \{a\}$ $\Gamma_1\{b\} = \{a, b\}$ $\Gamma_1\{c\} = \{a, c\}$
$\Gamma_1\{a, b\} = \{a, b\}$ $\Gamma_1\{a, c\} = E$ $\Gamma_1\{b, c\} = \{b, c\}$ $\Gamma_1 E = E$

$\delta_1 \varnothing = \varnothing$ $\delta_1 \{a\} = \{a\}$ $\delta_1 \{b\} = \varnothing$ $\delta_1 \{c\} = \{c\}$
$\delta_1 \{a, b\} = \{b\}$ $\delta_{1.}\{a, c\} = \{c\}$ $\delta_1 \{b, c\} = \{b, c\}$ $\delta_1 E = E$

Where:

$$A_b^{(1)} = \{\varnothing, \{a\}, \{c\}, \{b, c\}, E\}$$

$\Gamma_2 \varnothing = \varnothing$ $\Gamma_2\{a\} = \{a, b\}$ $\Gamma_2\{b\} = \{a, b\}$ $\Gamma_2\{c\} = \{a, c\}$
$\Gamma_2 \{a, b\} = E$ $\Gamma_2 \{a, c\} = E$ $\Gamma_2\{b, c\} = \{b, c\}$ $\Gamma_2 E = E$

$\delta_2 \varnothing = \varnothing$ $\delta_2 \{a\} = \{a\}$ $\delta_2 \{b\} = \varnothing$ $\delta_2 \{c\} = \varnothing$
$\delta_2\{a, b\} = \{b\}$ $\delta_{2.}\{a, c\} = \{c\}$ $\delta_2 \{b, c\} = \{c\}$ $\delta_2 E = E$

From which:

$$A_b^{(2)} = \{\varnothing, \{a\}, E\}$$

It can be seen that $A_b{}^{(2)} \subset A_b{}^{(1)}$, as $\forall A_j \in P(E) : \boldsymbol{\Gamma_1} A_j \subset \boldsymbol{\Gamma_2} A_j$.

It must be taken into account that:

$$A_b^{(2)} \subset A_b^{(1)} \not\Rightarrow (\forall A_j \in P(E): \Gamma_1 A_j \subset \Gamma_2 A_j)$$

We do not forget that the open set, also respectively to the closed, only takes in those groupings which are found related with themselves. Therefore, when the open set is a relationship between products or groups of financial products it is a constituent part of the open set of another relationship between products or groups of financial products, these last relationships do not always comprise those of the first group.

Also, even though $A_b{}^{(2)} \subset A_b{}^{(1)}$, one of the two pretopologies may well be strictly finer than the other.

In our efforts to give financial meaning to the representation through an example, by using a pretopology of the relationships between groups of financial products, we voluntarily limited the sense of the functional **application** to that of a simple **relationship**. It is evident that the possibilities of formalising phenomena are much wider. Many more aspects of the economic and financial life of businesses and institutions can be object to processing. Due to the necessary brevity, here we find ourselves obliged to close this interesting subject, to move on to another which presents itself to us no less suggestively: the construction of lattices of pretopologies.

6 Lattices of Pretopologies

The possibility of comparing related economic or management structures represented by means of pretopologies suggests the construction of **pretopology lattices**.

For this let us begin with an example. Let us assume a family of financial relationships, represented through pretopologies in E, defined by $\boldsymbol{\Gamma}_p$, $p = 1, 2..., m$. It is assumed that $(\boldsymbol{\Gamma}_p, \boldsymbol{\Gamma}_q)$, $p, q = 1, 2..., m$ has a lower extreme, which is to say a relationship between groupings of financial products, and an upper extreme, and therefore another relationship between groupings of financial products, for the relation of order $\leq$ defined in the previous epigraph, " $\boldsymbol{\Gamma}_1$ is finer than $\boldsymbol{\Gamma}_2$ ".

The upper extreme is represented by $\boldsymbol{\Gamma}_p$ and the lower extreme by $\boldsymbol{\Gamma}_q$ (the inverse representation may be taken). This family has the configuration of a lattice, if the upper and lower extremes are understood.

It is evident that a $\boldsymbol{\Gamma}_{pq}$ such that $\boldsymbol{\Gamma}_{pq} A_j = \boldsymbol{\Gamma}_p A_j \cap \boldsymbol{\Gamma}_q A_j$, $p, q = 1, 2,.... m$ is finer than $\boldsymbol{\Gamma}_p$ and $\boldsymbol{\Gamma}_q$.

Let us assume that $\boldsymbol{\Gamma}_r$ is a finer pretopology than $\boldsymbol{\Gamma}_p$ and $\boldsymbol{\Gamma}_q$. Therefore $\boldsymbol{\Gamma}_r A_j \subset \boldsymbol{\Gamma}_p A_j$ and $\boldsymbol{\Gamma}_r A_j \subset \boldsymbol{\Gamma}_q A_j$, where $\boldsymbol{\Gamma}_r A_j \subset \boldsymbol{\Gamma}_{pq} A_j$. Therefore $\boldsymbol{\Gamma}_r$ is finer than $\boldsymbol{\Gamma}_{pq}$.

We will provide an illustrated example by means of figures.

Let us assume three pretopologies which are given by $\boldsymbol{\Gamma}_1$, $\boldsymbol{\Gamma}_2$ and $\boldsymbol{\Gamma}_3$. We do not forget that in our example each pretopology shows a set of relationships amongst groups of financial products which fulfil certain requisites. In the following figure we represent $\boldsymbol{\Gamma}_1$, $\boldsymbol{\Gamma}_2$, $\boldsymbol{\Gamma}_3$, $\boldsymbol{\Gamma}_{1\cap 2}$,$\boldsymbol{\Gamma}_{1\cap 3}$,$\boldsymbol{\Gamma}_{2\cap 3}$,$\boldsymbol{\Gamma}_{1\cup 2}$,$\boldsymbol{\Gamma}_{1\cup 3}$,$\boldsymbol{\Gamma}_{2\cup 3}$,$\boldsymbol{\Gamma}_{1\cap 2\cap 3}$ and $\boldsymbol{\Gamma}_{1\cup 2\cup 3}$. It can be observed that it deals with pretopologies which are not all isotone. Therefore in $\boldsymbol{\Gamma}_3$ one has:

$$A_j = \{a\} \qquad \boldsymbol{\Gamma}_3\{a\} = E$$
$$A_k = \{a, b\} \qquad \boldsymbol{\Gamma}_3\{a, b\} = \{a, b\}$$

with $\boldsymbol{\Gamma}_3\{a\} \not\subset \boldsymbol{\Gamma}_3\{a, b\}$ the non-fulfillment of the corresponding axioms can be observed,

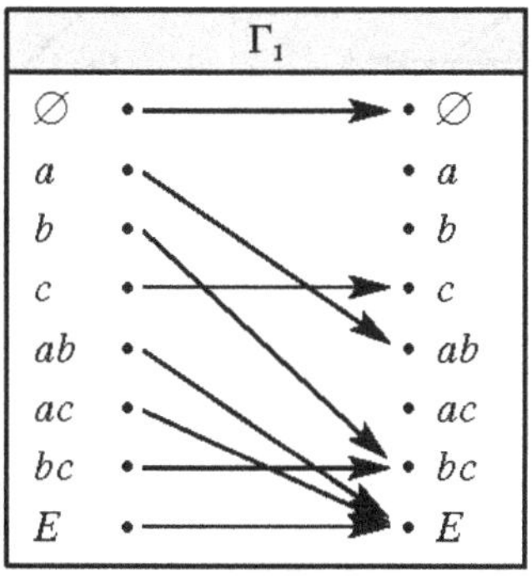

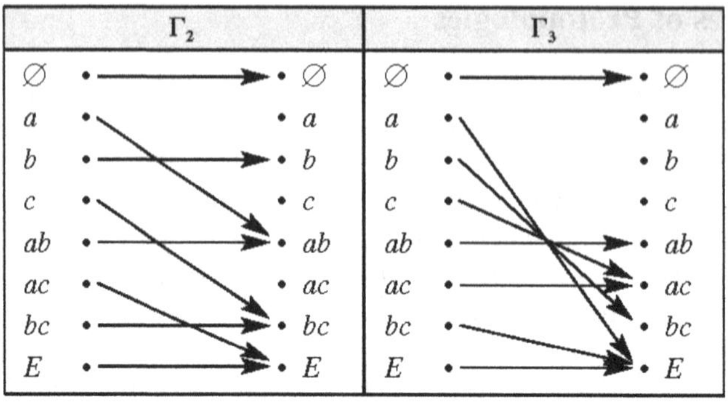
Γ2
Γ3
∅ a b c ab ac bc E
∅ a b c ab ac bc E

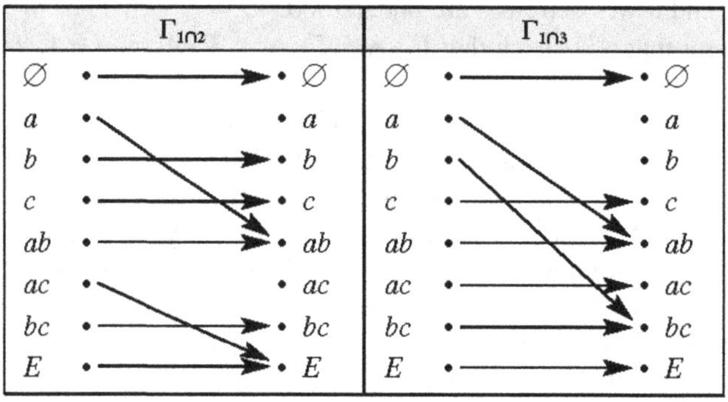
Γ1∩2
Γ1∩3
∅ a b c ab ac bc E
∅ a b c ab ac bc E

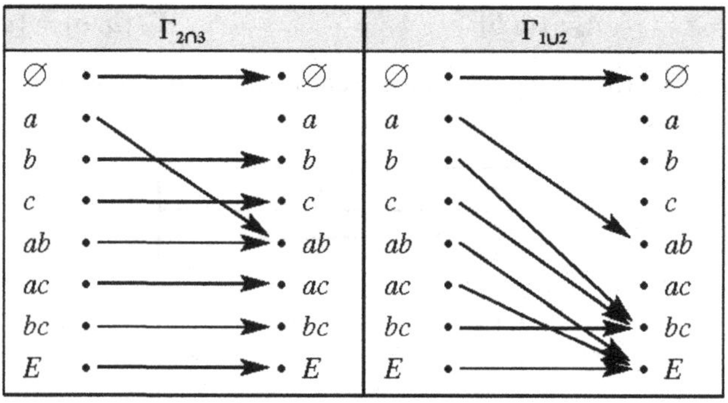
Γ2∩3
Γ1∪2
∅ a b c ab ac bc E
∅ a b c ab ac bc E

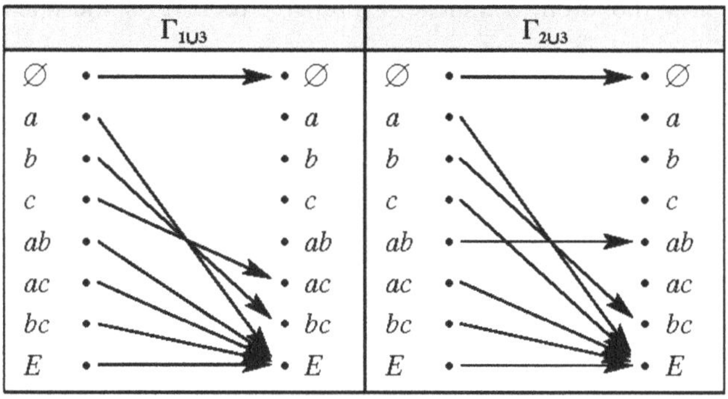

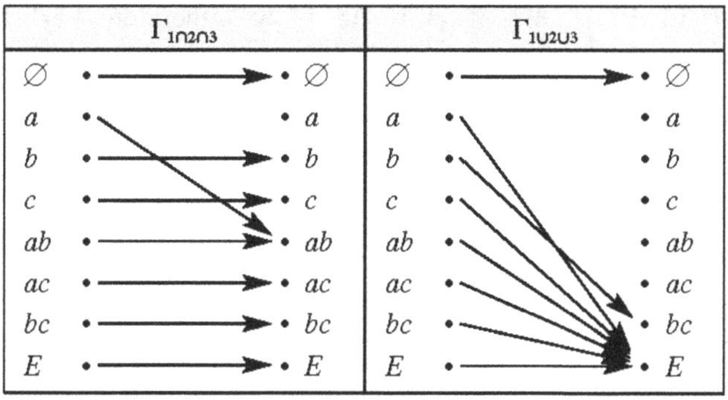

In the following we present the lattice corresponding to the previous pretopologies.

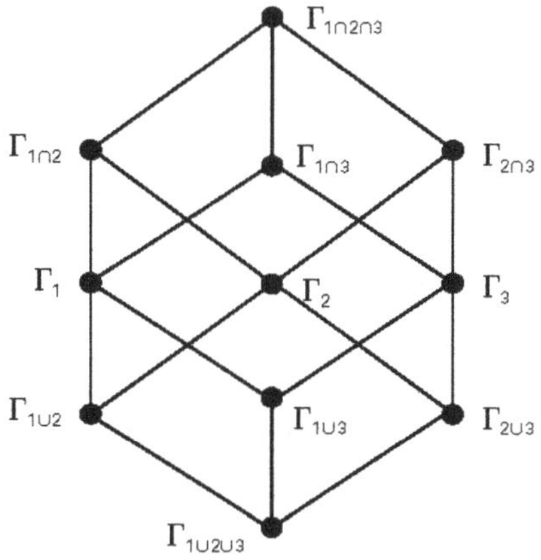

It is usual to name this lattice: **lattice with *n* generators**, in our case, $n = 3$.

We will now move on to indicate a property relative to the opens of isotone pretopologies.

Let $A_b^{(p)}$ be the set of **opens** of the pretopology $\boldsymbol{\Gamma_p}$ in E and $A_b^{(q)}$ the set of **opens** of the pretopology $\boldsymbol{\Gamma_q}$. Therefore the pretopology $\boldsymbol{\Gamma_\cup}$, the lower extreme of $\boldsymbol{\Gamma_p}$ and $\boldsymbol{\Gamma_q}$ has as an open:

$$A_b^{(\cup)} = A_b^{(p)} \cap A_b^{(q)}$$

and respectively the upper extreme has as an open:

$$A_b^{(\cap)} = A_b^{(p)} \cap A_b^{(q)}$$

One has $A_b^{(\cup)} \subset A_b^{(p)} \cap A_b^{(q)}$ as $\Gamma_\cup \preccurlyeq \Gamma_p$ and $\Gamma_\cup \preccurlyeq \Gamma_q$ where $A_b^{(\cup)} \subset A_b^{(p)}$ and $A_b^{(\cup)} \subset A_b^{(q)}$.

Also, $A_b^{(p)} \cap A_b^{(q)} \subset A_b^{(\cup)}$ if $A_j \subset A_b^{(p)} \cap A_b^{(q)}$, therefore $\Gamma_q \overline{A_j} = \overline{A_j}$; and $\Gamma_q \overline{A_j} = \overline{A_j}$; and, in this way $\Gamma_p \overline{A_j} \cup \Gamma_q \overline{A_j} = \overline{A_j}$ which entails that $A_j \subset A_b^{(\cup)}$.

$\Gamma_p \preccurlyeq \Gamma_\cap$ y $\Gamma_q \preccurlyeq \Gamma_\cap$, and therefore $A_b^{(p)} \subset A_b^{(\cap)}$ and $A_b^{(q)} \subset A_b^{(\cap)}$.

We will look at an example begining with the following figure in $\boldsymbol{\Gamma_1}$ and $\boldsymbol{\Gamma_2}$ which have been taken as isotones. The opens have been drawn in bold.

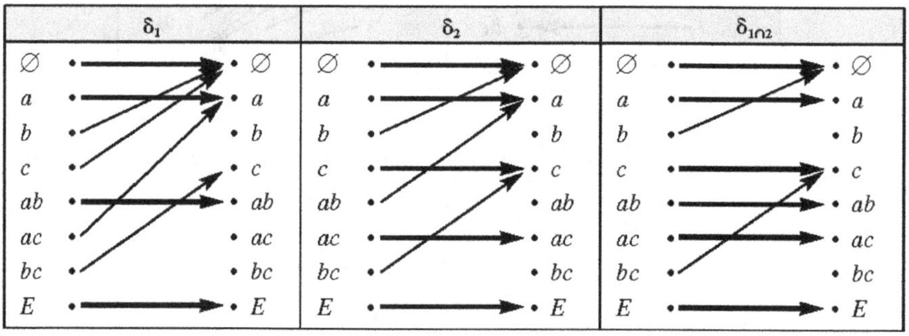

It has become evident that the union of the opens of δ_1 and of δ_2 coincides with the open set of $\delta_{1\cap 2}$. And this **having taken both topologies as isotone**. When this requisite is not fulfilled equality does not always take place. We will now see this through a counterexample with $\boldsymbol{\Gamma_2}$ and $\boldsymbol{\Gamma_3}$ where $\boldsymbol{\Gamma_3}$ **is not isotone**. It may be observed in the following figure that the union of the opens of δ_2 and of δ_3 does not give the open set of $\delta_{2\cap 3}$. Also in $\delta_{2\cap 3}$ the open $\delta\{a, b\} = \{a, b\}$ is obtained, which did not exist in the union of the opens δ_2 and δ_3.

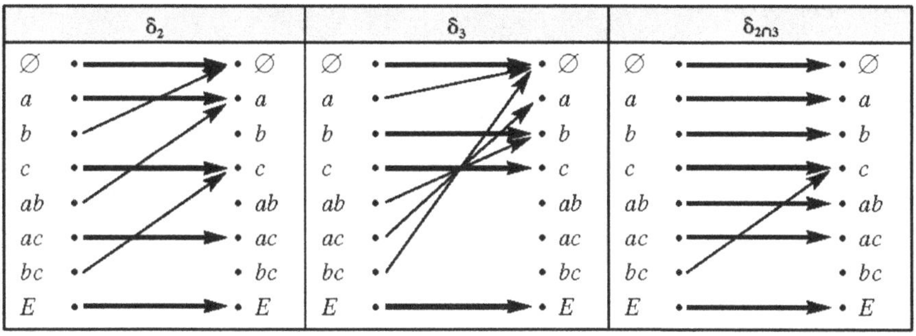

To conclude this epigraph we will advance a general observation relative to **distributive pretopologies**.

A **topology** may be defined by the opens which contain a distributive pretopology. In exchange, a pretopology **is defined by its adherent functional application Γ** (or what becomes the same for its interior functional application δ). As has been mentioned previously, two distributive pretopologies may be **discinct** and have the same open sets, which therefore gives rise to the **same** topology. In this way, except for some particular cases, bijection between the set of distributive pretopologies of *E* and the set of topologies of *E* does not exist.

7 Moore Closing

Let us assume a new finite referential[9] E and the set of its parts P (E). A functional application M of P(E) in P(E), which has the following properties is named Moore closing[10].

1. $\forall A_j \in P(E)$:	$A_j \subset M A_j$	Extensivity
2. $\forall A_j \in P(E)$:	$M(M A_j) = M A_j$	Idempotentcy
3. $\forall A_j, A_k \in P(E)$:		
	$(A_j \subset A_k) \Rightarrow (M A_j \subset M A_k)$	Isotonia

The first of these axioms: $A_j \subset M A_j$ implies that $M E = E$ is fulfilled.

We will look at a simple example, taking into account for greater precision our supposition of financial products, in which the referential is:

$$E = \{a, b, c\}$$

[9] The notion of Moore closing, as in those of pretopology and topology, is valid for all sets whether finite or not, but in the framework of this work we will only consider the finite case.

[10] In Moore closing the notation of the functional application **Γ** is ususally changed for the notation *M*.

and the application M:

$$M\varnothing = \varnothing \qquad M\{a\} = \{a, c\} \qquad M\{b\} = E \qquad M\{c\} = \{a, c\}$$
$$M\{a, b\} = E \qquad M\{a, c\} = \{a, c\} \qquad M\{b, c\} = E \qquad ME = E$$

The following figure is enough to visually show that the previous application constitutes a Moore closing, as the idempotency axiom (as well as the other two) is fulfilled:

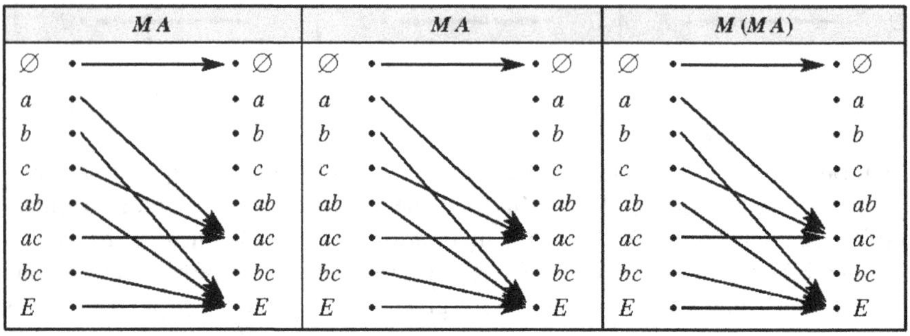

It is important to observe that in the axiomatic of the Moore closing the axiom M ∅ = ∅ **does not appear** as occurs in pretopologies. In the following figure a Moore closing has been represented in which M ∅ = $\{a\}$, and which therefore is **not a pretopology.**

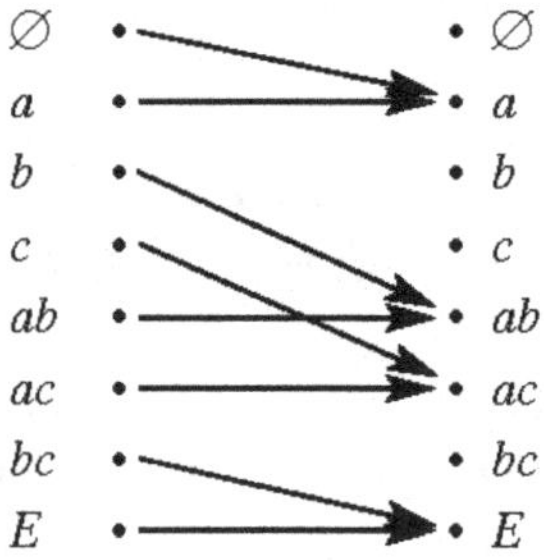

If $\boldsymbol{\Gamma}(\boldsymbol{\Gamma} A_j) = \boldsymbol{\Gamma} A_j$ is demanded in an isotone pretopology it may immediately be signalled that M ∅ = ∅.

$\boldsymbol{\Gamma}(\boldsymbol{\Gamma} A_j) = \boldsymbol{\Gamma} A_j$ can be reached by:

$$\hat{\Gamma} A_j = \Gamma^n A_j = \Gamma^{n+1} A_j$$

with n large enough. Therefore:

$$\hat{\Gamma}(\hat{\Gamma} A_j) = \Gamma^n (\Gamma^n A_j) = \Gamma^{2n} A_j = \Gamma^n A_j = \hat{\Gamma} A_j$$

In this way, the pretopology $\hat{\Gamma}$ which arose from the isotone pretopology $\mathbf{\Gamma}$ is an isotone pretopology which is a Moore closing at the same time.

It is important to know how to obtain a Moore closing[11]. One usually begins from an easily obtained element in the environment of the problem posed. One of these is the well-known Moore family. Let us take a look at a Moore family.

If $F(E) \in P(E)$ is a family which fulfils the two following properties:

$$E \in F(E)$$
$$(A_1 \in F(E), A_2 \in F(E)......, A_i \in F(E)...)$$
$$\Rightarrow ((A_1 \cap A_2 \cap \ldots \cap A_i \cap \ldots.) \in F(E))$$

it is said that $F(E)$ is a Moore family.

The concept of a Moore family merits an example.

Let us assume the referential:

$$E = \{a, b, c\}$$

The following families $F_1(E)$, $F_2(E)$:

$$F_1(E) = \{ \{a\}, \{a, b\}, \{a, c\}, E\}$$
$$F_2(E) = \{ \varnothing, \{a\}, \{b\}, \{a, b\}, \{a, c\}, E\}$$

are Moore families in E.

From a Moore family it is possible to construct only one Moore closing, performing:

$$M A_j = \bigcap_{F \in F_{A_j}(E)} F$$

in which $F_{A_j}(E)$ is a subset of the elements of $F(E)$ which contains A_j and F which contains all elements of $F_{A_j}(E)$.

We will take the family $F_1(E)$ as an example of the referential $E = \{a, b, c\}$ and we have:

$F_{\varnothing}(E) = F(E)$	$M\varnothing = \{a\}^{24}$
$F_{\{a\}}(E) = F(E)$	$M\{a\} = \{a\}$
$F_{\{b\}}(E) = \{\{a, b\}, E\}$	$M\{b\} = \{a, b\}$
$F_{\{c\}}(E) = \{\{a, c\}, E\}$	$M\{c\} = \{a, c\}$
$F_{\{a,b\}}(E) = \{\{a, b\}, E\}$	$M\{a, b\} = \{a, b\}$
$F_{\{a,c\}}(E) = \{\{a, c\}, E\}$	$M\{a, c\} = \{a, c\}$
$F_{\{b,c\}}(E) = E$	$M\{b, c\} = E$
$F_E(E) = E$	$ME = E$

[11] To this respect we recommend the work of Gil-Aluja, J.: Elementos para una teoría de la decisi6n en la incertidumbre. Ed. Milladoiro. Vigo, 1999, pages. 194-203.

In this way the Moore closing presented in the following graph is obtained:

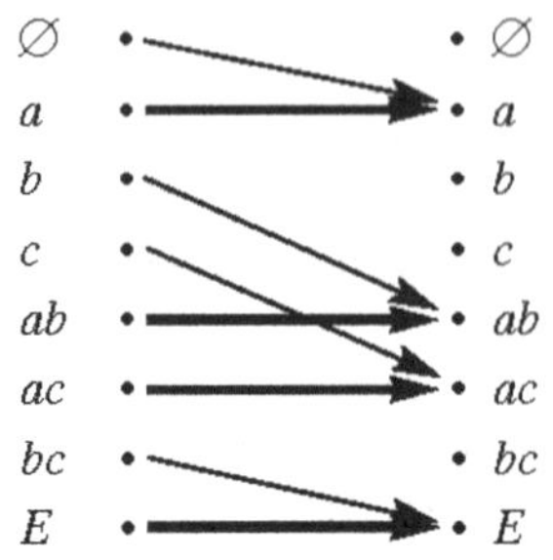

We will now move on to look at the verification of the extensive:

In the first place one has:

$\forall A_j \in P(E)$:

$$E \in F_{A_j}(E)$$

therefore:

$$F_{A_j}(E) \text{ is not empty}$$

and from here:

$$A_j \in M\,A_j$$

as:

$\forall F \in F_{A_j}(E)$:

$$A_j \in F$$

We stress that the condition of extension implies that $M\,E = E$, but in exchange it does not demand that $M\,\varnothing = \varnothing$ as is compulsory in pretopologies and topologies. In effect, there are Moore closings in which $M\,\varnothing = \varnothing$. We therefore insist that a Moore closing is a pretopology if $M\,\varnothing = \varnothing$ is fulfilled.

In continuation we will verify the idempotency.

Having taken into account that $F_{A_j}(E)$ is not empty for all of $A_j \in E$, $M\,A_j \in F_{A_j}(E)$ may be written and, therefore $M\,A_j \in F$ for all $F \in F_{A_j}(E)$ and $F \supset A_j$. Which also allows us to write that $M\,A_j$ is the smallest element of $F(E)$ which contains A_j. It may therefore be written as so:

$$M\,(M\,A_j) = \bigcap_{F \in F_{M\,A_j}(E)} F \subset M\,A_j$$

But as $M\ A_j \subset F_{M A_j}(E)$, it results that $M\ (M\ A_j) = M$. In this way the idempotency has been confirmed.

The isotone is easily verified. If $A_j \subset A_k$, all elements of $F(E)$ which contain A_k also contain A_j. We therefore have:

$$\mathrm{M}\ A_k \supset \mathrm{M}\ A_j$$

The existing uniqueness in the corresponding Moore family and Moore closing is therefore verified by the algorithm of its construction.

Another interesting property can also be observed. The elements of $P(E)$ which are images of another element of $P(E)$ are their own image themselves (this is necessary for idempotency). These elements are known as **closes**. Therefore in the previous figure: $\{\{a\}, \{a, b\}, \{a, c\}, E\}$ is the set of closes of M. We must remind ourselves that E is always a close.

Finally, we will state that the subset $F(E)$ of the closed of a Moore closing is a Moore family. On the other hand, the image of any element $A_j \in P\ (E)$ is the smallest close which it contains.

We have made a special mention to Moore closings in relation with pretopologies due to the interest which arises in their use for solving a great number of questions which revolve around economics and management. To check that which we have just stated it is enough for us to look towards the **theory of affinities**. But maybe it would not be prudent to advance events, above all taking into account that we are dealing with this interesting aspect of pretopological idea in the final part of this work. We will now move on to the heart of our work in that concerning pretopologies: **Pretopologies in uncertainty**.

Chapter 2
Pretopologies in Uncertainty

1 Axiomatic of an Uncertain Pretopology

In the sections which have made up the previous block of this first part each formal approach has been accompanied by a simple reference to the area of financial studies with the aim of trying to reduce the level of abstraction inherent to pretopological spaces. In this second chapter we will focus, preferentially but not exclusively, on the formal aspects which we consider fundamental, avoiding wherever possible deviation away from the main objective which we are pursuing. Even in this way, each axiomatic will be accompanied by a **didactic example**.

We will begin with a finite referential *E* and a set of **fuzzy parts** *P*(*E*) is considered . Taking into account that our attention is going to be centred on the treatment of finite and numerable structures and configurations, in our formal developments we will collect finite or *P*(*E*) families subsets. In first place we present the axioms of an uncertain pretopology.

Given a functional application $\mathbf{\Gamma}$ of *P*(*E*) in *P*(*E*), it is said that $\mathbf{\Gamma}$ is an **uncertain pretopology** of *E* if and only if the following axioms are fulfilled:

1. $\Gamma \varnothing = \varnothing$
2. $\forall \underset{\sim}{A}_j \in K(E)$:

$$\underset{\sim}{A}_j \subset \Gamma \underset{\sim}{A}_j$$

in which *K(E)* is a finite subset of *P*(*E*), which is includes of ∅ and E.

We should remember that the second of these axioms entails:

$$\mathbf{\Gamma} E = E$$

The pair (*E*,$\mathbf{\Gamma}$) is known as "uncertain pretopological space" and in reality represents a fuzzy pretopological space with a finite configuration.

The functional application $\mathbf{\Gamma}$ for all elements of *P*(*E*) is called "adherent application" or simply "adherence".

As in that which occurs in the "crisp" supposition, the "interior application" or simply "interior" my be associated to $\mathbf{\Gamma}$, defined by:

$\forall \underset{\sim}{A}_j \in K(E)$:

$$\delta \underset{\sim}{A}_j = \overline{\Gamma \overline{\underset{\sim}{A}_j}}$$

in which $\overline{\underset{\sim}{A}_j}$ is the complement of $\underset{\sim}{A}_j$.

J. Gil Aluja & A.M. Gil Lafuente: Towards an Advanced Modelling, STUDFUZZ 276, pp. 29–68.
springerlink.com

The established axiomatic leads to:

$$\delta \varnothing = \varnothing \text{ and } \delta E = E$$

in the same way:

$\forall \underset{\sim}{A}_j \in K(E)$:

$$\underset{\sim}{A}_j \supset \delta \underset{\sim}{A}_j$$

From these concepts the "open subsets" and the "closed subsets" can be defined.

It is said that $\underset{\sim}{A}_j$ is **closed** or **a close** of the uncertain pretopology when:

$$\underset{\sim}{A}_j = \Gamma \underset{\sim}{A}_j$$

It is also said that $\underset{\sim}{A}_j$ is **open** or **an open** of the uncertain pretopology when:

$$\underset{\sim}{A}_j = \delta \underset{\sim}{A}_j$$

It is easy to see that:

$$(\underset{\sim}{A}_j = \Gamma \underset{\sim}{A}_j) \Rightarrow (\overline{\underset{\sim}{A}}_j = \delta \overline{\underset{\sim}{A}}_j)$$

$$(\underset{\sim}{A}_j = \delta \underset{\sim}{A}_j) \Rightarrow (\overline{\underset{\sim}{A}}_j = \Gamma \overline{\underset{\sim}{A}}_j)$$

When these two conditions are met the following results:

$$(\underset{\sim}{A}_j \in K(E)) \Rightarrow (\overline{\underset{\sim}{A}}_j \in K(E))$$

which is to say, if the set of closed and opens contain $\underset{\sim}{A}_j$ they must also contain $\overline{\underset{\sim}{A}}_j$.

In an uncertain pretopology, the necessary and sufficient condition for a close to also be an open is that the closed set c or that the open set contains $\underset{\sim}{A}_j$ and $\overline{\underset{\sim}{A}}_j$, which is the same.

The demonstration is almost trivial. If $\underset{\sim}{A}_j$ is an open, then $\underset{\sim}{A}_j = \delta \underset{\sim}{A}_j$. If is a close then $\underset{\sim}{A}_j = \Gamma \underset{\sim}{A}_j = \overline{\delta \overline{\underset{\sim}{A}}_j}$, and therefore $\overline{\underset{\sim}{A}}_j = \delta \overline{\underset{\sim}{A}}_j$ and so $\underset{\sim}{A}_j$ is also open. The same reasoning is valid if $\underset{\sim}{A}_j$ is a close.

We will now move on to one of the fundamental aspects of uncertain pretopological conception: its **description**. And this, for how much that the referential E from which $K(E)$ is established has not always been considered in a uniform way. In effect, a first way consists in forming the referential E integrated by the **elements of the referential of the fuzzy subsets**. A second way would be that in whose origin the referential E would be formed by **fuzzy subsets** from which the set of its parts $P(E)$ is established.

a) We will move on to describe, in first place, a finite uncertain pretopology in E following the first of these ways. In general, it is not possible to proceed with an exhaustive enumeration, even for finite E, as L^E is not finite. A subset of closes C_e or a subset of opens A_b are possible to define, but this is not sufficient to define a unique uncertain pretopology either. One must therefore choose a $K(E)$, **respecting the previously stated axioms**. Let us move on to an example:

Let:

$$E = \{a, b, c\}$$

and

$$K(E) = \left\{ \varnothing,\ \underset{\sim}{A}_1 = \begin{array}{ccc} a & b & c \\ \hline .8 & 1 & .3 \end{array},\ \overline{\underset{\sim}{A}}_1 = \begin{array}{ccc} a & b & c \\ \hline .2 & 0 & .7 \end{array},\ \underset{\sim}{A}_2 = \begin{array}{ccc} a & b & c \\ \hline .2 & .4 & 1 \end{array},\ \overline{\underset{\sim}{A}}_2 = \begin{array}{ccc} a & b & c \\ \hline .8 & .6 & 0 \end{array},\ \underset{\sim}{A}_3 = \begin{array}{ccc} a & b & c \\ \hline .6 & .8 & 1 \end{array},\ \overline{\underset{\sim}{A}}_3 = \begin{array}{ccc} a & b & c \\ \hline .4 & .2 & 0 \end{array},\ \underset{\sim}{A}_4 = \begin{array}{ccc} a & b & c \\ \hline 0 & 0 & .3 \end{array},\ \overline{\underset{\sim}{A}}_4 = \begin{array}{ccc} a & b & c \\ \hline 1 & 1 & .7 \end{array},\ E \right\}$$

Let us show **one** uncertain pretopology constructed from $K(E)$.

		$\underset{\sim}{A}_1$	$\overline{\underset{\sim}{A}}_1$	$\underset{\sim}{A}_2$	$\overline{\underset{\sim}{A}}_2$	
$\underset{\sim}{A}_j$	∅	a b c .8 1 .3	a b c .2 0 .7	a b c .2 .4 1	a b c .8 .6 0	..
$\Gamma\,\underset{\sim}{A}_j$	∅ cerrado	a b c 1 1 .7	a b c .2 .4 1	E	a b c .8 .6 0 cerrado	..
$\delta\,\underset{\sim}{A}_j$	∅ abierto	a b c .8 .6 0	a b c 0 0 .3	a b c .2 .4 1 abierto	∅	..

	$\underset{\sim}{A}_3$	$\overline{\underset{\sim}{A}}_3$	$\underset{\sim}{A}_4$	$\overline{\underset{\sim}{A}}_4$	
...	a b c .6 8 .1	a b c .4 2 .0	a b c .0 .0 3	a b c 1 1 7	E
...	E	E	a b c .0 .0 3 cerrado	E	E cerrado
...	∅	∅	∅	a b c 1 1 7 abierto	E abierto

We will present these results in the following figures.

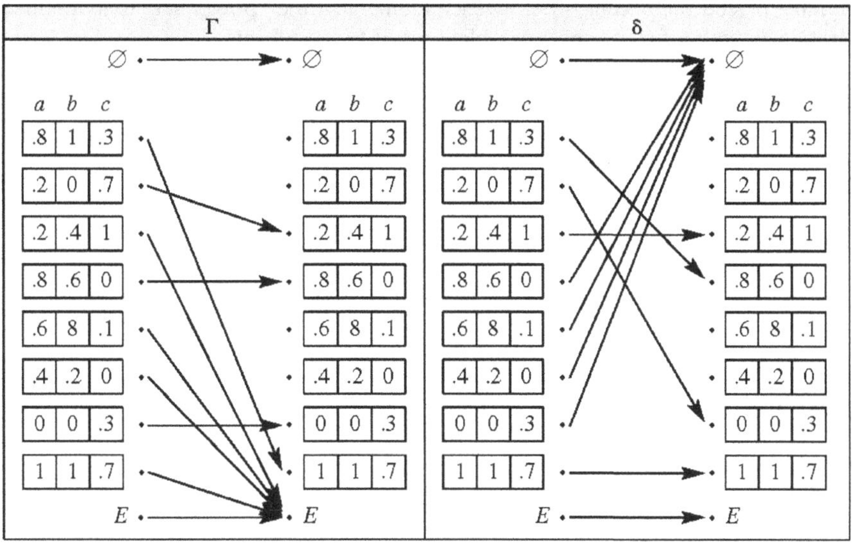

Having taken into account that the fuzzy subsets $\underset{\sim}{A}_1, \underset{\sim}{A}_2, \underset{\sim}{A}_3, \underset{\sim}{A}_4$ and their complements **are not ordered**, some arcs of δ can go (and come in this case) downwards.

It is necessary to point out that when a fuzzy subset is such that $\underset{\sim}{A}_j = \underset{\sim}{A}_j$, so for all $x : \mu_{\underset{\sim}{A}_j}(x) = 0{,}5$. When an uncertain pretopology of $K(E)$ contains an $\underset{\sim}{A}_j$ of this nature the cardinal of **Γ** and δ is odd.

b) We will now move on to develop the idea of uncertain pretopology, moving into the second way. In this for the movement towards uncertainty and with the aim of representing it, the "crisp" elements of the referential E may be substituted for fuzzy subsets. In this way we will have a referential E formed by fuzzy subsets:

$$E = \{\underset{\sim}{P}_1, \underset{\sim}{P}_2, \ldots, \underset{\sim}{P}_n\}$$

In this case, the set of its parts, $P(E)$, will have a cardinal equal to 2^n, the same as in the ordinary pretopology. Each element of the "power set", $P(E)$ will either be a fuzzy subset or a set of fuzzy subsets. In this way, therefore, we can write:

$$P(E) = \Big\{\{\underset{\sim}{P}_1\}, \{\underset{\sim}{P}_2\}, \ldots, \{\underset{\sim}{P}_n\}, \{\underset{\sim}{P}_1, \underset{\sim}{P}_2\}, \{\underset{\sim}{P}_1, \underset{\sim}{P}_3\}, \ldots, \{\underset{\sim}{P}_{n-1}, \underset{\sim}{P}_n\},$$
$$\{\underset{\sim}{P}_1, \underset{\sim}{P}_2, \underset{\sim}{P}_3\}, \ldots, \{\underset{\sim}{P}_{n-2}, \underset{\sim}{P}_{n-1}, \underset{\sim}{P}_n\}, \ldots, \{\underset{\sim}{P}_1, \underset{\sim}{P}_2, \ldots, \underset{\sim}{P}_n\} = E\Big\}$$

If, for a greater level of descriptive convenience, these elements of the "power set" are expressed through $\underset{\sim}{A}_j$, $j = 1, 2, \ldots 2^n$, we will have:

$$P(E) = \{\underset{\sim}{A}_1, \underset{\sim}{A}_2, \ldots, \underset{\sim}{A}_{2^n}\}$$

We have placed the accent ~ below each element of the "power set" to symbolise that it deals with a fuzzy subset or a clear set of fuzzy subsets.

Now we can propose how to choose those elements of the "power set" $P(E)$ which form $K(E) \in P(E)$, between which the application Γ will be established. It seems evident that the self nature of the problem studied should, in principle, give the way of determining the elements of the "power set" which give rise to $K(E)$.

Let us see a supposition:

A financial institution has a set of products:

$$E = \{\underset{\sim}{P}_1, \underset{\sim}{P}_2, \ldots, \underset{\sim}{P}_6\}$$

which they wish to group in a **sufficiently** homogenous way.

Each one of them comes described in a referential of its attributes (qualities, characteristics or singularities):

$$\{a, b, c\}$$

through the following fuzzy subsets:

$$\underset{\sim}{P}_1 = \begin{array}{ccc} a & b & c \\ \hline \multicolumn{1}{|c|}{.4} & \multicolumn{1}{c|}{1} & \multicolumn{1}{c|}{.8} \\ \hline \end{array} \qquad \underset{\sim}{P}_2 = \begin{array}{ccc} a & b & c \\ \hline \multicolumn{1}{|c|}{.8} & \multicolumn{1}{c|}{.3} & \multicolumn{1}{c|}{.7} \\ \hline \end{array} \qquad \underset{\sim}{P}_3 = \begin{array}{ccc} a & b & c \\ \hline \multicolumn{1}{|c|}{.6} & \multicolumn{1}{c|}{.9} & \multicolumn{1}{c|}{.4} \\ \hline \end{array}$$

$$\underset{\sim}{P}_4 = \begin{array}{ccc} a & b & c \\ \hline \multicolumn{1}{|c|}{.9} & \multicolumn{1}{c|}{.4} & \multicolumn{1}{c|}{.5} \\ \hline \end{array} \qquad \underset{\sim}{P}_5 = \begin{array}{ccc} a & b & c \\ \hline \multicolumn{1}{|c|}{1} & \multicolumn{1}{c|}{.3} & \multicolumn{1}{c|}{.2} \\ \hline \end{array} \qquad \underset{\sim}{P}_6 = \begin{array}{ccc} a & b & c \\ \hline \multicolumn{1}{|c|}{.3} & \multicolumn{1}{c|}{.9} & \multicolumn{1}{c|}{.8} \\ \hline \end{array}$$

To the pertinent effects it is estimated that the grouping is possible if for each attribute the corresponding product reaches the following levels:

$$Q_a \geq 0.8 \qquad Q_b \geq 0.9 \qquad Q_c \geq 0.8$$

In this way we may gather together the financial products **which have** each attribute to the demanded level in one family. The families would therefore be:

$$F_1 = \{\underset{\sim}{P}_2, \underset{\sim}{P}_4, \underset{\sim}{P}_5\} = \left\{ \begin{array}{ccc} a & b & c \\ \hline \multicolumn{1}{|c|}{.8} & \multicolumn{1}{c|}{.3} & \multicolumn{1}{c|}{.7} \\ \hline \end{array}, \begin{array}{ccc} a & b & c \\ \hline \multicolumn{1}{|c|}{.9} & \multicolumn{1}{c|}{.4} & \multicolumn{1}{c|}{.5} \\ \hline \end{array}, \begin{array}{ccc} a & b & c \\ \hline \multicolumn{1}{|c|}{1} & \multicolumn{1}{c|}{.3} & \multicolumn{1}{c|}{.2} \\ \hline \end{array} \right\}$$

$$F_2 = \{\underset{\sim}{P}_1, \underset{\sim}{P}_3, \underset{\sim}{P}_6\} = \left\{ \begin{array}{ccc} a & b & c \\ \hline \multicolumn{1}{|c|}{.4} & \multicolumn{1}{c|}{1} & \multicolumn{1}{c|}{.8} \\ \hline \end{array}, \begin{array}{ccc} a & b & c \\ \hline \multicolumn{1}{|c|}{.6} & \multicolumn{1}{c|}{.9} & \multicolumn{1}{c|}{.4} \\ \hline \end{array}, \begin{array}{ccc} a & b & c \\ \hline \multicolumn{1}{|c|}{.3} & \multicolumn{1}{c|}{.9} & \multicolumn{1}{c|}{.8} \\ \hline \end{array} \right\}$$

$$F_3 = \{\underset{\sim}{P}_1, \underset{\sim}{P}_6\} = \left\{ \begin{array}{ccc} a & b & c \\ \hline \multicolumn{1}{|c|}{.4} & \multicolumn{1}{c|}{1} & \multicolumn{1}{c|}{.8} \\ \hline \end{array}, \begin{array}{ccc} a & b & c \\ \hline \multicolumn{1}{|c|}{.3} & \multicolumn{1}{c|}{.9} & \multicolumn{1}{c|}{.8} \\ \hline \end{array} \right\}$$

At the same time one has those families which do not have each attribute to the demanded level. They are the following:

$$\overline{F}_1 = \{\underset{\sim}{P}_1, \underset{\sim}{P}_3, \underset{\sim}{P}_6\} = \{ \begin{array}{|c|c|c|} \multicolumn{1}{c}{a} & \multicolumn{1}{c}{b} & \multicolumn{1}{c}{c} \\ \hline .4 & 1 & .8 \\ \hline \end{array} , \begin{array}{|c|c|c|} \multicolumn{1}{c}{a} & \multicolumn{1}{c}{b} & \multicolumn{1}{c}{c} \\ \hline .6 & .9 & .4 \\ \hline \end{array} , \begin{array}{|c|c|c|} \multicolumn{1}{c}{a} & \multicolumn{1}{c}{b} & \multicolumn{1}{c}{c} \\ \hline .3 & .9 & .8 \\ \hline \end{array} \}$$

$$\overline{F}_2 = \{\underset{\sim}{P}_2, \underset{\sim}{P}_4, \underset{\sim}{P}_5\} = \{ \begin{array}{|c|c|c|} \multicolumn{1}{c}{a} & \multicolumn{1}{c}{b} & \multicolumn{1}{c}{c} \\ \hline .8 & .3 & .7 \\ \hline \end{array} , \begin{array}{|c|c|c|} \multicolumn{1}{c}{a} & \multicolumn{1}{c}{b} & \multicolumn{1}{c}{c} \\ \hline .9 & .4 & .5 \\ \hline \end{array} , \begin{array}{|c|c|c|} \multicolumn{1}{c}{a} & \multicolumn{1}{c}{b} & \multicolumn{1}{c}{c} \\ \hline 1 & .3 & .2 \\ \hline \end{array} \}$$

$$\overline{F}_3 = \{\underset{\sim}{P}_2, \underset{\sim}{P}_3, \underset{\sim}{P}_4, \underset{\sim}{P}_5\} = \{ \begin{array}{|c|c|c|} \multicolumn{1}{c}{a} & \multicolumn{1}{c}{b} & \multicolumn{1}{c}{c} \\ \hline .8 & .3 & .7 \\ \hline \end{array} , \begin{array}{|c|c|c|} \multicolumn{1}{c}{a} & \multicolumn{1}{c}{b} & \multicolumn{1}{c}{c} \\ \hline .6 & .9 & .4 \\ \hline \end{array} , \begin{array}{|c|c|c|} \multicolumn{1}{c}{a} & \multicolumn{1}{c}{b} & \multicolumn{1}{c}{c} \\ \hline .9 & .4 & .5 \\ \hline \end{array} , \begin{array}{|c|c|c|} \multicolumn{1}{c}{a} & \multicolumn{1}{c}{b} & \multicolumn{1}{c}{c} \\ \hline 1 & .3 & .2 \\ \hline \end{array} \}$$

If groups are formed with these six families in such a way that each one of the groups joins those products which fulfil the demanded levels of the three attributes, that they fulfil two of them and not the other... and that they fulfil none of them, a combined study will have been performed which gives the **homogenous grouping of products** by attributes. In our case three subsets with distinct products have been found. These subsets are disjunctive.

$$G_1 = F_1 \cap F_2 \cap F_3 = \varnothing \qquad G_2 = F_1 \cap F_2 \cap \overline{F}_3 = \varnothing \qquad G_3 = F_1 \cap \overline{F}_2 \cap F_3 = \varnothing$$

$$G_4 = F_1 \cap \overline{F}_2 \cap \overline{F}_3 = \{\underset{\sim}{P}_2, \underset{\sim}{P}_4, \underset{\sim}{P}_5\} = \{ \begin{array}{|c|c|c|} \multicolumn{1}{c}{a} & \multicolumn{1}{c}{b} & \multicolumn{1}{c}{c} \\ \hline .8 & .3 & .7 \\ \hline \end{array} , \begin{array}{|c|c|c|} \multicolumn{1}{c}{a} & \multicolumn{1}{c}{b} & \multicolumn{1}{c}{c} \\ \hline .9 & .4 & .5 \\ \hline \end{array} , \begin{array}{|c|c|c|} \multicolumn{1}{c}{a} & \multicolumn{1}{c}{b} & \multicolumn{1}{c}{c} \\ \hline 1 & .3 & .2 \\ \hline \end{array} \}$$

$$G_5 = \overline{F}_1 \cap F_2 \cap F_3 = \{\underset{\sim}{P}_1, \underset{\sim}{P}_6\} = \{ \begin{array}{|c|c|c|} \multicolumn{1}{c}{a} & \multicolumn{1}{c}{b} & \multicolumn{1}{c}{c} \\ \hline .4 & 1 & .8 \\ \hline \end{array} , \begin{array}{|c|c|c|} \multicolumn{1}{c}{a} & \multicolumn{1}{c}{b} & \multicolumn{1}{c}{c} \\ \hline .3 & .9 & .8 \\ \hline \end{array} \}$$

$$G_6 = \overline{F}_1 \cap F_2 \cap \overline{F}_3 = \{\underset{\sim}{P}_3\} = \{ \begin{array}{|c|c|c|} \multicolumn{1}{c}{a} & \multicolumn{1}{c}{b} & \multicolumn{1}{c}{c} \\ \hline .6 & .9 & .4 \\ \hline \end{array} \}$$

$$G_7 = \overline{F}_1 \cap \overline{F}_2 \cap F_3 = \varnothing \qquad G_8 = \overline{F}_1 \cap \overline{F}_2 \cap \overline{F}_3 = \varnothing$$

In effect, the only groupings possible are G_4, G_5, G_6, which will be respectively named $\underset{\sim}{A}_1$, $\underset{\sim}{A}_2$, $\underset{\sim}{A}_3$. Therefore:

$$\underset{\sim}{A}_1 = \{ \begin{array}{|c|c|c|} \multicolumn{1}{c}{a} & \multicolumn{1}{c}{b} & \multicolumn{1}{c}{c} \\ \hline .8 & .3 & .7 \\ \hline \end{array} , \begin{array}{|c|c|c|} \multicolumn{1}{c}{a} & \multicolumn{1}{c}{b} & \multicolumn{1}{c}{c} \\ \hline .9 & .4 & .5 \\ \hline \end{array} , \begin{array}{|c|c|c|} \multicolumn{1}{c}{a} & \multicolumn{1}{c}{b} & \multicolumn{1}{c}{c} \\ \hline 1 & .3 & .2 \\ \hline \end{array} \}$$

which comprises the three financial products which only have, to the demanded level, property **a**.

$$\underset{\sim}{A}_2 = \{ \begin{array}{|c|c|c|} \multicolumn{1}{c}{a} & \multicolumn{1}{c}{b} & \multicolumn{1}{c}{c} \\ \hline .4 & 1 & .8 \\ \hline \end{array} , \begin{array}{|c|c|c|} \multicolumn{1}{c}{a} & \multicolumn{1}{c}{b} & \multicolumn{1}{c}{c} \\ \hline .3 & .9 & .8 \\ \hline \end{array} \}$$

which groups the two financial products with only qualities **b** and **c**.

$$\underset{\sim}{A}_3 = \left\{ \begin{array}{|c|c|c|} \multicolumn{1}{c}{a} & \multicolumn{1}{c}{b} & \multicolumn{1}{c}{c} \\ \hline .6 & .9 & .4 \\ \hline \end{array} \right\}$$

which shows the single financial product which only has property **b**.

Now, if we **join** the groups of two and three together, the union of $\underset{\sim}{A}_1$ and $\underset{\sim}{A}_2$ is found:

$$\underset{\sim}{A}_4 = \left\{ \begin{array}{|c|c|c|} \multicolumn{1}{c}{a} & \multicolumn{1}{c}{b} & \multicolumn{1}{c}{c} \\ \hline .4 & 1 & .8 \\ \hline \end{array}, \begin{array}{|c|c|c|} \multicolumn{1}{c}{a} & \multicolumn{1}{c}{b} & \multicolumn{1}{c}{c} \\ \hline .8 & .3 & .7 \\ \hline \end{array}, \begin{array}{|c|c|c|} \multicolumn{1}{c}{a} & \multicolumn{1}{c}{b} & \multicolumn{1}{c}{c} \\ \hline .9 & .4 & .5 \\ \hline \end{array}, \begin{array}{|c|c|c|} \multicolumn{1}{c}{a} & \multicolumn{1}{c}{b} & \multicolumn{1}{c}{c} \\ \hline 1 & .3 & .2 \\ \hline \end{array}, \begin{array}{|c|c|c|} \multicolumn{1}{c}{a} & \multicolumn{1}{c}{b} & \multicolumn{1}{c}{c} \\ \hline .3 & .9 & .8 \\ \hline \end{array} \right\}$$

This includes the three financial products with quality **a** and the two with qualities **b** and **c**. In this way, among the five they **have all** of the properties.

The union of $\underset{\sim}{A}_1$ and $\underset{\sim}{A}_3$:

$$\underset{\sim}{A}_5 = \left\{ \begin{array}{|c|c|c|} \multicolumn{1}{c}{a} & \multicolumn{1}{c}{b} & \multicolumn{1}{c}{c} \\ \hline .8 & .3 & .7 \\ \hline \end{array}, \begin{array}{|c|c|c|} \multicolumn{1}{c}{a} & \multicolumn{1}{c}{b} & \multicolumn{1}{c}{c} \\ \hline .6 & .9 & .4 \\ \hline \end{array}, \begin{array}{|c|c|c|} \multicolumn{1}{c}{a} & \multicolumn{1}{c}{b} & \multicolumn{1}{c}{c} \\ \hline .9 & .4 & .5 \\ \hline \end{array}, \begin{array}{|c|c|c|} \multicolumn{1}{c}{a} & \multicolumn{1}{c}{b} & \multicolumn{1}{c}{c} \\ \hline 1 & .3 & .2 \\ \hline \end{array} \right\}$$

In this the single financial product which only has characteristic **b** and the three which have **a** are found. Therefore, between the four only these two properties **a** and **b** are held.

The union of $\underset{\sim}{A}_2$ and $\underset{\sim}{A}_3$:

$$\underset{\sim}{A}_6 = \left\{ \begin{array}{|c|c|c|} \multicolumn{1}{c}{a} & \multicolumn{1}{c}{b} & \multicolumn{1}{c}{c} \\ \hline .4 & 1 & .8 \\ \hline \end{array}, \begin{array}{|c|c|c|} \multicolumn{1}{c}{a} & \multicolumn{1}{c}{b} & \multicolumn{1}{c}{c} \\ \hline .6 & .9 & .4 \\ \hline \end{array}, \begin{array}{|c|c|c|} \multicolumn{1}{c}{a} & \multicolumn{1}{c}{b} & \multicolumn{1}{c}{c} \\ \hline .3 & .9 & .8 \\ \hline \end{array} \right\}$$

In this element we find the single financial product which has characteristic **b** to which the two that only have properties **b** and **c** are incorporated.

The union of $\underset{\sim}{A}_1$, $\underset{\sim}{A}_2$ and $\underset{\sim}{A}_3$:

$$\underset{\sim}{A}_7 = \left\{ \begin{array}{|c|c|c|} \multicolumn{1}{c}{a} & \multicolumn{1}{c}{b} & \multicolumn{1}{c}{c} \\ \hline .4 & 1 & .8 \\ \hline \end{array}, \begin{array}{|c|c|c|} \multicolumn{1}{c}{a} & \multicolumn{1}{c}{b} & \multicolumn{1}{c}{c} \\ \hline .8 & .3 & .7 \\ \hline \end{array}, \begin{array}{|c|c|c|} \multicolumn{1}{c}{a} & \multicolumn{1}{c}{b} & \multicolumn{1}{c}{c} \\ \hline .6 & .9 & .4 \\ \hline \end{array}, \right.$$

$$\left. \begin{array}{|c|c|c|} \multicolumn{1}{c}{a} & \multicolumn{1}{c}{b} & \multicolumn{1}{c}{c} \\ \hline .9 & .4 & .5 \\ \hline \end{array}, \begin{array}{|c|c|c|} \multicolumn{1}{c}{a} & \multicolumn{1}{c}{b} & \multicolumn{1}{c}{c} \\ \hline 1 & .3 & .2 \\ \hline \end{array}, \begin{array}{|c|c|c|} \multicolumn{1}{c}{a} & \multicolumn{1}{c}{b} & \multicolumn{1}{c}{c} \\ \hline .3 & .9 & .8 \\ \hline \end{array} \right\}$$

To this set of fuzzy subsets we will only add the empty $\varnothing$ which represents the non-existence of financial products as the referential E coincides with $\underset{\sim}{A}_7$.

Perhaps a brief summary, adequately ordered, would help to achieve an overall perspective:

- Three financial products only have quality **a**.
- One financial product only has quality **b**.
- Two financial products only have qualities **b** and **c** together.
- Four financial products have qualities **a** and **b** together, one of quality **a** and the others of **b**.
- Three financial products have **b** and **c** together, one of quality **b** and the others **b** and **c**.
- Five financial products have all of the qualities **a**, **b** and **c** together, three of quality **a** and two of **b** and **c**.

We have therefore found a subset of the "power set" $K(E)$.

$$K(E) = \{\emptyset, G_4, G_5, G_6, G_4 \cup G_5. G_4 \cup G_6, G_5 \cup G_6, G_4 \cup G_5 \cup G_6 = E\}$$
$$= \{\emptyset, \{\underset{\sim}{P}_2, \underset{\sim}{P}_4, \underset{\sim}{P}_5\}, \{\underset{\sim}{P}_1, \underset{\sim}{P}_6\}, \{\underset{\sim}{P}_3\}, \{\underset{\sim}{P}_1, \underset{\sim}{P}_2, \underset{\sim}{P}_4, \underset{\sim}{P}_5, \underset{\sim}{P}_6\},$$
$$\{\underset{\sim}{P}_2, \underset{\sim}{P}_3, \underset{\sim}{P}_4, \underset{\sim}{P}_5\}, \{\underset{\sim}{P}_1, \underset{\sim}{P}_3, \underset{\sim}{P}_6\}, \{\underset{\sim}{P}_1, \underset{\sim}{P}_2, \underset{\sim}{P}_3, \underset{\sim}{P}_4, \underset{\sim}{P}_5, \underset{\sim}{P}_6\} = E\}$$

It can be seen that, in this case, together with the whole group a complement is found. In effect, if:

$$\underset{\sim}{A}_1 = \{\underset{\sim}{P}_2, \underset{\sim}{P}_4, \underset{\sim}{P}_5\} \qquad \underset{\sim}{A}_2 = \{\underset{\sim}{P}_1, \underset{\sim}{P}_6\} \qquad \underset{\sim}{A}_3 = \{\underset{\sim}{P}_3\}$$

therefore:

$$\underset{\sim}{A}_4 = \{\underset{\sim}{P}_1, \underset{\sim}{P}_2, \underset{\sim}{P}_4, \underset{\sim}{P}_5, \underset{\sim}{P}_6\} = \overline{\underset{\sim}{A}_3}$$
$$\underset{\sim}{A}_5 = \{\underset{\sim}{P}_2, \underset{\sim}{P}_3, \underset{\sim}{P}_4, \underset{\sim}{P}_5\} = \overline{\underset{\sim}{A}_2}$$
$$\underset{\sim}{A}_6 = \{\underset{\sim}{P}_1, \underset{\sim}{P}_3, \underset{\sim}{P}_6\} = \overline{\underset{\sim}{A}_1}$$

The referential E is evidently the complement of the empty $\emptyset$.

Having obtained $K(E)$, we will establish an adherent application $\mathbf{\Gamma}$ in such a way that it fulfils the second axiom of the uncertain pretopology which demands:

$\forall \underset{\sim}{A}_j \in K(E)$:

$$\underset{\sim}{A}_j \subset \Gamma \underset{\sim}{A}_j$$

In our example we have aimed to form a relationship of each group of financial products with the reviewed characteristics with another group or the same group which, at least, also has its qualities, also specified.

Among the possible applications, we propose the following refering to the original example, in which $Q_a \geq 0.8$, $Q_b \geq 0.9$, $Q_c \geq 0.8$.

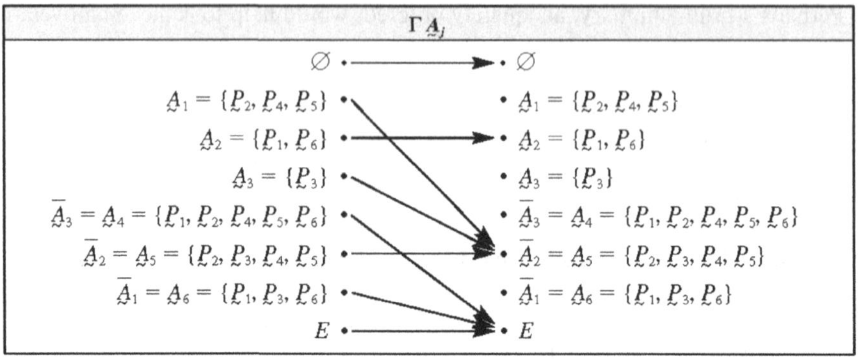

In this way each grouping of financial products A_j is related to another grouping, which as well as being composed of the same financial products may (or may not) include other products with qualities not held by the first. In this way, for example, A_1, which includes financial products with quality **a**, is related to A_2, in which that which has quality **b** is also found.

Having taken into account that, as we have indicated, the "interior application" is obtained in the following way:

$$\delta A_j = \overline{\Gamma \overline{A}_j}$$

we will have:

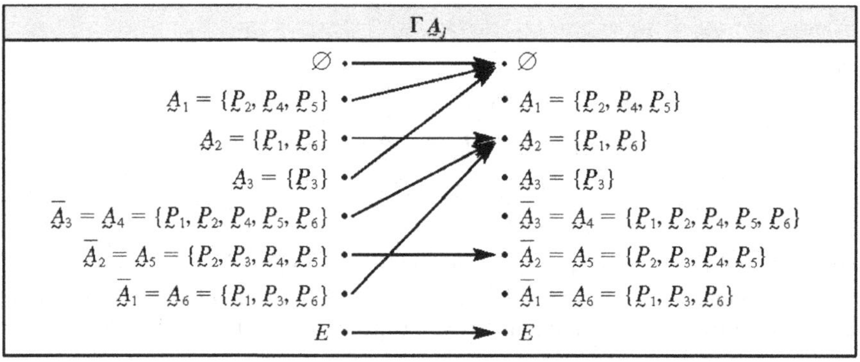

In the case of the **interior application**, the relationships are automatically determined by the adherent application and each grouping relates with the other formed by the same financial products or by less than those which it has. Therefore, the qualities covered are the same or less in the application than in the source.

In this way, for example, the element A_6 which is composed of products P_1, P_3, P_6 relates with the element A_2 with two of these products, specifically P_1 and P_6. If those three had the qualities **b** and **c** they now also have the same qualities **b** and **c**, but not one more.

In this supposition and given the process followed, we reach the situation that the **closes** are the same as the **opens**.

2 Isotonia in Uncertain Pretopologies

Given an uncertain pretoplogy, it is said to be "isotone" if:

$$\forall\, \underset{\sim}{A}_j, \underset{\sim}{A}_k \in K(E):$$

$$(\underset{\sim}{A}_j \subset A_k) \;\Rightarrow\; (\Gamma \underset{\sim}{A}_j \subset \Gamma \underset{\sim}{A}_k)$$

or, in the same way:

$$(\underset{\sim}{A}_j \subset A_k) \;\Rightarrow\; (\delta \underset{\sim}{A}_j \subset \delta \underset{\sim}{A}_k)$$

a) We will move on to develop a simple example following the first of the proposed ways for the establishment of the set $K(E)$. For this we will begin with the referential:

$$E = \{a, b, c\}$$

choosing a set $K(E)$ from which an adherent application **will be established which fulfils** the axiomatic of isotone uncertain pretopology:

$K(E) = \{$

$\varnothing$			$\underset{\sim}{A}_1$			$\overline{\underset{\sim}{A}_1}$			$\underset{\sim}{A}_2$			$\overline{\underset{\sim}{A}_2}$		
a	b	c	a	b	c	a	b	c	a	b	c	a	b	c
0	0	0	0	0	.1	1	1	.9	1	.2	.8	0	.8	.2

,

$\underset{\sim}{A}_3$			$\overline{\underset{\sim}{A}_3}$			$\underset{\sim}{A}_4$			$\overline{\underset{\sim}{A}_4}$			$\underset{\sim}{A}_5$		
a	b	c	a	b	c	a	b	c	a	b	c	a	b	c
.1	.8	.2	.9	.2	.8	.6	.1	.8	.4	.9	.2	1	.3	.8

,

$\overline{\underset{\sim}{A}_5}$			$\underset{\sim}{A}_6$			$\overline{\underset{\sim}{A}_6}$			E		
a	b	c	a	b	c	a	b	c	a	b	c
0	.7	.2	.5	1	1	.5	0	0	1	1	1

$\}$

In continuation we present this set $K(E)$ in the form of a lattice structure:

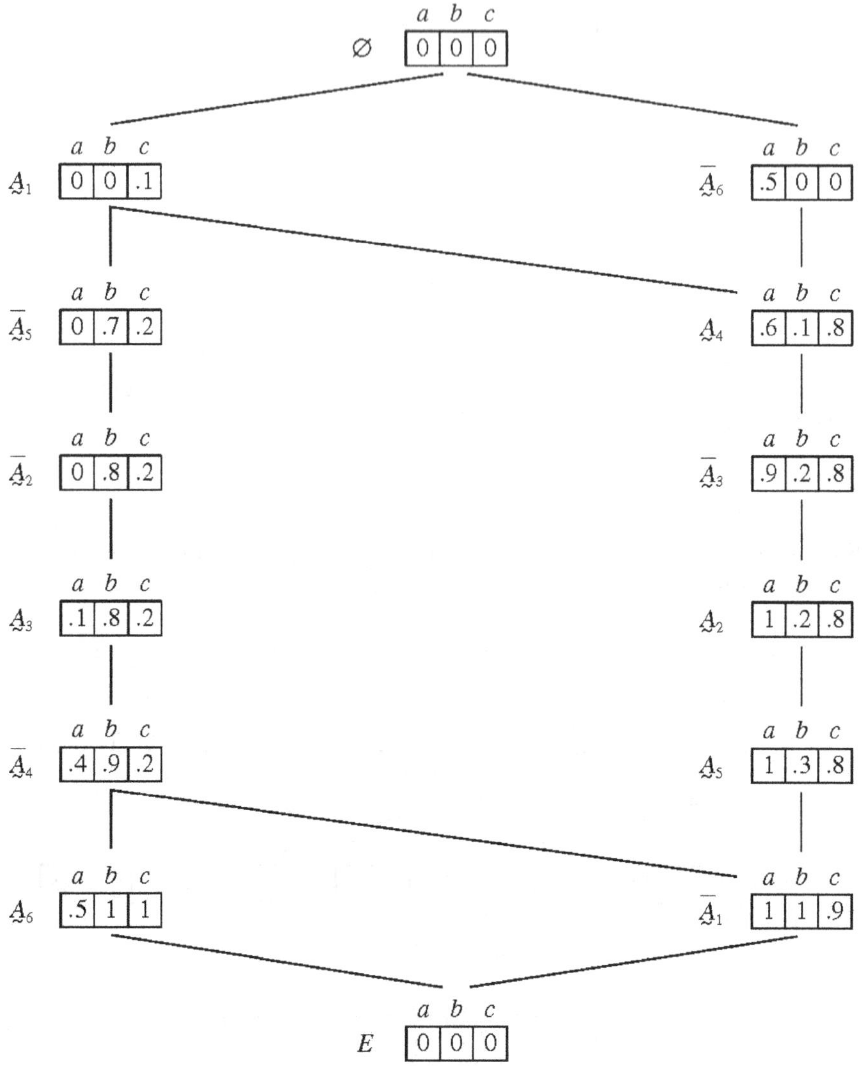

This structure shows that the fuzzy subsets which compose the set $K(E)$ form a partial order. This is important at the moment of obtaining the "opens" and "closes" in a Moore closing. But now we will continue to show an uncertain pre-topology formed in $K(E)$, firstly by means of the **adherence application** and later by the **interior application**.

Let the following be an adherence application:

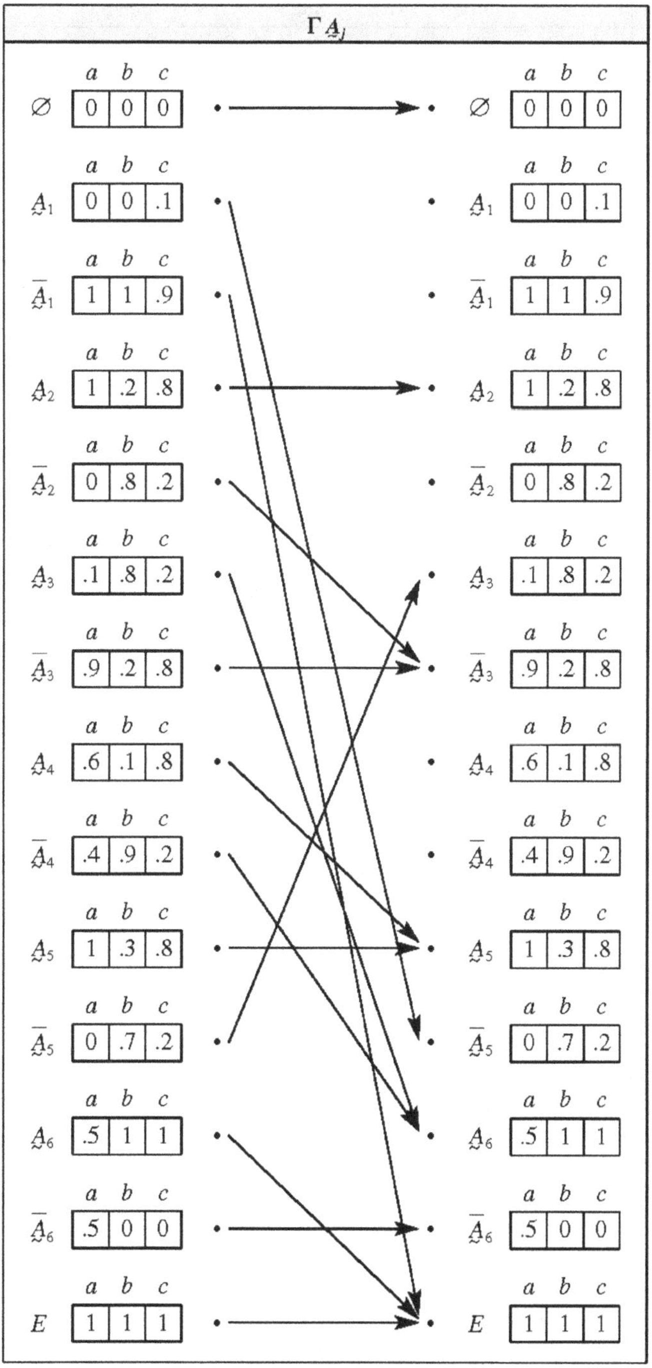

Likewise we assume the interior application:

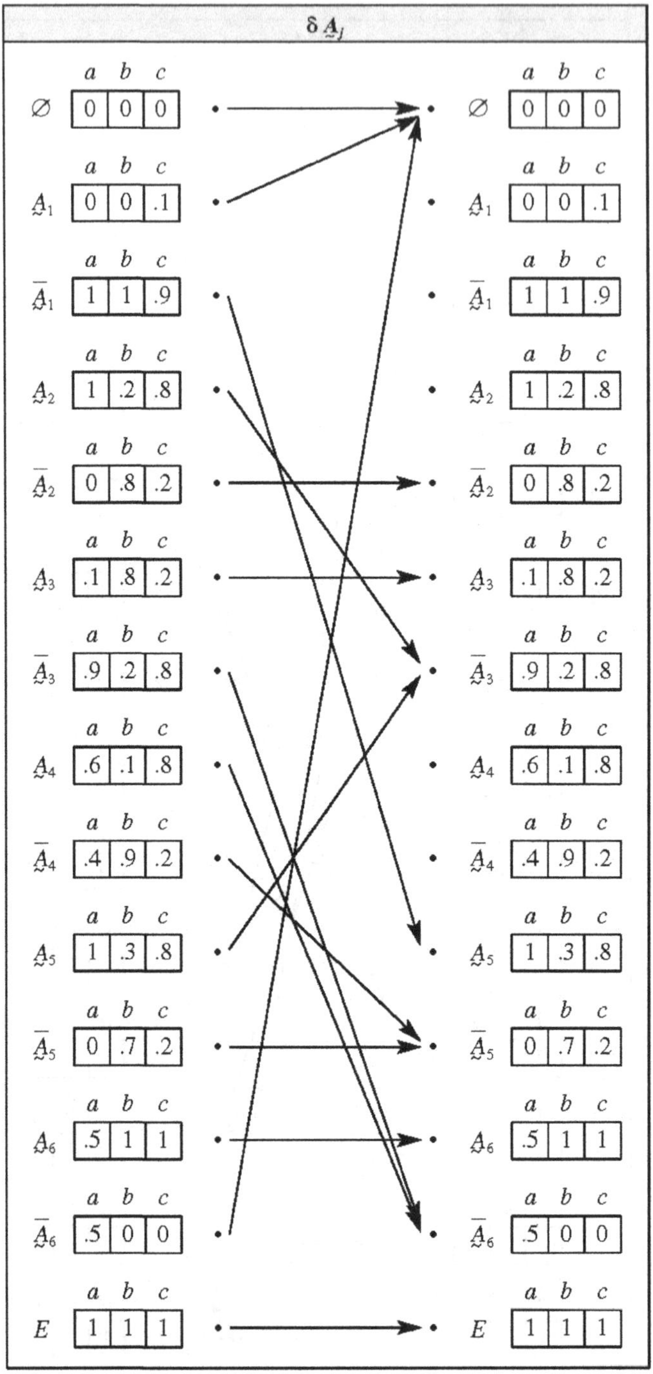

In the following table we summarise the isotone uncertain pretopology, in which we clearly show the closes and the opens.

$\underset{\sim}{A}_j$	a	b	c	$\Gamma \underset{\sim}{A}_j$	a	b	c	$\delta \underset{\sim}{A}_j$	a	b	c	
$\varnothing$	0	0	0	$\Gamma \varnothing$	0	0	0	$\delta \varnothing$	0	0	0	open closed
$\underset{\sim}{A}_1$	0	0	.1	$\Gamma \underset{\sim}{A}_1$	0	.7	.2	$\delta \underset{\sim}{A}_1$	0	0	0	
$\overline{\underset{\sim}{A}}_1$	1	1	.9	$\Gamma \overline{\underset{\sim}{A}}_1$	1	1	1	$\delta \overline{\underset{\sim}{A}}_1$	1	.3	.8	
$\underset{\sim}{A}_2$	1	.2	.8	$\Gamma \underset{\sim}{A}_2$	1	.2	.8	$\delta \underset{\sim}{A}_2$	.9	.2	.8	closed
$\overline{\underset{\sim}{A}}_2$	0	.8	.2	$\Gamma \overline{\underset{\sim}{A}}_2$	.1	.8	.2	$\delta \overline{\underset{\sim}{A}}_2$	0	.8	.2	open
$\underset{\sim}{A}_3$	.1	.8	.2	$\Gamma \underset{\sim}{A}_3$	.5	1	1	$\delta \underset{\sim}{A}_3$	.1	.8	.2	open
$\overline{\underset{\sim}{A}}_3$	.9	.2	.8	$\Gamma \overline{\underset{\sim}{A}}_3$	.9	.2	.8	$\delta \overline{\underset{\sim}{A}}_3$	.5	0	0	closed
$\underset{\sim}{A}_4$	.6	.1	.8	$\Gamma \underset{\sim}{A}_4$	1	.3	.8	$\delta \underset{\sim}{A}_4$	.5	0	0	
$\overline{\underset{\sim}{A}}_4$	.4	.9	.2	$\Gamma \overline{\underset{\sim}{A}}_4$	.5	1	1	$\delta \overline{\underset{\sim}{A}}_4$	0	.7	.2	
$\underset{\sim}{A}_5$	1	.3	.8	$\Gamma \underset{\sim}{A}_5$	1	.3	.8	$\delta \underset{\sim}{A}_5$	.9	.2	.8	closed
$\overline{\underset{\sim}{A}}_5$	0	.7	.2	$\Gamma \overline{\underset{\sim}{A}}_5$	.1	.8	.2	$\delta \overline{\underset{\sim}{A}}_5$	0	.7	.2	open
$\underset{\sim}{A}_6$	.5	1	1	$\Gamma \underset{\sim}{A}_6$	1	1	1	$\delta \underset{\sim}{A}_6$	.5	1	1	open
$\overline{\underset{\sim}{A}}_6$	.5	0	0	$\Gamma \overline{\underset{\sim}{A}}_6$	.5	0	0	$\delta \overline{\underset{\sim}{A}}_6$	0	0	0	closed
E	1	1	1	ΓE	1	1	1	δE	1	1	1	open closed

In this way it is shown that the subsets of opens and closes are respectively:

$$A_b = \{\varnothing, \overline{\underset{\sim}{A}}_2, \underset{\sim}{A}_3, \overline{\underset{\sim}{A}}_5, \underset{\sim}{A}_6, E\}$$
$$C_e = \{\varnothing, \underset{\sim}{A}_2, \overline{\underset{\sim}{A}}_3, \underset{\sim}{A}_5, \overline{\underset{\sim}{A}}_6, E\}$$

The fuzzy subsets which are at the same time open and closed are:

$$A_b \cap C_e = \{\varnothing, E\}$$

It will not have escaped the attention of our readers that an isotone uncertain pretopology may contain, as well as $\varnothing$ and E, fuzzy subsets which are open and closed at the same time. In the simple example which follows this can be confirmed.

$$E = \{a, b\}$$

$$K(E) = \{\varnothing, \underset{\sim}{A}_1, \overline{\underset{\sim}{A}}_1, \underset{\sim}{E}\}$$

With:

$\underset{\sim}{A}_1 =$

a	b
.1	.8

$\overline{\underset{\sim}{A}}_1 =$

a	b
.9	.2

If:

$$\Gamma \underset{\sim}{A}_1 = \underset{\sim}{A}_1 \quad \text{y} \quad \delta \underset{\sim}{A}_1 = \underset{\sim}{A}_1$$
$$\Gamma \overline{\underset{\sim}{A}}_1 = \overline{\underset{\sim}{A}}_1 \quad \text{y} \quad \delta \overline{\underset{\sim}{A}}_1 = \overline{\underset{\sim}{A}}_1$$

In this case it is confirmed that $\underset{\sim}{A}_1$ and $\overline{\underset{\sim}{A}}_1$ are both opens and closes at the same time.

b) When we move to the second of the proposed ways the process to follow can be parallel. In our specific case it can be confirmed that the example presented for the uncertain pretopology also satisfies the requisite for isotonia. In effect, we will now see that it fulfils the axiom:

$\forall \underset{\sim}{A}_j, \underset{\sim}{A}_k \in K(E)$:

$$(\underset{\sim}{A}_j \subset \underset{\sim}{A}_k) \Rightarrow (\Gamma \underset{\sim}{A}_j \subset \Gamma \underset{\sim}{A}_k)$$

For the adherent application:

$$
\begin{array}{ll}
\underset{\sim}{A}_1 \subset \{\underset{\sim}{A}_4, \underset{\sim}{A}_5, E\} & \Gamma \underset{\sim}{A}_1 = \underset{\sim}{A}_5 \subset \{\Gamma \underset{\sim}{A}_4, \Gamma \underset{\sim}{A}_5, \Gamma E\} \\
\underset{\sim}{A}_2 \subset \{\underset{\sim}{A}_4, \underset{\sim}{A}_6, E\} & \Gamma \underset{\sim}{A}_2 = \underset{\sim}{A}_2 \subset \{\Gamma \underset{\sim}{A}_4, \Gamma \underset{\sim}{A}_6, \Gamma E\} \\
\underset{\sim}{A}_3 \subset \{\underset{\sim}{A}_5, \underset{\sim}{A}_6, E\} & \Gamma \underset{\sim}{A}_3 = \underset{\sim}{A}_5 \subset \{\Gamma \underset{\sim}{A}_5, \Gamma \underset{\sim}{A}_6, \Gamma E\} \\
\underset{\sim}{A}_4 \subset \{E\} & \Gamma \underset{\sim}{A}_4 = E \subset \{\Gamma E\} \\
\underset{\sim}{A}_5 \subset \{E\} & \Gamma \underset{\sim}{A}_5 = \underset{\sim}{A}_5 \subset \{\Gamma E\} \\
\underset{\sim}{A}_6 \subset \{E\} & \Gamma \underset{\sim}{A}_6 = E \subset \{\Gamma E\}
\end{array}
$$

For the interior application:

$$
\begin{array}{ll}
\underset{\sim}{A}_1 \subset \{\underset{\sim}{A}_4, \underset{\sim}{A}_5, E\} & \delta \underset{\sim}{A}_4 = \varnothing \subset \{\delta \underset{\sim}{A}_4, \delta \underset{\sim}{A}_5, \delta E\} \\
\underset{\sim}{A}_2 \subset \{\underset{\sim}{A}_4, \underset{\sim}{A}_6, E\} & \delta \underset{\sim}{A}_2 = \underset{\sim}{A}_2 \subset \{\delta \underset{\sim}{A}_4, \delta \underset{\sim}{A}_6, \delta E\} \\
\underset{\sim}{A}_3 \subset \{\underset{\sim}{A}_5, \underset{\sim}{A}_6, E\} & \delta \underset{\sim}{A}_3 = \varnothing \subset \{\delta \underset{\sim}{A}_5, \delta \underset{\sim}{A}_6, \delta E\} \\
\underset{\sim}{A}_4 \subset \{E\} & \delta \underset{\sim}{A}_4 = \underset{\sim}{A}_2 \subset \{\delta E\} \\
\underset{\sim}{A}_5 \subset \{E\} & \delta \underset{\sim}{A}_5 = \underset{\sim}{A}_5 \subset \{\delta E\} \\
\underset{\sim}{A}_6 \subset \{E\} & \delta \underset{\sim}{A}_6 = \underset{\sim}{A}_2 \subset \{\delta E\}
\end{array}
$$

In this way it has been confirmed that we find ourselves before an isotone uncertain pretopology with the following axiomatic:

1.° $\Gamma \varnothing = \varnothing$
2.° $\forall \underset{\sim}{A}_j \in K(E)$:

$$\underset{\sim}{A}_j \subset \Gamma \underset{\sim}{A}_j$$

3.° $\forall \underset{\sim}{A}_j, \underset{\sim}{A}_k \in K(E)$:

$$(\underset{\sim}{A}_j \subset \underset{\sim}{A}_k) \Rightarrow (\Gamma \underset{\sim}{A}_j \subset \Gamma \underset{\sim}{A}_k)$$

Returning to our specific case as a set of closes and opens we respectively have:

$$C_e = \{\varnothing, \underset{\sim}{A}_2, \overline{\underset{\sim}{A}}_2 = \underset{\sim}{A}_5, E\} = \left\{\varnothing, \{\underset{\sim}{P}_1, \underset{\sim}{P}_6\}, \{\underset{\sim}{P}_2, \underset{\sim}{P}_3, \underset{\sim}{P}_4, \underset{\sim}{P}_5\}, E\right\}$$

$$A_b = \{\varnothing, \underset{\sim}{A}_2, \overline{\underset{\sim}{A}}_2 = \underset{\sim}{A}_5, E\} = \left\{\varnothing, \{\underset{\sim}{P}_1, \underset{\sim}{P}_6\}, \{\underset{\sim}{P}_2, \underset{\sim}{P}_3, \underset{\sim}{P}_4, \underset{\sim}{P}_5\}, E\right\}$$

It can be observed that when $\underset{\sim}{A}_j$ appears in the closed or in the open then its compliment $\overline{\underset{\sim}{A}}_j$ is also found. Here, as we have already pointed out, the set of opens coincides with the set of closes.

3 Moore's Closing in Uncertainty

One of the most used concepts in the area of economics and management of uncertainty for the treatment of groupings is the "Moore closing". We have seen that in a deterministic context for the existence of a Moore closing it is necessary that the functional application **Γ** of a set of parts $K(E)$ over $K(E)$ has the properties of **extensivity**, **idempotency** and **isotonia**[1]. So, given that all isotone uncertain pretopology fulfils the conditions of extensivity and isotonia, if it also has idempotency then we find ourselves before an **uncertain Moore closing**.

Let us once again remind ourselves that in uncertain Moore closings $\mathbf{\Gamma}\, E = E$, as a consequence of extension. However, it is not necessary that $\mathbf{\Gamma}\, \varnothing = \varnothing$, as occurs in pretopologies. Therefore, not all uncertain Moore closings are uncertain pretopologies. But when in an uncertain Moore closing one has:

$\varnothing \in K(E)$:

$$\mathbf{\Gamma}\, \varnothing = \varnothing$$

then this uncertain Moore closing is an isotone uncertain pretopology with idempotency.

With these explanations made we will move on to expound the axiomatic of Moore closings.

1. $\forall\, \underset{\sim}{A}_j \in K(E)$:

$$\underset{\sim}{A}_j \subset \Gamma\, \underset{\sim}{A}_j \qquad \text{extensio}$$

2. $\forall\, \underset{\sim}{A}_j \in K(E)$:

$$\Gamma(\Gamma\, \underset{\sim}{A}_j) = \Gamma\, \underset{\sim}{A}_j \qquad \text{idempotency}$$

3. $\forall\, \underset{\sim}{A}_j, \underset{\sim}{A}_k \in K(E)$:

$$(\underset{\sim}{A}_j \subset \underset{\sim}{A}_k) \Rightarrow (\Gamma\, \underset{\sim}{A}_j \subset \Gamma\, \underset{\sim}{A}_k) \qquad \text{isotonia}$$

Given a set $K(E)$ and a functional application **Γ** the property of idempotency evidently does not always exist. However, in all cases it fulfils in its transitive close **Γ**. It is this way as if **Γ** is convolved then once again **Γ** is obtained. For all of this, if **Γ** is taken as functional application then the idempotency has been fulfilled.

When the interior application δ is considered in place of the axioms 1, 2 and 3, we will have:

[1] Kaufmann, A., y Gil Aluja, J.: *Técnicas de gestión de empresa. Previsiones, decisiones y estrategias*. Ed. Pirámide. Madrid, 1992, page 347.

1. $\forall\, \underset{\sim}{A}_j \in K(E)$:

$$\underset{\sim}{A}_j \supset \delta\, \underset{\sim}{A}_j$$

2. $\forall\, \underset{\sim}{A}_j \in K(E)$:

$$\delta(\delta\, \underset{\sim}{A}_j) = \delta\, \underset{\sim}{A}_j$$

3. $\forall\, \underset{\sim}{A}_j, \underset{\sim}{A}_k \in K(E)$:

$$(\underset{\sim}{A}_j \subset \underset{\sim}{A}_k) \Rightarrow (\delta\, \underset{\sim}{A}_j \subset \delta\, \underset{\sim}{A}_k)$$

Here that which has been expounded for the adherent application, with respect to the property of idempotency, is also valid. The interior application δ fulfils this axiom through its own construction.

In an exemplified way we will see this in **both ways**, with the respective isotone uncertain pretopologies taken as an operative base.

a) In the first way, with the aim of obtaining $\hat{\Gamma}\, \underset{\sim}{A}_j$ and $\delta\, \underset{\sim}{A}_j$ are expressed $\Gamma\, \underset{\sim}{A}_j$ and $\delta\, \underset{\sim}{A}_j$ in matrix form. We will begin with $\Gamma\, \underset{\sim}{A}_j$:

$\Gamma\, \underset{\sim}{A}_j =$

	$\varnothing$	$\underset{\sim}{A}_1$	$\overline{\underset{\sim}{A}}_1$	$\underset{\sim}{A}_2$	$\overline{\underset{\sim}{A}}_2$	$\underset{\sim}{A}_3$	$\overline{\underset{\sim}{A}}_3$	$\underset{\sim}{A}_4$	$\overline{\underset{\sim}{A}}_4$	$\underset{\sim}{A}_5$	$\overline{\underset{\sim}{A}}_5$	$\underset{\sim}{A}_6$	$\overline{\underset{\sim}{A}}_6$	E
$\varnothing$	1													
$\underset{\sim}{A}_1$											1			
$\overline{\underset{\sim}{A}}_1$														1
$\underset{\sim}{A}_2$				1										
$\overline{\underset{\sim}{A}}_2$						1								
$\underset{\sim}{A}_3$												1		
$\overline{\underset{\sim}{A}}_3$							1							
$\underset{\sim}{A}_4$										1				
$\overline{\underset{\sim}{A}}_4$												1		
$\underset{\sim}{A}_5$										1				
$\overline{\underset{\sim}{A}}_5$						1								
$\underset{\sim}{A}_6$														1
$\overline{\underset{\sim}{A}}_6$													1	
E														1

The successive compositions $\mathbf{\Gamma}^2\underset{\sim}{A}_j$, $\mathbf{\Gamma}^3\underset{\sim}{A}_j$,..., are found until a matrix the same as the first is reached. We will therefore have the matrix $\hat{\Gamma}\underset{\sim}{A}_j$. In effect:

$\Gamma^2 \underset{\sim}{A}_j =$

	$\varnothing$	$\underset{\sim}{A}_1$	$\overline{\underset{\sim}{A}}_1$	$\underset{\sim}{A}_2$	$\overline{\underset{\sim}{A}}_2$	$\underset{\sim}{A}_3$	$\overline{\underset{\sim}{A}}_3$	$\underset{\sim}{A}_4$	$\overline{\underset{\sim}{A}}_4$	$\underset{\sim}{A}_5$	$\overline{\underset{\sim}{A}}_5$	$\underset{\sim}{A}_6$	$\overline{\underset{\sim}{A}}_6$	E
$\varnothing$	1													
$\underset{\sim}{A}_1$						1								
$\overline{\underset{\sim}{A}}_1$														1
$\underset{\sim}{A}_2$				1										
$\overline{\underset{\sim}{A}}_2$												1		
$\underset{\sim}{A}_3$														1
$\overline{\underset{\sim}{A}}_3$							1							
$\underset{\sim}{A}_4$										1				
$\overline{\underset{\sim}{A}}_4$														1
$\underset{\sim}{A}_5$										1				
$\overline{\underset{\sim}{A}}_5$												1		
$\underset{\sim}{A}_6$														1
$\overline{\underset{\sim}{A}}_6$													1	
E														1

$\Gamma^3 \underset{\sim}{A}_j =$

	$\varnothing$	$\underset{\sim}{A}_1$	$\overline{\underset{\sim}{A}}_1$	$\underset{\sim}{A}_2$	$\overline{\underset{\sim}{A}}_2$	$\underset{\sim}{A}_3$	$\overline{\underset{\sim}{A}}_3$	$\underset{\sim}{A}_4$	$\overline{\underset{\sim}{A}}_4$	$\underset{\sim}{A}_5$	$\overline{\underset{\sim}{A}}_5$	$\underset{\sim}{A}_6$	$\overline{\underset{\sim}{A}}_6$	E
$\varnothing$	1													
$\underset{\sim}{A}_1$												1		
$\overline{\underset{\sim}{A}}_1$														1
$\underset{\sim}{A}_2$				1										
$\overline{\underset{\sim}{A}}_2$												1		
$\underset{\sim}{A}_3$														1
$\overline{\underset{\sim}{A}}_3$							1							
$\underset{\sim}{A}_4$										1				
$\overline{\underset{\sim}{A}}_4$														1
$\underset{\sim}{A}_5$										1				
$\overline{\underset{\sim}{A}}_5$														1
$\underset{\sim}{A}_6$														1
$\overline{\underset{\sim}{A}}_6$													1	
E														1

$\Gamma^4 \underset{\sim}{A}_j =$

	$\varnothing$	$\underset{\sim}{A}_1$	$\overline{\underset{\sim}{A}}_1$	$\underset{\sim}{A}_2$	$\overline{\underset{\sim}{A}}_2$	$\underset{\sim}{A}_3$	$\overline{\underset{\sim}{A}}_3$	$\underset{\sim}{A}_4$	$\overline{\underset{\sim}{A}}_4$	$\underset{\sim}{A}_5$	$\overline{\underset{\sim}{A}}_5$	$\underset{\sim}{A}_6$	$\overline{\underset{\sim}{A}}_6$	E
$\varnothing$	1													
$\underset{\sim}{A}_1$														1
$\overline{\underset{\sim}{A}}_1$														1
$\underset{\sim}{A}_2$				1										
$\overline{\underset{\sim}{A}}_2$														1
$\underset{\sim}{A}_3$														1
$\overline{\underset{\sim}{A}}_3$							1							
$\underset{\sim}{A}_4$										1				
$\overline{\underset{\sim}{A}}_4$														1
$\underset{\sim}{A}_5$										1				
$\overline{\underset{\sim}{A}}_5$														1
$\underset{\sim}{A}_6$														1
$\overline{\underset{\sim}{A}}_6$													1	
E														1

$\Gamma^5 \underset{\sim}{A}_j =$

	$\varnothing$	$\underset{\sim}{A}_1$	$\overline{\underset{\sim}{A}}_1$	$\underset{\sim}{A}_2$	$\overline{\underset{\sim}{A}}_2$	$\underset{\sim}{A}_3$	$\overline{\underset{\sim}{A}}_3$	$\underset{\sim}{A}_4$	$\overline{\underset{\sim}{A}}_4$	$\underset{\sim}{A}_5$	$\overline{\underset{\sim}{A}}_5$	$\underset{\sim}{A}_6$	$\overline{\underset{\sim}{A}}_6$	E
$\varnothing$	1													
$\underset{\sim}{A}_1$														1
$\overline{\underset{\sim}{A}}_1$														1
$\underset{\sim}{A}_2$				1										
$\overline{\underset{\sim}{A}}_2$														1
$\underset{\sim}{A}_3$														1
$\overline{\underset{\sim}{A}}_3$							1							
$\underset{\sim}{A}_4$										1				
$\overline{\underset{\sim}{A}}_4$														1
$\underset{\sim}{A}_5$										1				
$\overline{\underset{\sim}{A}}_5$														1
$\underset{\sim}{A}_6$														1
$\overline{\underset{\sim}{A}}_6$													1	
E														1

We stop here as $\Gamma^5 \underset{\sim}{A}_j = \Gamma^4 \underset{\sim}{A}_j$ and, therefore:

$$\hat{\Gamma} \underset{\sim}{A}_j = \Gamma^5 \underset{\sim}{A}_j = \Gamma^4 \underset{\sim}{A}_j$$

We have, in short.

$\Gamma \underset{\sim}{A}_j =$

	$\emptyset$	$\underset{\sim}{A}_1$	$\overline{\underset{\sim}{A}}_1$	$\underset{\sim}{A}_2$	$\overline{\underset{\sim}{A}}_2$	$\underset{\sim}{A}_3$	$\overline{\underset{\sim}{A}}_3$	$\underset{\sim}{A}_4$	$\overline{\underset{\sim}{A}}_4$	$\underset{\sim}{A}_5$	$\overline{\underset{\sim}{A}}_5$	$\underset{\sim}{A}_6$	$\overline{\underset{\sim}{A}}_6$	E
$\emptyset$	1													
$\underset{\sim}{A}_1$														1
$\overline{\underset{\sim}{A}}_1$														1
$\underset{\sim}{A}_2$				1										
$\overline{\underset{\sim}{A}}_2$														1
$\underset{\sim}{A}_3$														1
$\overline{\underset{\sim}{A}}_3$							1							
$\underset{\sim}{A}_4$										1				
$\overline{\underset{\sim}{A}}_4$														1
$\underset{\sim}{A}_5$										1				
$\overline{\underset{\sim}{A}}_5$														1
$\underset{\sim}{A}_6$														1
$\overline{\underset{\sim}{A}}_6$													1	
E														1

For a clearer picture we present this process in sagittate form.

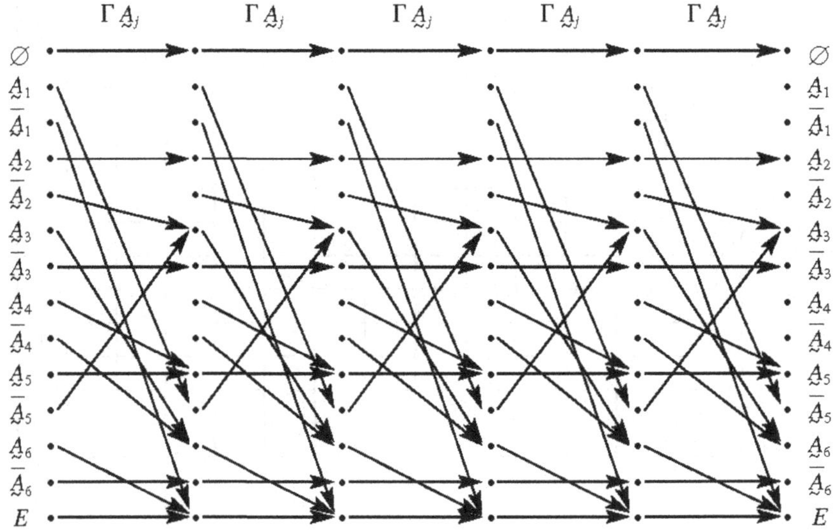

Which gives, in short:

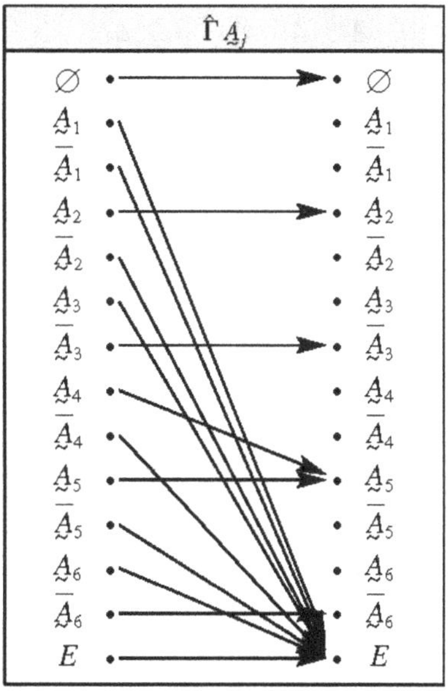

We will now continue with $\hat{\delta}\underset{\sim}{A}_j$, beginning from $\delta\underset{\sim}{A}_j$:

$\delta\underset{\sim}{A}_j =$

	∅	$\underset{\sim}{A}_1$	$\overline{\underset{\sim}{A}}_1$	$\underset{\sim}{A}_2$	$\overline{\underset{\sim}{A}}_2$	$\underset{\sim}{A}_3$	$\overline{\underset{\sim}{A}}_3$	$\underset{\sim}{A}_4$	$\overline{\underset{\sim}{A}}_4$	$\underset{\sim}{A}_5$	$\overline{\underset{\sim}{A}}_5$	$\underset{\sim}{A}_6$	$\overline{\underset{\sim}{A}}_6$	E
∅	1													
$\underset{\sim}{A}_1$	1													
$\overline{\underset{\sim}{A}}_1$										1				
$\underset{\sim}{A}_2$							1							
$\overline{\underset{\sim}{A}}_2$					1									
$\underset{\sim}{A}_3$						1								
$\overline{\underset{\sim}{A}}_3$													1	
$\underset{\sim}{A}_4$													1	
$\overline{\underset{\sim}{A}}_4$											1			
$\underset{\sim}{A}_5$							1							
$\overline{\underset{\sim}{A}}_5$											1			
$\underset{\sim}{A}_6$												1		
$\overline{\underset{\sim}{A}}_6$	1													
E														1

One successively finds:

$\delta^2 \underset{\sim}{A}_j =$

	$\varnothing$	$\underset{\sim}{A}_1$	$\overline{\underset{\sim}{A}}_1$	$\underset{\sim}{A}_2$	$\overline{\underset{\sim}{A}}_2$	$\underset{\sim}{A}_3$	$\overline{\underset{\sim}{A}}_3$	$\underset{\sim}{A}_4$	$\overline{\underset{\sim}{A}}_4$	$\underset{\sim}{A}_5$	$\overline{\underset{\sim}{A}}_5$	$\underset{\sim}{A}_6$	$\overline{\underset{\sim}{A}}_6$	E
$\varnothing$	1													
$\underset{\sim}{A}_1$	1													
$\overline{\underset{\sim}{A}}_1$							1							
$\underset{\sim}{A}_2$													1	
$\overline{\underset{\sim}{A}}_2$					1									
$\underset{\sim}{A}_3$						1								
$\overline{\underset{\sim}{A}}_3$	1													
$\underset{\sim}{A}_4$	1													
$\overline{\underset{\sim}{A}}_4$											1			
$\underset{\sim}{A}_5$													1	
$\overline{\underset{\sim}{A}}_5$											1			
$\underset{\sim}{A}_6$												1		
$\overline{\underset{\sim}{A}}_6$	1													
E														1

$\delta^3 \underset{\sim}{A}_j =$

	$\varnothing$	$\underset{\sim}{A}_1$	$\overline{\underset{\sim}{A}}_1$	$\underset{\sim}{A}_2$	$\overline{\underset{\sim}{A}}_2$	$\underset{\sim}{A}_3$	$\overline{\underset{\sim}{A}}_3$	$\underset{\sim}{A}_4$	$\overline{\underset{\sim}{A}}_4$	$\underset{\sim}{A}_5$	$\overline{\underset{\sim}{A}}_5$	$\underset{\sim}{A}_6$	$\overline{\underset{\sim}{A}}_6$	E
$\varnothing$	1													
$\underset{\sim}{A}_1$	1													
$\overline{\underset{\sim}{A}}_1$													1	
$\underset{\sim}{A}_2$	1													
$\overline{\underset{\sim}{A}}_2$					1									
$\underset{\sim}{A}_3$						1								
$\overline{\underset{\sim}{A}}_3$	1													
$\underset{\sim}{A}_4$	1													
$\overline{\underset{\sim}{A}}_4$											1			
$\underset{\sim}{A}_5$	1													
$\overline{\underset{\sim}{A}}_5$											1			
$\underset{\sim}{A}_6$												1		
$\overline{\underset{\sim}{A}}_6$	1													
E														1

$\delta^4 \underset{\sim}{A}_j =$

	$\emptyset$	$\underset{\sim}{A}_1$	$\overline{\underset{\sim}{A}}_1$	$\underset{\sim}{A}_2$	$\overline{\underset{\sim}{A}}_2$	$\underset{\sim}{A}_3$	$\overline{\underset{\sim}{A}}_3$	$\underset{\sim}{A}_4$	$\overline{\underset{\sim}{A}}_4$	$\underset{\sim}{A}_5$	$\overline{\underset{\sim}{A}}_5$	$\underset{\sim}{A}_6$	$\overline{\underset{\sim}{A}}_6$	E
$\emptyset$	1													
$\underset{\sim}{A}_1$	1													
$\overline{\underset{\sim}{A}}_1$	1													
$\underset{\sim}{A}_2$	1													
$\overline{\underset{\sim}{A}}_2$					1									
$\underset{\sim}{A}_3$						1								
$\overline{\underset{\sim}{A}}_3$	1													
$\underset{\sim}{A}_4$	1													
$\overline{\underset{\sim}{A}}_4$											1			
$\underset{\sim}{A}_5$	1													
$\overline{\underset{\sim}{A}}_5$											1			
$\underset{\sim}{A}_6$												1		
$\overline{\underset{\sim}{A}}_6$	1													
E														1

$\delta^5 \underset{\sim}{A}_j =$

	$\emptyset$	$\underset{\sim}{A}_1$	$\overline{\underset{\sim}{A}}_1$	$\underset{\sim}{A}_2$	$\overline{\underset{\sim}{A}}_2$	$\underset{\sim}{A}_3$	$\overline{\underset{\sim}{A}}_3$	$\underset{\sim}{A}_4$	$\overline{\underset{\sim}{A}}_4$	$\underset{\sim}{A}_5$	$\overline{\underset{\sim}{A}}_5$	$\underset{\sim}{A}_6$	$\overline{\underset{\sim}{A}}_6$	E
$\emptyset$	1													
$\underset{\sim}{A}_1$	1													
$\overline{\underset{\sim}{A}}_1$	1													
$\underset{\sim}{A}_2$	1													
$\overline{\underset{\sim}{A}}_2$					1									
$\underset{\sim}{A}_3$						1								
$\overline{\underset{\sim}{A}}_3$	1													
$\underset{\sim}{A}_4$	1													
$\overline{\underset{\sim}{A}}_4$											1			
$\underset{\sim}{A}_5$	1													
$\overline{\underset{\sim}{A}}_5$											1			
$\underset{\sim}{A}_6$												1		
$\overline{\underset{\sim}{A}}_6$	1													
E														1

It therefore results that $\delta^4 \underset{\sim}{A}_j = \delta^5 \underset{\sim}{A}_j = \hat{\delta} \underset{\sim}{A}_j$.

Sagittate presentation gives:

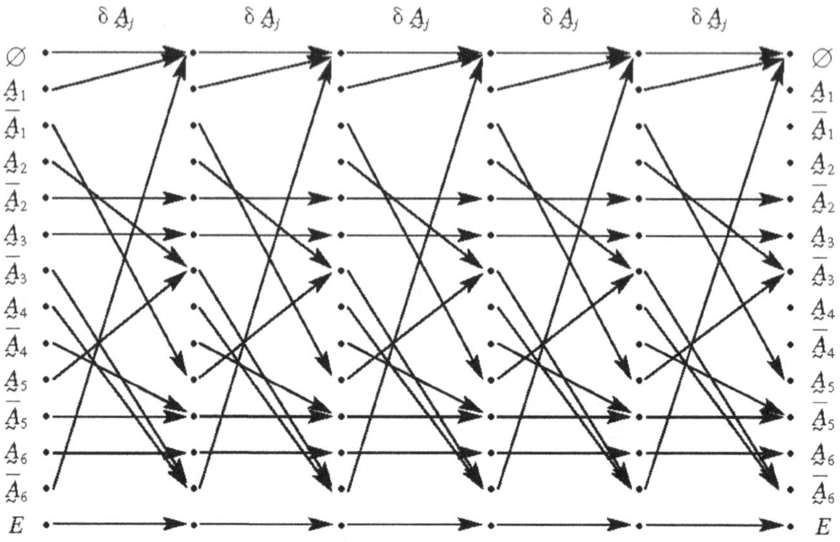

Which we can call a "Moore opening", as opposed to a "Moore closing, which can be expressed in the following way:

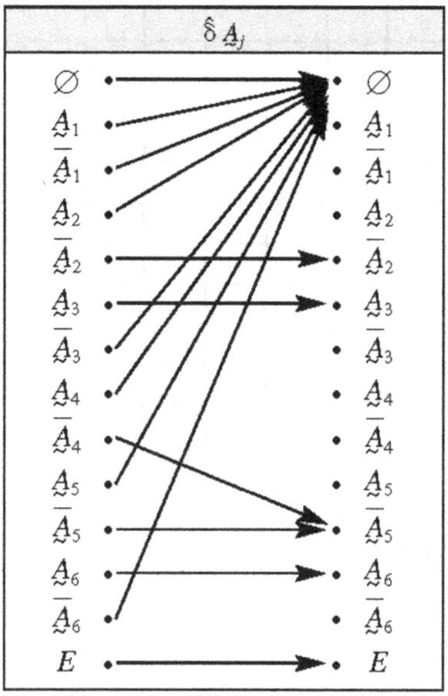

Upon seeing the previous graph we can ask about the cause which gives rise to the existence of an arc, the descending $(\overline{\underset{\sim}{A}}_4, \overline{\underset{\sim}{A}}_5)$, when they should all usually be **non-descendent**. We have advanced the answer when, upon constructing the lattice of $K(E)$, the existence of a partial order was shown. If adequate ordering is established the following results:

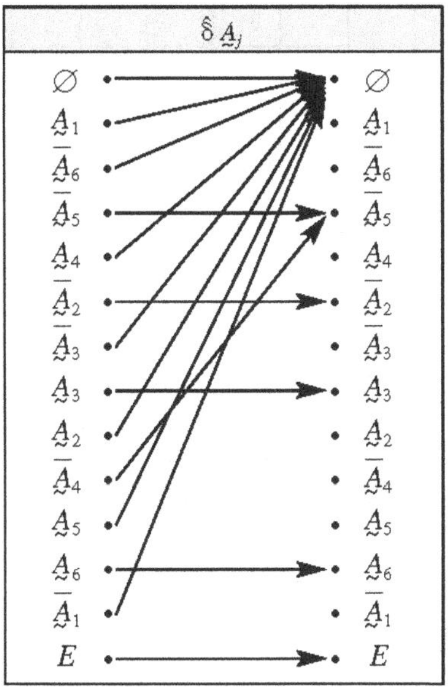

in which the arcs are now **non-descending**.

As it could not be in any other way, the sets of opens and closes are the same as those initially written, which is to say:

$$A_b = \{\varnothing, \overline{\underset{\sim}{A}}_2, \underset{\sim}{A}_3, \overline{\underset{\sim}{A}}_5, \underset{\sim}{A}_6, E\}$$

$$C_e = \{\varnothing, \underset{\sim}{A}_2, \overline{\underset{\sim}{A}}_3, \underset{\sim}{A}_5, \overline{\underset{\sim}{A}}_6, E\}$$

b) When we move further into the second way everything is easier because, in our case, as the functional application is its own transitive close it automatically has this property. To confirm this we will express $\boldsymbol{\Gamma}$ in its matrix form and perform the maxmin composition of the corresponding matrix with itself:

$\Gamma =$

	$\varnothing$	$\underset{\sim}{A}_1$	$\underset{\sim}{A}_2$	$\underset{\sim}{A}_3$	$\underset{\sim}{A}_4$	$\underset{\sim}{A}_5$	$\underset{\sim}{A}_6$	E
$\varnothing$	1							
$\underset{\sim}{A}_1$						1		
$\underset{\sim}{A}_2$			1					
$\underset{\sim}{A}_3$						1		
$\underset{\sim}{A}_4$								1
$\underset{\sim}{A}_5$						1		
$\underset{\sim}{A}_6$								1
E								1

$\Gamma^2 =$

	$\varnothing$	$\underset{\sim}{A}_1$	$\underset{\sim}{A}_2$	$\underset{\sim}{A}_3$	$\underset{\sim}{A}_4$	$\underset{\sim}{A}_5$	$\underset{\sim}{A}_6$	E
$\varnothing$	1							
$\underset{\sim}{A}_1$						1		
$\underset{\sim}{A}_2$			1					
$\underset{\sim}{A}_3$						1		
$\underset{\sim}{A}_4$								1
$\underset{\sim}{A}_5$						1		
$\underset{\sim}{A}_6$								1
E								1

It can be seen that $\mathbf{\Gamma}^2$ is the same as $\mathbf{\Gamma}$ and therefore no matter how much it continues to convolve, the same matrix will always be obtained, which is to say $\hat{\Gamma}$.

We will now see in detail that the pretopological space of our example is idempotent.

$$\Gamma(\Gamma\varnothing) = \Gamma\varnothing = \varnothing \qquad \Gamma\varnothing = \varnothing$$
$$\Gamma(\Gamma\underset{\sim}{A}_1) = \Gamma\underset{\sim}{A}_5 = \underset{\sim}{A}_5 \qquad \Gamma\underset{\sim}{A}_1 = \underset{\sim}{A}_5$$
$$\Gamma(\Gamma\underset{\sim}{A}_2) = \Gamma\underset{\sim}{A}_2 = \underset{\sim}{A}_2 \qquad \Gamma\underset{\sim}{A}_2 = \underset{\sim}{A}_2$$
$$\Gamma(\Gamma\underset{\sim}{A}_3) = \Gamma\underset{\sim}{A}_5 = \underset{\sim}{A}_5 \qquad \Gamma\underset{\sim}{A}_3 = \underset{\sim}{A}_5$$
$$\Gamma(\Gamma\underset{\sim}{A}_4) = \Gamma E = E \qquad \Gamma\underset{\sim}{A}_4 = E$$
$$\Gamma(\Gamma\underset{\sim}{A}_5) = \Gamma\underset{\sim}{A}_5 = \underset{\sim}{A}_5 \qquad \Gamma\underset{\sim}{A}_5 = \underset{\sim}{A}_5$$
$$\Gamma(\Gamma\underset{\sim}{A}_6) = \Gamma E = E \qquad \Gamma\underset{\sim}{A}_6 = E$$
$$\Gamma(\Gamma E) = \Gamma E = E \qquad \Gamma E = E$$

As it could not be in another way, the fulfillment of the condition of idempotency has been clearly shown. Given that it was previously seen that uncertain pretopology has the properties of extension and isotonia, we find ourselves before a **Moore closing**, which also has the property $\mathbf{\Gamma}\,\varnothing = \varnothing$.

4 Distributive Uncertain Pretopology

We will now move on to define a distributive uncertain pretopological space. If the following property is demanded from an isotone pretopology:

$\forall\, \underset{\sim}{A}_j, \underset{\sim}{A}_k \in K(E)$:

$$\Gamma(\underset{\sim}{A}_j \cup \underset{\sim}{A}_k) = \Gamma\underset{\sim}{A}_j \cup \Gamma\underset{\sim}{A}_k$$

in which if $K(E) \subset P(E)$ we find ourselves before a **partially distributive uncertain pretopology** and if $K(E) = P(E)$ it is said that the uncertain pretopology is distributive. We can also express:

$\forall \underset{\sim}{A}_j, \underset{\sim}{A}_k \in K(E)$:

$$\delta (\underset{\sim}{A}_j \cap \underset{\sim}{A}_k) = \delta \underset{\sim}{A}_j \cap \delta \underset{\sim}{A}_k$$

In this way we have the following axiomatic:

1. $\Gamma \varnothing = \varnothing$
2. $\forall \underset{\sim}{A}_j \in K(E)$:

$$\underset{\sim}{A}_j \subset \Gamma \underset{\sim}{A}_j$$

3. $\forall \underset{\sim}{A}_j, \underset{\sim}{A}_k \subset K(E)$:

$$(\underset{\sim}{A}_j \subset \underset{\sim}{A}_k) \Rightarrow (\Gamma \underset{\sim}{A}_j \subset \Gamma \underset{\sim}{A}_k)$$

4. $\forall \underset{\sim}{A}_j, \underset{\sim}{A}_k \in K(E)$:

$$\Gamma (\underset{\sim}{A}_j \cup \underset{\sim}{A}_k) \Rightarrow (\Gamma \underset{\sim}{A}_j \cup \Gamma \underset{\sim}{A}_k)$$

All distributive uncertain pretopological space is isotone, but not all of the isotones have to be distributive due to the necessity of fulfillment of this last axiom.

a) Let us immediately look at an example in the first way. Let:

$$E = \{a, b, c\}$$

In the following table a distributive uncertain pretopology is described[2].

[2] Kaufmann, A.: «Prétopologie ordinaire et prétopologie floue». *Note de travail 115*. La Tronche,1983, pages. 54-56.

$K(E)$

	$\underset{\sim}{A}_j$			$\Gamma\underset{\sim}{A}_j$			$\delta\underset{\sim}{A}_j$			
	a	*b*	*c*	*a*	*b*	*c*	*a*	*b*	*c*	
$\varnothing$	0	0	0	0	0	0	0	0	0	A_b, C_e
$\underset{\sim}{A}_1$	.7	.6	1	.7	.6	1	.5	.6	1	C_e
$\underset{\sim}{A}_2$	.3	.6	1	.5	.6	1	.3	.6	1	A_b
$\underset{\sim}{A}_3$	.7	.5	.5	.7	.5	.5	.5	.5	.5	C_e
$\overline{\underset{\sim}{A}}_1 = \underset{\sim}{A}_4$	.3	.4	0	.5	.4	0	.3	.4	0	A_b
$\overline{\underset{\sim}{A}}_2 = \underset{\sim}{A}_5$	.7	.4	0	.7	.4	0	.5	.4	0	C_e
$\underset{\sim}{A}_6$	.5	.5	.5	.5	.5	.5	.5	.5	.5	A_b, C_e
$\overline{\underset{\sim}{A}}_3 = \underset{\sim}{A}_7$	.3	.5	.5	.5	.5	.5	.3	.5	.5	A_b
$\underset{\sim}{A}_8$	.5	.6	1	.5	.6	1	.5	.6	.1	A_b, C_e
$\overline{\underset{\sim}{A}}_8 = \underset{\sim}{A}_9$	.5	.4	0	.5	.4	0	.5	.4	0	A_b, C_e
E	1	1	1	1	1	1	1	1	1	A_b, C_e

Let us confirm that the condition of distributivity is met:

	(1)			(2)			(3)			(4)			(5)			(6)			(7)		
	$\underset{\sim}{A}_j$			$\underset{\sim}{A}_k$			$\underset{\sim}{A}_j \cup \underset{\sim}{A}_k$			$\Gamma(\underset{\sim}{A}_j \cup \underset{\sim}{A}_k)$			$\Gamma\underset{\sim}{A}_j$			$\Gamma\underset{\sim}{A}_k$			$\Gamma\underset{\sim}{A}_j \cup \underset{\sim}{A}_k$		
	a	*b*	*c*	*a*	*b*	*c*	*a*	*b*	*c*	*a*	*b*	*c*	*a*	*b*	*c*	*a*	*b*	*c*	*a*	*b*	*c*
$\underset{\sim}{A}_1, \underset{\sim}{A}_2$	.7	.6	1	.3	.6	1	.7	.6	1	.7	.6	1	.7	.6	1	.5	.6	1	.7	.6	1
$\underset{\sim}{A}_1, \underset{\sim}{A}_3$	.7	.6	1	.7	.5	.5	.7	.6	1	.7	.6	1	.7	.6	1	.7	.5	.5	.7	.6	1
$\underset{\sim}{A}_1, \underset{\sim}{A}_4$	.7	.6	1	.3	.4	0	.7	.6	1	.7	.6	1	.7	.6	1	.5	.4	0	.7	.6	1
$\underset{\sim}{A}_1, \underset{\sim}{A}_5$	.7	.6	1	.7	.4	0	.7	.6	1	.7	.6	1	.7	.6	1	.7	.4	0	.7	.6	1
$\underset{\sim}{A}_1, \underset{\sim}{A}_6$	.7	.6	1	.5	.5	.5	.7	.6	1	.7	.6	1	.7	.6	1	.5	.5	.5	.7	.6	1
$\underset{\sim}{A}_1, \underset{\sim}{A}_7$	.7	.6	1	.3	.5	.5	.7	.6	1	.7	.6	1	.7	.6	1	.5	.5	.5	.7	.6	1
$\underset{\sim}{A}_1, \underset{\sim}{A}_8$	.7	.6	1	.5	.6	1	.7	.6	1	.7	.6	1	.7	.6	1	.5	.6	1	.7	.6	1
$\underset{\sim}{A}_1, \underset{\sim}{A}_9$	.7	.6	1	.5	.4	0	.7	.6	1	.7	.6	1	.7	.6	1	.5	.4	0	.7	.6	1
$\underset{\sim}{A}_2, \underset{\sim}{A}_3$	.3	.6	1	.7	.5	.5	.7	.6	1	.7	.6	1	.5	.6	1	.7	.5	.5	.7	.6	1

	(1)			(2)			(3)			(4)			(5)			(6)			(7)		
	$\underset{\sim}{A}_j$			$\underset{\sim}{A}_k$			$\underset{\sim}{A}_j \cup \underset{\sim}{A}_k$			$\Gamma(\underset{\sim}{A}_j \cup \underset{\sim}{A}_k)$			$\Gamma \underset{\sim}{A}_j$			$\Gamma \underset{\sim}{A}_k$			$\Gamma \underset{\sim}{A}_j \cup \underset{\sim}{A}_k$		
	a	*b*	*c*	*a*	*b*	*c*	*a*	*b*	*c*	*a*	*b*	*c*	*a*	*b*	*c*	*a*	*b*	*c*	*a*	*b*	*c*
$\underset{\sim}{A}_2, \underset{\sim}{A}_4$	.3	.6	1	.3	.4	0	.3	.6	1	.5	.6	1	.5	.6	1	.5	.4	0	.5	.6	1
$\underset{\sim}{A}_2, \underset{\sim}{A}_5$	.3	.6	1	.7	.4	0	.7	.6	1	.7	.6	1	.5	.6	1	.7	.4	0	.7	.6	1
$\underset{\sim}{A}_2, \underset{\sim}{A}_6$	.3	.6	1	.5	.5	.5	.5	.6	1	.5	.6	1	.5	.6	1	.5	.5	.5	.5	.6	1
$\underset{\sim}{A}_2, \underset{\sim}{A}_7$	.3	.6	1	.3	.5	.5	.3	.6	1	.5	.6	1	.5	.6	1	.5	.5	.5	.5	.6	1
$\underset{\sim}{A}_2, \underset{\sim}{A}_8$	.3	.6	1	.5	.6	.1	.5	.6	1	.5	.6	1	.5	.6	1	.5	.6	.1	.5	.6	1
$\underset{\sim}{A}_2, \underset{\sim}{A}_9$	.3	.6	1	.5	.4	0	.5	.6	1	.5	.6	1	.5	.6	1	.5	.4	0	.5	.6	1
$\underset{\sim}{A}_3, \underset{\sim}{A}_4$	.7	.5	.5	.3	.4	0	.7	.5	.5	.7	.5	.5	.7	.5	.5	.5	.4	0	.7	.5	.5
$\underset{\sim}{A}_3, \underset{\sim}{A}_5$	.7	.5	.5	.7	.4	0	.7	.5	.5	.7	.5	.5	.7	.5	.5	.7	.4	0	.7	.5	.5
$\underset{\sim}{A}_3, \underset{\sim}{A}_6$	.7	.5	.5	.5	.5	.5	.7	.5	.5	.7	.5	.5	.7	.5	.5	.5	.5	.5	.7	.5	.5
$\underset{\sim}{A}_3, \underset{\sim}{A}_7$	.7	.5	.5	.3	.5	.5	.7	.5	.5	.7	.5	.5	.7	.5	.5	.5	.5	.5	.7	.5	.5
$\underset{\sim}{A}_3, \underset{\sim}{A}_8$	.7	.5	.5	.5	.6	1	.7	.6	1	.7	.6	1	.7	.5	.5	.5	.6	1	.7	.6	1
$\underset{\sim}{A}_3, \underset{\sim}{A}_9$	.7	.5	.5	.5	.4	0	.7	.5	.5	.7	.5	.5	.7	.5	.5	.5	.4	0	.7	.5	.5
$\underset{\sim}{A}_4, \underset{\sim}{A}_5$	.3	.4	0	.7	.4	0	.7	.4	0	.7	.4	0	.5	.4	0	.7	.4	0	.7	.4	0
$\underset{\sim}{A}_4, \underset{\sim}{A}_6$	.3	.4	0	.5	.5	.5	.5	.5	.5	.5	.5	.5	.5	.4	0	.5	.5	.5	.5	.5	.5
$\underset{\sim}{A}_4, \underset{\sim}{A}_7$	.3	.4	0	.3	.5	.5	.3	.5	.5	.5	.5	.5	.5	.4	0	.5	.5	.5	.5	.5	.5
$\underset{\sim}{A}_4, \underset{\sim}{A}_8$	.3	.4	0	.5	.6	1	.5	.6	1	.5	.6	1	.5	.4	0	.5	.6	.1	.5	.6	1
$\underset{\sim}{A}_4, \underset{\sim}{A}_9$	.3	.4	0	.5	.4	0	.5	.4	0	.5	.4	0	.5	.4	0	.5	.4	0	.5	.4	0
$\underset{\sim}{A}_5, \underset{\sim}{A}_6$	.7	.4	0	.5	.5	.5	.7	.5	.5	.7	.5	.5	.7	.4	0	.5	.5	.5	.7	.5	.5
$\underset{\sim}{A}_5, \underset{\sim}{A}_7$	.7	.4	0	.3	.5	.5	.7	.5	.5	.7	.5	.5	.7	.4	0	.5	.5	.5	.7	.5	.5
$\underset{\sim}{A}_5, \underset{\sim}{A}_8$	.7	.4	0	.5	.6	.1	.7	.6	1	.7	.6	1	.7	.4	0	.5	.6	1	.7	.6	1
$\underset{\sim}{A}_5, \underset{\sim}{A}_9$	.7	.4	0	.5	.4	0	.7	.4	0	.7	.4	0	.7	.4	0	.5	.4	0	.7	.4	0
$\underset{\sim}{A}_6, \underset{\sim}{A}_7$	.5	.5	.5	.3	.5	.5	.5	.5	.5	.5	.5	.5	.5	.5	.5	.5	.5	.5	.5	.5	.5
$\underset{\sim}{A}_6, \underset{\sim}{A}_8$	.5	.5	.5	.5	.6	1	.5	.6	1	.5	.6	1	.5	.5	.5	.5	.6	.1	.5	.6	1
$\underset{\sim}{A}_6, \underset{\sim}{A}_9$	.5	.5	.5	.5	.4	0	.5	.5	.5	.5	.5	.5	.5	.5	.5	.5	.4	.0	.5	.5	.5
$\underset{\sim}{A}_7, \underset{\sim}{A}_8$	.3	.5	.5	.5	.6	1	.5	.6	1	.5	.6	1	.5	.5	.5	.5	.6	1	.5	.6	1

	(1)	(2)	(3)	(4)	(5)	(6)	(7)
	$\underset{\sim}{A}_j$	$\underset{\sim}{A}_k$	$\underset{\sim}{A}_j \cup \underset{\sim}{A}_k$	$\Gamma(\underset{\sim}{A}_j \cup \underset{\sim}{A}_k)$	$\Gamma \underset{\sim}{A}_j$	$\Gamma \underset{\sim}{A}_k$	$\Gamma \underset{\sim}{A}_j \cup \underset{\sim}{A}_k$
	a b c	a b c	a b c	a b c	a b c	a b c	a b c
$\underset{\sim}{A}_7, \underset{\sim}{A}_9$	.3 .5 .5	.5 .4 0	.5 .5 .5	.5 .5 .5	.5 .5 .5	.5 .4 0	.5 .5 .5
$\underset{\sim}{A}_8, \underset{\sim}{A}_9$	.5 .6 1	.5 .4 0	.5 .6 1	.5 .6 1	.5 .6 1	.5 .4 0	.5 .6 1

The verification of $\Gamma\,(\underset{\sim}{A}_j \cup \underset{\sim}{A}_k) \Rightarrow \Gamma\,\underset{\sim}{A}_j \cup \Gamma\,\underset{\sim}{A}_k$ also brings the verification of $\delta\,(\underset{\sim}{A}_j \cap \underset{\sim}{A}_k) \Rightarrow \delta\,\underset{\sim}{A}_j \cap \delta\,\underset{\sim}{A}_k$ In this way we obviously find ourselves before a partially distributive uncertain pretopology of E. The opens of this pretopology come given by:

$$A_b = \left\{ \varnothing\ ,\ \begin{array}{ccc} a & b & c \\ \hline .3 & .6 & .1 \end{array}\ ,\ \begin{array}{ccc} a & b & c \\ \hline .3 & .4 & 0 \end{array}\ ,\ \begin{array}{ccc} a & b & c \\ \hline .5 & .5 & .5 \end{array}\ ,\right.$$

$$\left. \begin{array}{ccc} a & b & c \\ \hline .3 & .5 & .5 \end{array}\ ,\ \begin{array}{ccc} a & b & c \\ \hline .5 & .6 & 1 \end{array}\ ,\ \begin{array}{ccc} a & b & c \\ \hline .5 & .4 & 0 \end{array}\ ,\ E \right\}$$

The following figure shows the lattice of these opens, which form a topology.

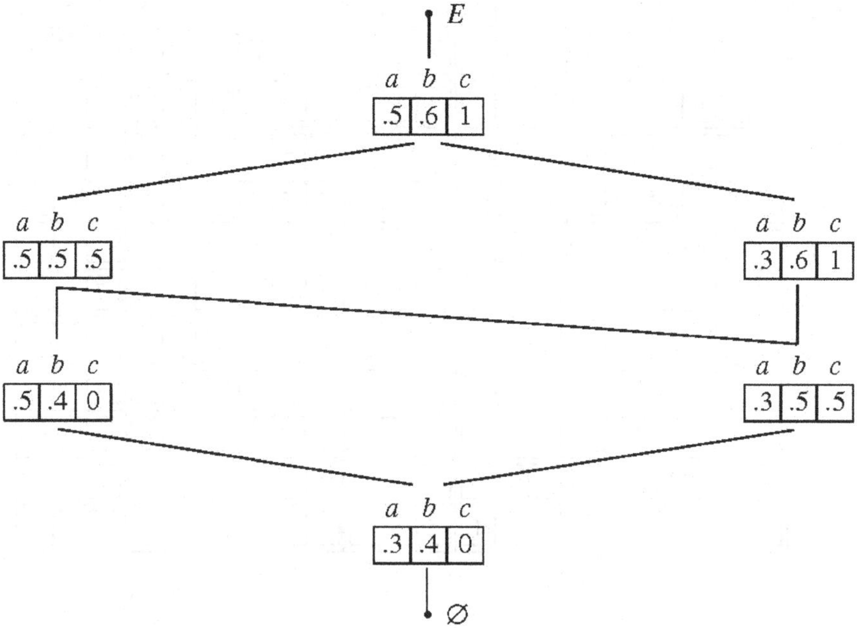

The closes of this distributive uncertain pretopology:

$$C_e = \left\{ \emptyset , \begin{array}{ccc} a & b & c \\ \hline \multicolumn{1}{|c|}{.7} & \multicolumn{1}{c|}{.6} & \multicolumn{1}{c|}{.1} \\ \hline \end{array} , \begin{array}{ccc} a & b & c \\ \hline \multicolumn{1}{|c|}{.7} & \multicolumn{1}{c|}{.5} & \multicolumn{1}{c|}{.5} \\ \hline \end{array} , \begin{array}{ccc} a & b & c \\ \hline \multicolumn{1}{|c|}{.7} & \multicolumn{1}{c|}{.4} & \multicolumn{1}{c|}{0} \\ \hline \end{array} , \begin{array}{ccc} a & b & c \\ \hline \multicolumn{1}{|c|}{.5} & \multicolumn{1}{c|}{.5} & \multicolumn{1}{c|}{.5} \\ \hline \end{array} , \begin{array}{ccc} a & b & c \\ \hline \multicolumn{1}{|c|}{.5} & \multicolumn{1}{c|}{.6} & \multicolumn{1}{c|}{1} \\ \hline \end{array} , \begin{array}{ccc} a & b & c \\ \hline \multicolumn{1}{|c|}{.5} & \multicolumn{1}{c|}{.4} & \multicolumn{1}{c|}{0} \\ \hline \end{array} , E \right\}$$

also form a lattice which we present in the following figure.

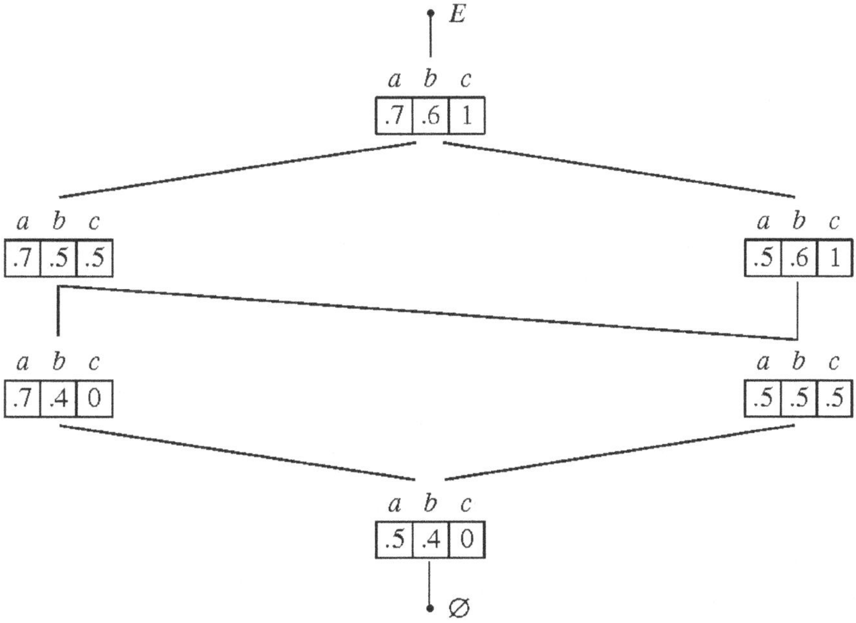

The lattices A_b and C_e are, evidently, isomorphic.
The subset of the opens and closes is:

$$A_b \cap C_e = \left\{ \emptyset , \begin{array}{ccc} a & b & c \\ \hline \multicolumn{1}{|c|}{.5} & \multicolumn{1}{c|}{.5} & \multicolumn{1}{c|}{.5} \\ \hline \end{array} , \begin{array}{ccc} a & b & c \\ \hline \multicolumn{1}{|c|}{.5} & \multicolumn{1}{c|}{.6} & \multicolumn{1}{c|}{1} \\ \hline \end{array} , \begin{array}{ccc} a & b & c \\ \hline \multicolumn{1}{|c|}{.5} & \multicolumn{1}{c|}{.4} & \multicolumn{1}{c|}{0} \\ \hline \end{array} , E \right\}$$

b) We now propose to confirm if the reiterated example of isotone uncertain pretopology of the second way fulfils the axiom corresponding to distributivity. We will begin its verification in the following table.

$\underset{\sim}{A}_j$	$\underset{\sim}{A}_k$	$\underset{\sim}{A}_j \cup \underset{\sim}{A}_k$	$\Gamma(\underset{\sim}{A}_j \cup \underset{\sim}{A}_k)$	$\Gamma\underset{\sim}{A}_j$	$\Gamma\underset{\sim}{A}_k$	$\Gamma\underset{\sim}{A}_j \cup \Gamma\underset{\sim}{A}_k$
$\underset{\sim}{A}_1$	$\underset{\sim}{A}_2$	$\underset{\sim}{A}_4$	E	$\underset{\sim}{A}_5$	$\underset{\sim}{A}_2$	E
$\underset{\sim}{A}_1$	$\underset{\sim}{A}_3$	$\underset{\sim}{A}_5$	$\underset{\sim}{A}_5$	$\underset{\sim}{A}_5$	$\underset{\sim}{A}_5$	$\underset{\sim}{A}_5$
$\underset{\sim}{A}_1$	$\underset{\sim}{A}_4$	$\underset{\sim}{A}_4$	E	$\underset{\sim}{A}_5$	E	E
$\underset{\sim}{A}_1$	$\underset{\sim}{A}_5$	$\underset{\sim}{A}_5$	$\underset{\sim}{A}_5$	$\underset{\sim}{A}_5$	$\underset{\sim}{A}_5$	$\underset{\sim}{A}_5$
$\underset{\sim}{A}_1$	$\underset{\sim}{A}_6$	E	E	$\underset{\sim}{A}_5$	E	E
$\underset{\sim}{A}_1$	E	E	E	$\underset{\sim}{A}_5$	E	E
$\underset{\sim}{A}_2$	$\underset{\sim}{A}_3$	$\underset{\sim}{A}_6$	E	$\underset{\sim}{A}_2$	$\underset{\sim}{A}_5$	E
$\underset{\sim}{A}_2$	$\underset{\sim}{A}_4$	$\underset{\sim}{A}_4$	E	$\underset{\sim}{A}_2$	E	E
$\underset{\sim}{A}_2$	$\underset{\sim}{A}_5$	E	E	$\underset{\sim}{A}_2$	$\underset{\sim}{A}_5$	E
$\underset{\sim}{A}_2$	$\underset{\sim}{A}_6$	$\underset{\sim}{A}_6$	E	$\underset{\sim}{A}_2$	E	E
$\underset{\sim}{A}_2$	E	E	E	$\underset{\sim}{A}_2$	E	E
$\underset{\sim}{A}_3$	$\underset{\sim}{A}_4$	E	E	$\underset{\sim}{A}_5$	E	E
$\underset{\sim}{A}_3$	$\underset{\sim}{A}_5$	$\underset{\sim}{A}_5$	$\underset{\sim}{A}_5$	$\underset{\sim}{A}_5$	$\underset{\sim}{A}_5$	$\underset{\sim}{A}_5$
$\underset{\sim}{A}_3$	$\underset{\sim}{A}_6$	$\underset{\sim}{A}_6$	E	$\underset{\sim}{A}_5$	E	E
$\underset{\sim}{A}_3$	E	E	E	$\underset{\sim}{A}_5$	E	E
$\underset{\sim}{A}_4$	$\underset{\sim}{A}_5$	E	E	$\underset{\sim}{E}$	$\underset{\sim}{A}_5$	E
$\underset{\sim}{A}_4$	$\underset{\sim}{A}_6$	E	E	E	E	E
$\underset{\sim}{A}_4$	E	E	E	$\underset{\sim}{E}$	E	E
$\underset{\sim}{A}_5$	$\underset{\sim}{A}_6$	E	E	$\underset{\sim}{A}_5$	E	E
$\underset{\sim}{A}_5$	E	E	E	$\underset{\sim}{A}_5$	E	E
$\underset{\sim}{A}_6$	E	E	E	E	E	E

In effect, our pretopology is also distributive.
But now, in this way the open set which coincides with the close set:

$$A_b = C_e = \{\varnothing, \{\underset{\sim}{P}_1, \underset{\sim}{P}_6\}, \{\underset{\sim}{P}_2, \underset{\sim}{P}_3, \underset{\sim}{P}_4, \underset{\sim}{P}_5\}, E\}$$

has the structure of a boolean lattice (distributive and complemented) and its representation, here very simple, is:

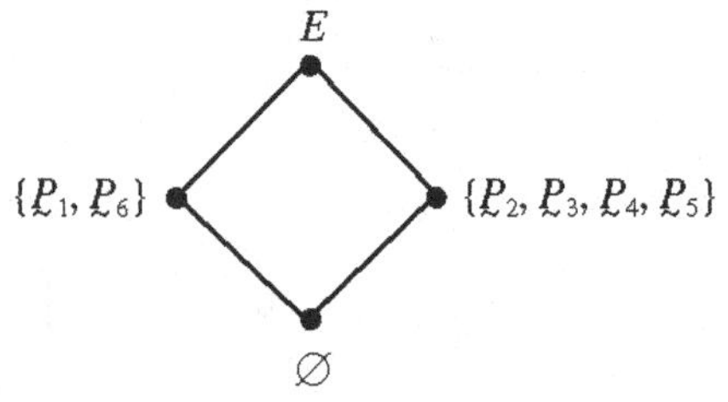

Let us move a little more into pretopological precision.

5 Uncertain Moore Pretopology

It is convenient to repeat that which we have previously stated; a **Moore closing is not always** a pretopology, since $\boldsymbol{\Gamma}\, \varnothing = \varnothing$ is not demanded, a necessary condition in this. This gives us a reason to present the axiomatic of an "Moore's uncertain pretopology". For this we will add this requisite to those demanded in the close, to which we will add the property of distributivity. Therefore, Moore's uncertain pretopology is a distributive uncertain pretopology with idempotency.

The axiomatic of an uncertain Moore pretopology is the following:

1. $\Gamma \varnothing = \varnothing$
2. $\forall \underset{\sim}{A}_j \in K(E)$:

$$\underset{\sim}{A}_j \subset \Gamma\, \underset{\sim}{A}_j \qquad \text{extension}$$

3. $\forall \underset{\sim}{A}_j \in K(E)$:

$$\Gamma(\Gamma\, \underset{\sim}{A}_j) = \Gamma\, \underset{\sim}{A}_j \qquad \text{idempotency}$$

4. $\forall \underset{\sim}{A}_j, \underset{\sim}{A}_k \in K(E)$:

$$(\underset{\sim}{A}_j \subset \underset{\sim}{A}_k) \Rightarrow (\Gamma\, \underset{\sim}{A}_j \subset \Gamma\, \underset{\sim}{A}_k) \qquad \text{isotone}$$

5. $\forall \underset{\sim}{A}_j, \underset{\sim}{A}_k \in K(E)$:

$$\Gamma\,(\underset{\sim}{A}_j \cup \underset{\sim}{A}_k) \Rightarrow (\Gamma\, \underset{\sim}{A}_j \cup \Gamma\, \underset{\sim}{A}_k) \qquad \text{distributivity}$$

As a consequence of the second axiom, the following also results:

$$\boldsymbol{\Gamma}\ E = E$$

In Moore closings we have seen how to confirm or in which way the idempotency is obtained. Distributivity has also been addressed upon making reference to distributive uncertain pretopology. It may result as superfluous to insist on both properties. However, we feel that we must include some brief considerations aided by examples in this part. We will begin with a general and straightforward consideration, starting with a referential E and the finite set $K(E) \subset P(E)$:

$$K(E) = \{\varnothing, \underset{\sim}{A}_1, \underset{\sim}{A}_2 = \overline{\underset{\sim}{A}_1}, \underset{\sim}{A}_3, \underset{\sim}{A}_4 = \overline{\underset{\sim}{A}_3}, \underset{\sim}{A}_5, \underset{\sim}{A}_6 = \overline{\underset{\sim}{A}_5}, E\}$$

Let us assume that:

$$\underset{\sim}{A}_1 \subset \underset{\sim}{A}_3 \subset \underset{\sim}{A}_5$$

For this:

$$\underset{\sim}{A}_6 \subset \underset{\sim}{A}_4 \subset \underset{\sim}{A}_2$$

We establish a Moore pretopology, given by the following adherent and interior applications:

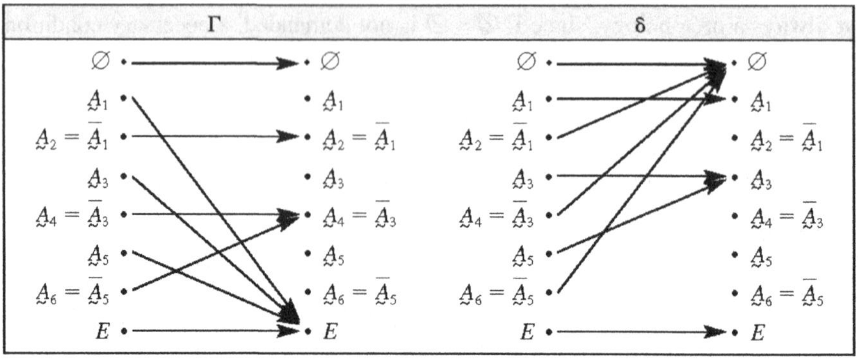

We visually confirm the idempotency:

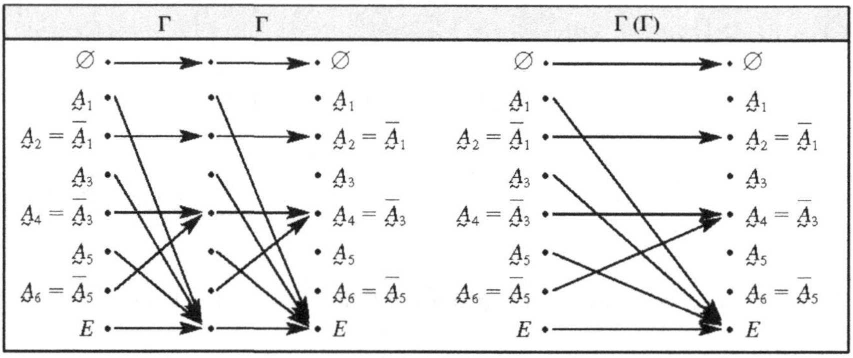

The closes are:

$$C_e = \{\emptyset, \underset{\sim}{A}_2, \underset{\sim}{A}_4, E\}$$

and the opens:

$$A_b = \{\emptyset, \underset{\sim}{A}_1, \underset{\sim}{A}_3, E\}$$

Concerning $\underset{\sim}{A}_5$ and $\underset{\sim}{A}_6$ they are neither open nor closed.

We can now move on to some more specific suppositions. For this we will adopt the examples used in distributive uncertain pretopologies. If the property of **idempotency** is fulfilled we can say that we find ourselves before an uncertain pretopology of Moore.

a) In the first of the ways the adherent application was:

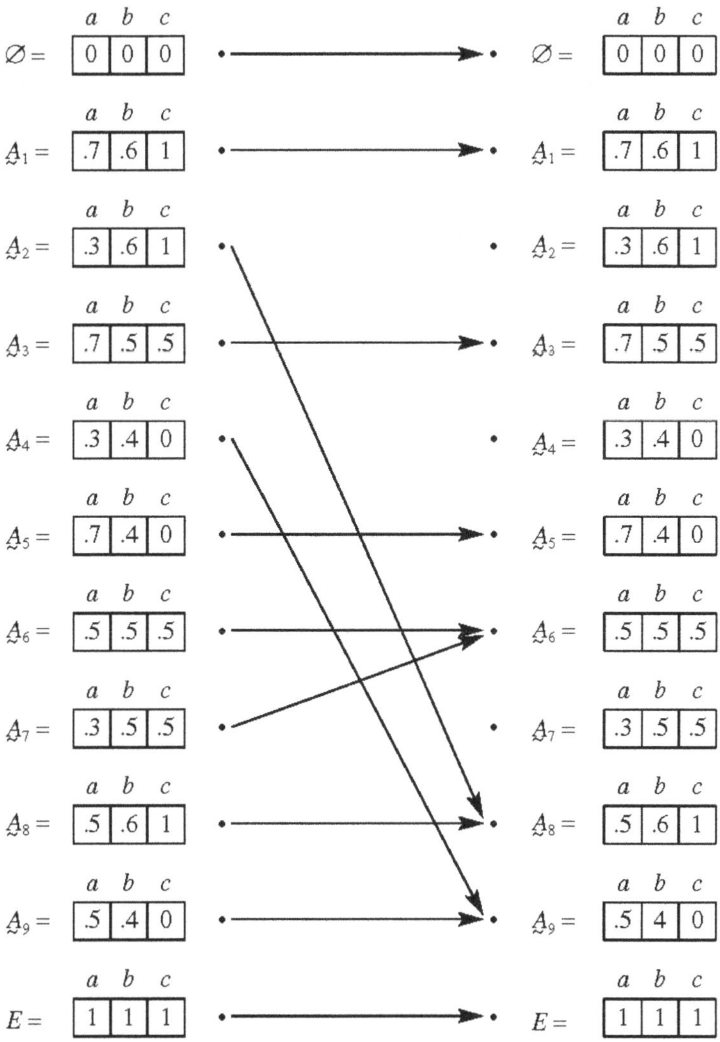

We perform the confirming of the idempotency:

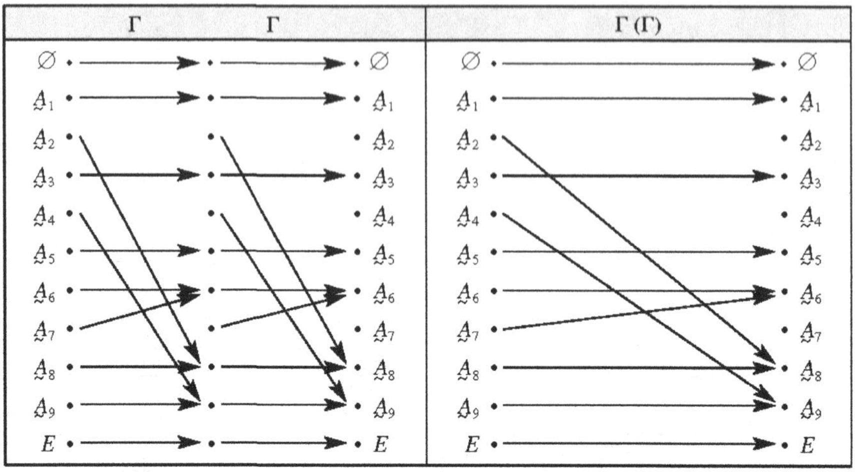

Even though it stands out that **Γ (Γ) = Γ** , we have wanted to once again separate the process which demonstrates that the distributive uncertain pretopology of our example is a Moore's uncertain pretopology.

b) The second of the ways of distributive uncertain pretopology had as an adherent application:

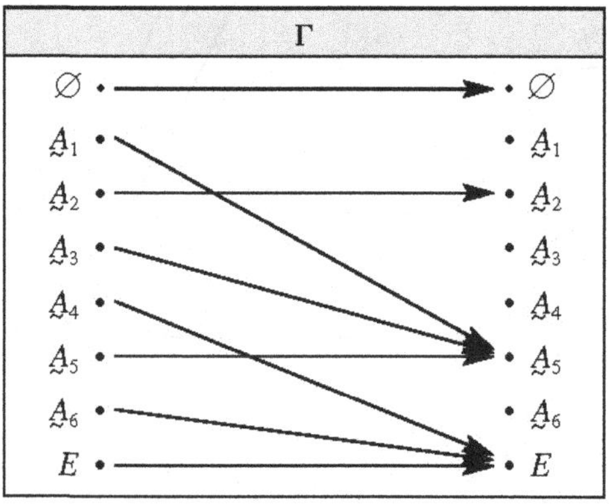

We move on to confirm if the axiom of idempotency is fulfilled:

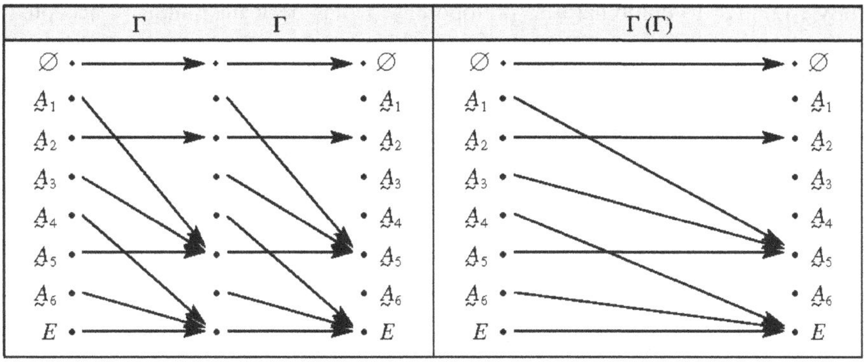

Throughout this exposition we have confirmed that our examples of pretopological space have the properties of extension, idempotency, isotonia and distributivity. We therefore find ourselves before some **uncertain Moore pretopologies.**

Let us remember, once again, the existing difference between a Moore closing and a Moore pretopology, such as we have defined them: in the first $\mathbf{\Gamma}\, \varnothing = \varnothing$ is not demanded, while this condition is necessary in the second.

6 Brief Final Discussion on Pretopological Spaces

Throughout the development of our work it has been possible to observe that as the pretopological space assumed new axioms the cohesion among the elements of the subset $K(E)$ was reinforced. Our central, very singular, examples have passed all of the "tests" until arriving at this end. This does not normally occur in this way. We have wanted to develop this outline in an attempt to cover the widest range of phenomena which reality may present. Fortunately, it will not always be necessary to use uncertain Moore pretopologies. In many cases the isotone uncertain pretopology or distributive uncertain pretopology will be enough (idempotency is not required). The Moore closing has been a basic piece for many studies and, thanks to this, it has been possible to develop the "theory of affinities"[3] which we have used on many occasions in order to solve problems of grouping in economics and management relationships. The susceptible connections to be performed among the distinct common groups now signify the passing to a distinct sphere. Here, in its different axiomatics, uncertain pretopology is providing wide possibilities of developing formal methods capable of giving a complete answer to economics and management problems which arise in a world immersed in uncertainty, and which may provide yet more in the future.

---

[3] Kaufmann, A., y Gil Aluja, J.: «Selection of affinities by means of fuzzy relations and Galois latices ». Proceedings of the XI Euro O. R. Congress. Aachen, 16-19 July 1991.: «Prétopologie ordinaire et prétopologie floue». Note de travail 115. La Tronche, 1983, pages. 54-56.

In this part of the study which we propose, we have incorporated some suggestions received from our research group, "Economics and Management of Uncertainty", relative to combinatorial pretopology. It has been unanimously accepted by the members that our proposal of considering pretopology as a starting point and basic support for a modelling sustained by non-numerical mathematics. It is a task which we have personally imposed in the last lustrum in order to offer **a united body** to those whose worries drive them towards the new twists and turns in which the proposals which **disagree with the geometrical conception** of the economics and management universe still find themselves today.

Finally, an elemental idea of synthesis which is unfortunately forgotten and whose oversight produces more than a few headaches for those, like ourselves, immersed in the area of economics application who wish to use structures typical of spaces for the problems which reality poses: **pretopologies** have their reason for being like **functional applications**. As we will see in continuation, the same does not occur with **topologies** in which the spaces are presented as **sets** taken from a subset of a referential which may acquire a different nature depending on the intentions of those who must give solutions to complex problems.

Part II

Towards an Idea of Uncertain Topological Space

Chapter 3
Topology

1 Genesis of a Topology

In the introductory part of our work we have attempted to provide a visual idea of **topological** space. For this concepts such as **environment**, **proximity**, **adherence**, **accumulation**, **isolation** and **border** have been turned to, all of them taken from the texts which may today be considered classic in topological literature. It now corresponds to place these notions in vocabulary which is easily assimilable by the necessary algorithms for the treatment of economics and management problems. Although, and above all, we hope that our work gets as close as possible to a study whose base is found in the strictest axiomatics. It will be seen that some words have remained, resisting our expected demands. Others have left their protagonism to new terms, supposedly more suitable to express the thinking which they wish to represent. From these parameters we will begin our journey making reference to the notion of **filter base** as an elemental concept capable of generating other concepts.

Given a set E and a non-empty subset $B(E)$ of the "power set" $P(E)$, which is to say:

$$B(E) \subset P(E)$$
$$B(E) = \varnothing$$

it is said that $B(E)$ is a **filter-base** in E if:

1. $\forall \underset{\sim}{A}_j, \underset{\sim}{A}_k \in B(E)$:

$$\exists \underset{\sim}{A}_q \in B(E) \text{ and } \underset{\sim}{A}_q \subset \underset{\sim}{A}_j \cap \underset{\sim}{A}_k$$

The intersection of two subsets of $B(E)$ contain a subset of $P(E)$ which belongs to $B(E)$.

$$2. \ \varnothing \notin B(E)$$

Of the intersection of a finite number of elements of $P(E)$ which belong to $B(E)$ an empty cannot result.

We immediately move on to a referential example.

$$E = \{a, b, c, d\}$$

J. Gil Aluja & A.M. Gil Lafuente: Towards an Advanced Modelling, STUDFUZZ 276, pp. 71–106.
springerlink.com

such that:

$$B(E) = \{\ \{d\}, \{b, d\}, \{c, d\}, \{a, b, d\}, \{b, c, d\}\ \}$$

whose representation in a Boole lattice is the following:

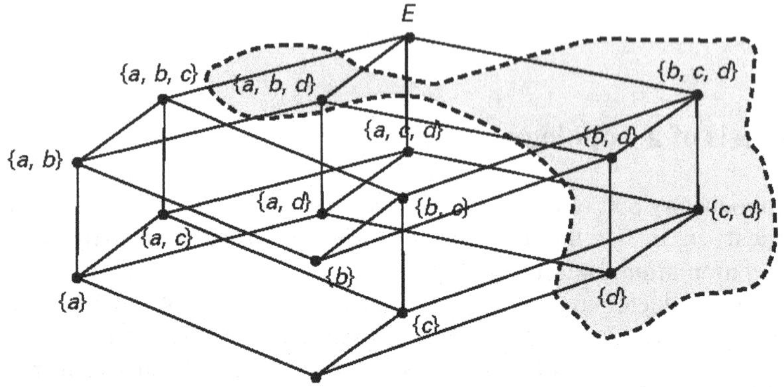

Let us see if this subset fulfils the conditions to be a filter-base

$$\{d\} \cap \{b, d\} = \{d\}$$
$$\{d\} \cap \{c, d\} = \{d\}$$
$$\{d\} \cap \{a, b, d\} = \{d\}$$
$$\{d\} \cap \{b, c, d\} = \{d\}$$

$$\{b, d\} \cap \{c, d\} = \{d\}$$
$$\{b, d\} \cap \{a, b, d\} = \{b, d\}$$
$$\{b, d\} \cap \{b, c, d\} = \{b, d\}$$

$$\{c, d\} \cap \{a, b, d\} = \{d\}$$
$$\{c, d\} \cap \{b, c, d\} = \{c, d\}$$

$$\{a, b, d\} \cap \{b, c, d\} = \{b, d\}$$

As all of the intersections contain a subset of $P(E)$ which belongs to $B(E)$, it fulfils the first of the **filter-base** axioms.

In the same way, given that the intersections of all the elements of $B(E)$ give non-empty elements of $P(E)$, it also fulfils the second axiom and this subset is therefore a **filter-base**.

An interesting aspect of filter-bases is their possibility of equivalence. So, two filter-bases $B_1(E)$ and $B_2(E)$ are equivalent if all elements of $B_1(E)$ contain at least one element of $B_2(E)$ and vice versa.

Let us assume two filter-bases:

$$B_1(E) = \{\ \{d\}, \{b, d\}, \{c, d\}, \{a, b, d\}, \{b, c, d\}\ \}$$
$$B_2(E) = \{\ \{d\}, \{\mathrm{a}, d\}, \{\mathrm{b}, d\}, \{\mathrm{a}, c, d\}\ \}$$

We perform:

$$\{d\} \supset \{d\} \qquad \{b, d\} \supset \{b, d\} \qquad \{c, d\} \supset \{d\}$$
$$\{a, b, d\} \supset \{b, d\} \qquad \{b, c, d\} \supset \{b, d\}$$

$$\{d\} \supset \{d\} \qquad \{a, d\} \supset \{d\} \qquad \{b, d\} \supset \{b, d\}$$
$$\{a, c, d\} \supset \{c, d\}$$

It has been shown that we are before **two equivalent filter-bases**
Let us advance in our process to move on to the notion of **filter**.
The **filter given rise to by a base** $B(E)$ is the subset of all the elements of $P(E)$ which contain an element of $B(E)$. Therefore we may write:

$$F_B(E) = \{A_j \in P(E) / \exists\, A_k \in B(E), A_k \subset A_j\}$$

The previous filter-base $B(E)$:

$$B(E) = \big\{ \{d\}, \{b, d\}, \{c, d\}, \{a, b, d\}, \{b, c, d\} \big\}$$

gives rise to the filter:

$$F(E) = \big\{ \{d\}, \{a, d\}, \{b, d\}, \{c, d\}, \{a, b, d\}, \{a, c, d\}, \{b, c, d\}, E \big\}$$

The elements of a filter form a **Boole lattice**. Let us see this in our example:

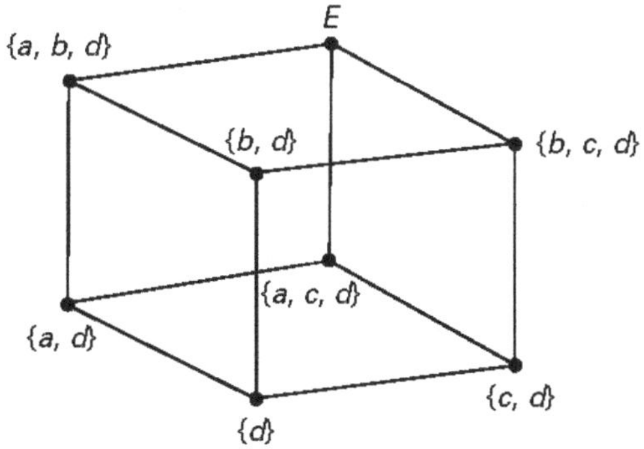

Two equivalent filter-bases give rise to the **same filter**.
We will confirm this with the bases $B_1(E)$, $B_2(E)$.

$$B_1(E) = \{\ \{d\}, \{b, d\}, \{c, d\}, \{a, b, d\}, \{b, c, d\}\ \}$$

As we have seen the **engendered filter** is:

$$F_1(E) = \{ \{d\}, \{a, d\}, \{b, d\}, \{c, d\}, \{a, b, d\}, \{a, c, d\}, \{b, c, d\}, E \}$$

For $B_2(E)$:

$$B_2(E) = \{ \{d\}, \{a, d\}, \{b, d\}, \{\text{a}, c, d\} \}$$

the **engendered filter** is:

$$F_2(E) = \{ \{d\}, \{a, d\}, \{b, d\}, \{c, d\}, \{a, b, d\}, \{\text{a}, c, d\}, \{b, c, d\}, E \}$$

Without a doubt the two filters **coincide**.

We will now move on to briefly present two interesting aspects relative to the filters which, later, will be of use in the treatment of topologies.

Let two filters be $F_1(E)$ and $F_2(E)$. It is said that one of them, $F_1(E)$, **is finer** than the other, $F_2(E)$, if all elements which belong to $F_1(E)$ also belong to $F_2(E)$. It would be in this way:

$$\left(\forall A_j \in P(E), A_j \in F_1(E) \Rightarrow A_j \in F_2(E)\right) \Leftrightarrow (F_1(E) \text{ finer than } F_2(E))$$

Let us see an example:

$$F_1(E) = \{ \{a, d\}, \{a, b, d\}, \{a, c, d\}, E \}$$

$$F_2(E) = \{ \{d\}, \{\text{a}, d\}, \{\text{b}, d\}, \{c, d\}, \{a, b, d\}, \{\text{a}, c, d\}, \{b, c, d\}, E \}$$

It is easily confirmed that $F_1(E)$ is finer than $F_2(E)$.

Let us assume, finally, that one has a non-empty element $A_j \in P(E)$ and another element $A_k \in F(E)$, **the intersections** $A_j \cap A_k$ form a set $F(A_j)$which is a filter of A_j, if all elements of F(E) give rise to a non-empty element upon intersecting with A_j . We are therefore dealing with an " A_j filter induced by $F(E)$ over A_j".

Let us consider the filter:

$$F_1(E) = \{ \{a, d\}, \{a, b, d\}, \{a, c, d\}, E \}$$

and the element:

$$A_j = \{b, c, d\}$$

It is confirmed that its intersection with the elements of are not empty:

$$\{b, c, d\} \cap \{a, d\} = \{d\} \qquad \{b, c, d\} \cap \{a, b, d\} = \{b, d\}$$

$$\{b, c, d\} \cap \{a, c, d\} = \{c, d\} \qquad \{b, c, d\} \cap E = \{b, c, d\}$$

It results that:

$$F(A_j) = \{ \{d\}, \{b, d\}, \{c, d\}, \{b, c, d\} \}$$

If we advance with the filter idea to place it in the area of the relationships which the **functional application** may be considered as a way of connection between the "power set" of the two sets. In effect, we suppose two finite sets E_1, E_2 and the corresponding "power set" $P(E_1)$, $Q(E_2)$. Given a functional application of E_1 in E_2 it is confirmed that the subset of images of a filter-base $B(E_1)$ constitutes a filter-base $B(E_2)$ in E_2. We move on to the verification with a simple example. Let:

$$E_1 = \{a, b, c, d\}$$

$$E_2 = \{\alpha, \beta, \gamma\}$$

and a functional application as in the following:

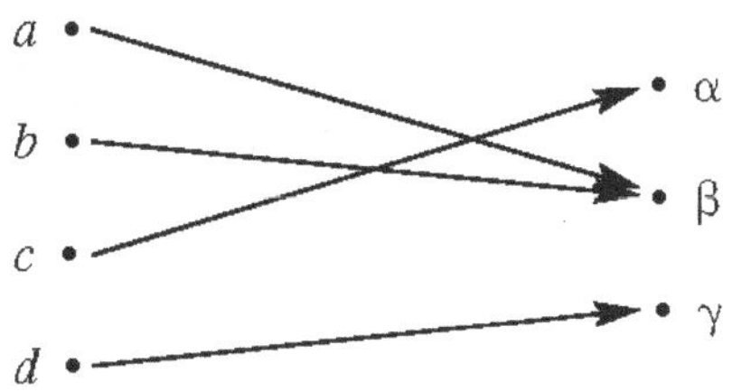

We consider the already used filter-base:

$$B(E_1) = \{\ \{d\}, \{b, d\}, \{c, d\}, \{a, b\ d\}, \{b, c, d\}\ \}$$

The images of this filter-base are:

$$\Gamma\,(d) = \{\gamma\}$$
$$\Gamma\,\{b, d\} = \Gamma\,\{b\} \cup \Gamma\,\{d\} = \{\beta, \gamma\}$$
$$\Gamma\,\{c, d\} = \Gamma\,\{c\} \cup \Gamma\,\{d\} = \{\alpha, \gamma\}$$
$$\Gamma\,\{a, b, d\} = \Gamma\,\{a\} \cup \Gamma\,\{b\} \cup \Gamma\,\{d\} = \{\beta, \gamma\}$$
$$\Gamma\,\{b, c, d\} = \Gamma\,\{b\} \cup \Gamma\,\{c\} \cup \Gamma\,\{d\} = \{\alpha, \beta, \gamma\}$$

We therefore have:

$$\{\{\gamma\}, \{\alpha, \gamma\}, \{\beta, \gamma\}, \{\alpha, \beta, \gamma\}\}$$

which is, in effect, a filter-base E_2, $B(E_2)$.

Let us illustrate that which we have just expounded through the following figure:

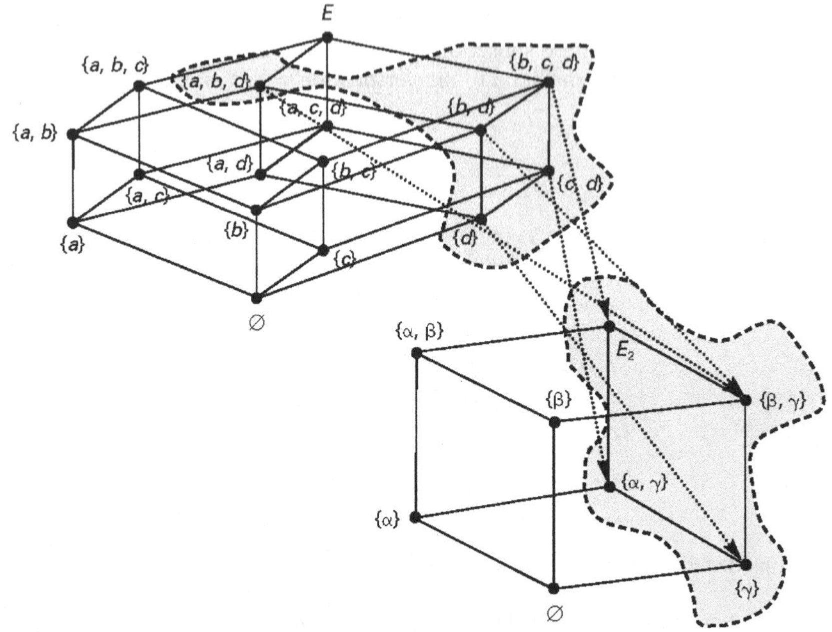

We believe it prudent to make an observation, as obvious as it my seem: if the image of **a filter** $F(E_1)$ is **a filter-base** $B(E_2)$, it is not always a filter as a consequence of the nature of the functional application. Let it be that a filter therefore depends on the specific functional application established.

2 Topological Space

We believe that the time has arrived to fully enter the study of topological spaces.

A pretopology, as we have seen, comes defined by either its adherent functional application $\boldsymbol{\Gamma}$ or by its interior application δ, which is the same. A topology may be considered as a particular case of isotone pretopology defined as much by $\boldsymbol{\Gamma}$ as by δ. Therefore, all topology is an isotone pretopology, but an isotone pretopology is not always a topology. The custom exists of designating a topology by means of a subset $T(E) \in P(E)$ whose elements are called " open-T subsets" or simply "opens". This subset $T(E)$ should have the following properties for all its elements:

1. $\varnothing \in T(E)$
2. $E \in T(E)$
3. $(A_j \in T(E) \text{ y } A_k \in T(E)) \Rightarrow (A_j \cap A_k \in T(E))$
4. $(A_j \in T(E) \text{ y } A_k \in T(E)) \Rightarrow (A_j \cup A_k \in T(E))$

The pair (E, $T(E)$) is named **Topological Space**, although the pair (E, T) is also frequently used.

Now that the axioms of a topological **space** have been established we will see if its generation starting from a **filter-base** is possible.

Let:

$$E = \{a, b, c, d\}$$

We establish a **filter-base** such as:

$$B(E) = \{ \ d\}, \{b, d\}, \{c, d\}, \{a, b, d\}, \{b, c, d\} \ \}$$

From which the **filter** is obtained:

$$F_B(E) = \{ \ \{d\}, \{a, d\}, \{b, d\}, \{c, d\}, \{a, b, d\}, \{a, c, d\}, \{b, c, d\}, E \ \}$$

Given that the filter $F_B(E)$ already **includes** the referential E, it is only necessary to **add** the empty $\emptyset$. We have:

$$F_B(E) = \{ \ \emptyset, \{d\}, \{a, d\}, \{b, d\}, \{c, d\}, \{a, b, d\}, \{a, c, d\}, \{b, c, d\}, E \ \}$$

Does this constitute a topology?

We do not believe it necessary to confirm the first three axioms, as their fulfillment is necessary by construction. Even when it may seem superfluous we will verify some unions of the fourth:

$$\{d\} \cup \{a, d\} = \{a, d\}$$
$$\{d\} \cup \{b, d\} = \{b, d\}$$
....
$$\{d\} \cup E = E$$

$$\{a, d\} \cup \{b, d\} = \{a, b, d\}$$
$$\{a, d\} \cup \{c, d\} = \{a, c, d\}$$
$$\{a, d\} \cup \{a, b, d\} = \{a, b, d\}$$
$$\{a, d\} \cup \{a, c, d\} = \{a, c, d\}$$
$$\{a, d\} \cup \{b, c, d\} = E$$
$$\{a, d\} \cup E = E$$

....

....

$$\{d\} \cup \{a, d\} \cup \{b, d\} \cup \{c, d\} \cup \{a, b, d\} \cup \{a, c, d\} = E$$
$$\{d\} \cup \{a, d\} \cup \{b, d\} \cup \{c, d\} \cup \{a, b, d\} \cup \{b, c, d\} = E$$
$$\{d\} \cup \{a, d\} \cup \{b, d\} \cup \{c, d\} \cup \{a, b, d\} \cup E = E$$

$\{d\} \cup \{a, d\} \cup \{b, d\} \cup \{c, d\} \cup \{a, c, d\} \cup \{b, c, d\} = E$

$\{d\} \cup \{a, d\} \cup \{b, d\} \cup \{c, d\} \cup \{a, c, d\} \cup E = E$

$\{d\} \cup \{a, d\} \cup \{b, d\} \cup \{c, d\} \cup \{b, c, d\} \cup E = E$

$\{d\} \cup \{a, d\} \cup \{b, d\} \cup \{c, d\} \cup \{a, b, d\} \cup \{a, c, d\} \cup \{b, c, d\} = E$

$\{d\} \cup \{a, d\} \cup \{b, d\} \cup \{c, d\} \cup \{a, b, d\} \cup \{a, c, d\} \cup E = E$

$\{d\} \cup \{a, d\} \cup \{b, d\} \cup \{c, d\} \cup \{a, b, d\} \cup \{a, c, d\} \cup \{b, c, d\} \cup E = E$

There is no doubt that the filter to which the empty ∅ has been added is a topology. Therefore it is possible to confirm that the **filter-base** is found in the genesis of a **topology**.

In continuation we propose the performance of a more extensive study of the notion of topology, but first we will present a simple example[1]:

$$E = \{a, b, c, d\}$$

it is said that:

$$T(E) = \{ \varnothing, \{a\}, \{a, c\}, \{a, b, c\}, \{a, c, d\}, E \}$$

is a topology of E if it fulfils the established axiomatic.

The axioms 3 and 4 can be confirmed in $T(E)$:

$\varnothing \cap \{a\} = \varnothing$

$\varnothing \cap \{a, c\} = \varnothing$

$\varnothing \cap \{a, b, c\} = \varnothing$

$\varnothing \cap \{a, c, d\} = \varnothing$

$\varnothing \cap E = \varnothing$

$\{a\} \cap \{a, c\} = \{a\}$

$\{a\} \cap \{a, b, c\} = \{a\}$

$\{a\} \cap \{a, c, d\} = \{a\}$

$\{a\} \cap E = \{a\}$

$\{a, c\} \cap \{a, b, c\} = \{a, c\}$

$\{a, c\} \cap \{a, c, d\} = \{a, c\}$

$\{a, c\} \cap E = \{a, c\}$

$\{a, b, c\} \cap \{a, c, d\} = \{a, c\}$

$\{a, b, c\} \cap E = \{a, b, c\}$

$\{a, c, d\} \cap E = \{a, c, d\}$

$\varnothing \cup \{a\} = \{a\}$

$\varnothing \cup \{a, c\} = \{a, c\}$

$\varnothing \cup \{a, b, c\} = \{a, b, c\}$

$\varnothing \cup \{a, c, d\} = \{a, c, d\}$

$\varnothing \cup E = E$

$\{a, c\} \cup \{a, b, c\} = \{a, b, c\}$

$\{a, c\} \cup \{a, c, d\} = \{a, c, d\}$

$\{a, c\} \cup E = E$

$\{a, b, c\} \cup \{a, c, d\} = E$

$\{a, b, c\} \cup E = E$

[1] Kaufmann, A.: «Prétopologie ordinaire et prétopologie floue». Note de Travail 115. La Tronche 1983, pages. 17-19.

$\{a\} \cup \{a, c\} = \{a, c\}$ $\{a, c, d\} \cup E = E$

$\{a\} \cup \{a, b, c\} = \{a, b, c\}$

$\{a\} \cup \{a, c, d\} = \{a, c, d\}$

$\{a\} \cup E = E$

Therefore, T (E) is a topology and its elements are the **open-*T* 's** or simply **opens**.

The complementary subsets of the open-T 's of a topology are called **closed-*T* subsets** or, in a simpler way, **closes.**

In our example the closes are:

$$\{\varnothing, \{b\}, \{d\}, \{b, d\}, \{b, c, d\}, E\}$$

In the following lattice we show the open-T 's and the closed-T 's .

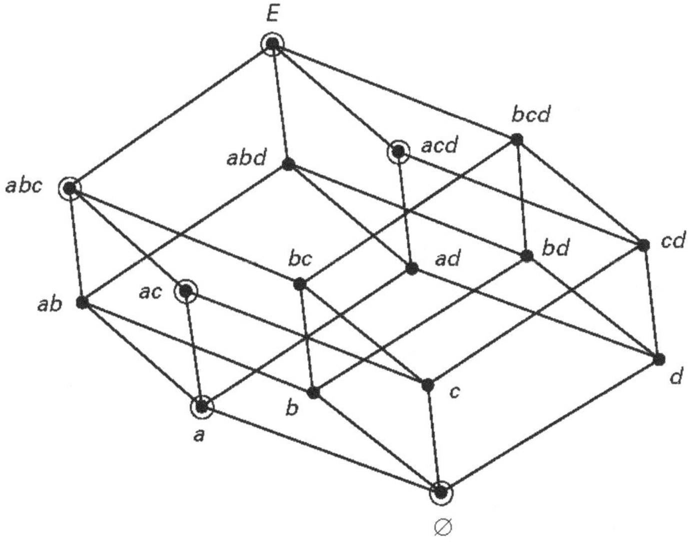

⊙ open-T 's of the topology $T(E)$

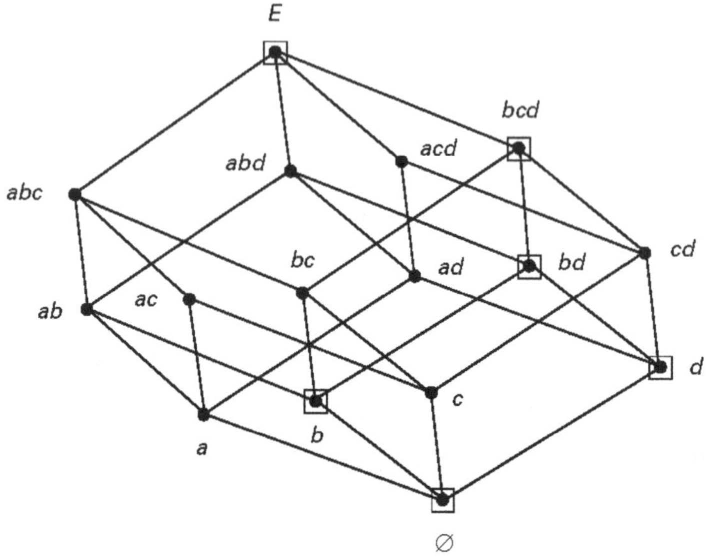

closed-T 's of the topology $T(E)$

Certain elements may be neither opens nor closes. This occurs with $\{c\}$, $\{a, b\}$, $\{a, d\}$, $\{b, c\}$, $\{c, d\}$, $\{a, b, d\}$, depending on how the previous figures are interpreted.

Another way of defining a topology T(E) comes given by the use of the **interior application**. The interior of $A_j \in \mathbf{\Gamma}(E)$ **is the greatest open-*T* contained in** $\boldsymbol{A_j}$. In this way, taking the example of the first of the previous figures we have:

$\delta\varnothing = \varnothing$	$\delta\,\{a\} = \{a\}$	$\delta\,\{b\} = \varnothing$	$\delta\,\{c\} = \varnothing$
$\delta\,\{d\} = \varnothing$	$\delta\,\{a, b\} = \{a\}$	$\delta\,\{a, c\} = \{a, c\}$	$\delta\,\{a, d\} = \{a\}$
$\delta\,\{b, c\} = \varnothing$	$\delta\,\{b, d\} = \varnothing$	$\delta\,\{c, d\} = \varnothing$	$\delta\,\{a, b, c\} = \{a, b, c\}$
$\delta\,\{a, b, d\} = \{a\}$	$\delta\,\{a, c, d\} = \{a,c,d\}$	$\delta\,\{b, c, d\} = \varnothing$	$\delta E = E$

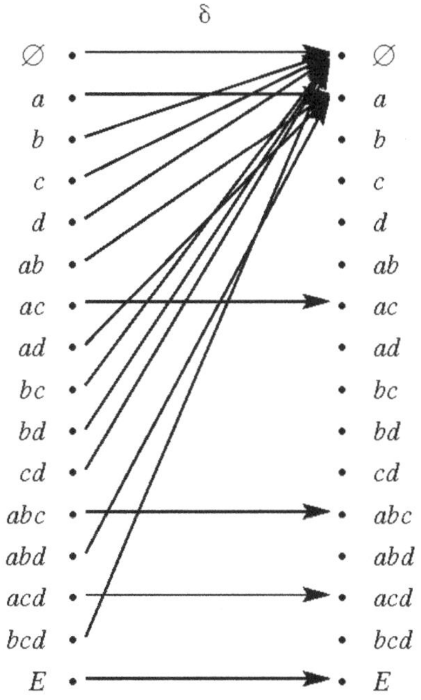

T(*E*) can also be defined by its adherent application **Γ**, in which the closes intervene. The adherence of $A_j \in \mathbf{\Gamma}(E)$ **is the smallest** closed-*T* which contains A_j. In this way, considering the second of the previous figures we have:

$\Gamma \varnothing = \varnothing$	$\Gamma\{a\} = E$	$\Gamma\{b\} = \{b\}$	$\Gamma\{c\} = \{b, c, d\}$
$\Gamma\{d\} = \{d\}$	$\Gamma\{a, b\} = E$	$\Gamma\{a, c\} = E$	$\Gamma\{a, d\} = E$
$\Gamma\{b, c\} = \{b, c, d\}$	$\Gamma\{b, d\} = \{b, d\}$	$\Gamma\{c, d\} = \{b, c, d\}$	$\Gamma\{a, b, c\} = E$
$\Gamma\{a, b, d\} = E$	$\Gamma\{a ,c, d\} = E$	$\Gamma\{b, c, d\} = \{b, c, d\}$	$\Gamma E = E$

The following figure represents the adherent application $\mathbf{\Gamma} A_j$.

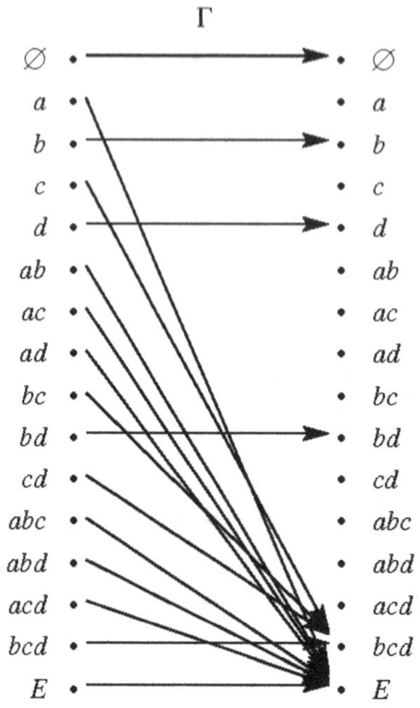

defined closes by the adherent application

If a topology $T(E)$ is defined by its interior application δ or by its adherent application **Γ** (which is practically the same), it is seen that a topology $T(E)$ of E fulfils the axioms:

$\varnothing \in \mathbf{\Gamma}(E)$
$\mathrm{E} \in \mathbf{\Gamma}(E)$
$\mathbf{\Gamma}\varnothing = \varnothing$
$\forall A_j, A_k \in T(E)$

$$(A_j \subset A_k) \Rightarrow (\mathbf{\Gamma} A_j \subset \mathbf{\Gamma} A_k)$$

We are therefore dealing with an **isotone pretopology**. The reciprocal is not always certain: an isotone pretopology is not necessarily a topology. As we have previously pointed out, **isotone pretopology** contains a **topology** as a particular case. It should therefore not be deduced that all isotone pretopology is a topology. To this respect let us take a look at a counterexample.

$$E = \{a, b, c\}$$

and the isotone pretopology:

$\Gamma\varnothing = \varnothing$ $\quad \Gamma\{a\} = \{a\}$ $\quad \Gamma\{b\} = \{b\}$ $\quad \Gamma\{c\} = \{c\}$
$\Gamma\{a, b\} = E$ $\quad \Gamma\{a, c\} = \{a, c\}$ $\quad \Gamma\{b, c\} = \{b, c\}$ $\quad \Gamma E = E$

we have:

$\delta\varnothing = \varnothing$ $\quad \delta\{a\} = \{a\}$ $\quad d\{b\} = \{b\}$ $\quad d\{c\} = \varnothing$
$\delta\{a, b\} = \{a, b\}$ $\quad \delta\{a, c\} = \{a, c\}$ $\quad \delta\{b, c\} = \{b, c\}$ $\quad \delta E = E$

Which, despite being an isotone pretopology is not however a topology.
In effect, axiom 3 demands:

$$(A_j \in T(E), A_k \in T(E)) \Rightarrow (A_j \cap A_k \in T(E))$$

To confirm that the isotone pretopology presented **does not fulfil** this property we begin by pointing out that the opens are:

$$A_b = \{\varnothing, \{a\}, \{b\}, \{a, b\}, \{a, c\}, \{b, c\}, E\}$$

It can be seen that:

$$\{a, c\} \cap \{b, c\} = \{c\}$$
$$\{c\} \notin T(E)$$

Although A_j fulfils all of the other axioms, the non-fulfillment of one of them is enough, as pointed out, for this isotone pretopology to not be a topology.
We now move on to compare the idea of topology with some pretopologies.

3 A Comparative Look at Some Spaces

Before moving into the nucleus of uncertain topological studies, the object of the next block of this second part, we are going to present a comparative table of some **pretopologies**, **of the Moore closing** and **of topology**, all in the area of finite boolean lattices. In the previous epigraph we have put forward a short introduction to this subject with the aim of carrying out a first outline of comparison between pretopology and topology. We have used the symbol **Γ** to refer to the adherent applications and the symbol δ to represent the interior applications.
We have also seen that it is possible to define a topology in E.

1. For the subset $T(E) \subset P(E)$ which makes up the "opens"
2. For the functional application δ, called the **interior application**. The interior of $A_j \in P(E)$ is the largest open contained in A_j.

3. For the functional application Γ, called the **adherent application**. The adherence of $A_j \in P(E)$ is the smallest close contained in A_j.

In effect, if we have a subset $T(E)$, Γ or δ are obtained in a univocal way. From here one can speak of a **Topological Space** from any of the pairs $(E, T(E))$, (E, δ) and (E, Γ), although, as we have repeatedly seen, the habit of designating the topological space by $(E, T(E))$ exists.

With this reminder finished we move on to the comparative table:

Main Properties / Espaces	$\Gamma\varnothing = \varnothing$	$\forall A_j \subset \Gamma A_j$ $\Gamma E = E$ Extensitivity	$\forall A_j, A_k \in P(E)$ $(A_j \subset A_k)$ $\Rightarrow (\Gamma A_j \subset \Gamma A_k)$ $\Rightarrow (\delta A_j \subset \delta A_k)$ Isotonia	$\forall A_j, A_k \in P(E)$ $\Gamma(A_j \cup A_k) = \Gamma A_j \cup \Gamma A_k$ $\Gamma(A_j \cap A_k) = \Gamma A_j \cap \Gamma A_k$ Distributivity	$\forall A_j \in P(E)$ $\Gamma(\Gamma A_j) = \Gamma A_j$ $\delta(\delta A_j) = \delta A_j$ Idempotency	$\forall A_j, A_k \in T(E)$ $(A_j \subset A_k) \in T(E)$ $\Rightarrow (A_j \cap A_k \in T(E))$ $\Rightarrow (A_j \cup A_k \in T(E))$
Pretopology	Yes	Yes				
Isotone Pretopology	Yes	Yes	Yes			
Distributive Pretopology	Yes	Yes	Yes	Yes		
Moore Closing		Yes	Yes		Yes	
Moore Pretopology	Yes	Yes	Yes	Yes	Yes	
Topology	Yes	Yes	Yes			Yes

With the concept of topology in E already presented and with the axiomatics corresponding to the distinct pretopologies and topology summarised in the previous table, in continuation we will present another example situated within a Boole lattice. For this we will take the referential:

$$E = \{a, b, c, d\}$$

The subset:

$$T(E) = \{\varnothing, \{c\}, \{a, c\}, \{b, c\}, \{a, b, c\}, E\}$$

is a topology in E.

In effect, the corresponding interior application δ is:

$\delta\varnothing = \varnothing$	$\delta\{a\} = \varnothing$	$\delta\{b\} = \varnothing$	$\delta\{c\} = \{c\}$
$\delta\{d\} = \varnothing$	$\delta\{a,b\} = \varnothing$	$\delta\{a, c\} = \{a, c\}$	$\delta\{a, d\} = \varnothing$
$\delta\{b, c\} = \{b, c\}$	$\delta\{b, d\} = \varnothing$	$\delta\{c, d\} = \{c\}$	$\delta\{a, b, c\} = \{a, b, c\}$
$\delta\{a, b, d\} = \varnothing$	$\delta\{a, c, d\} = \{a, c\}$	$\delta\{b, c, d\} = \{b, c\}$	$\delta E = E$

In the following figure we represent the topology defined by the opens of this interior application:

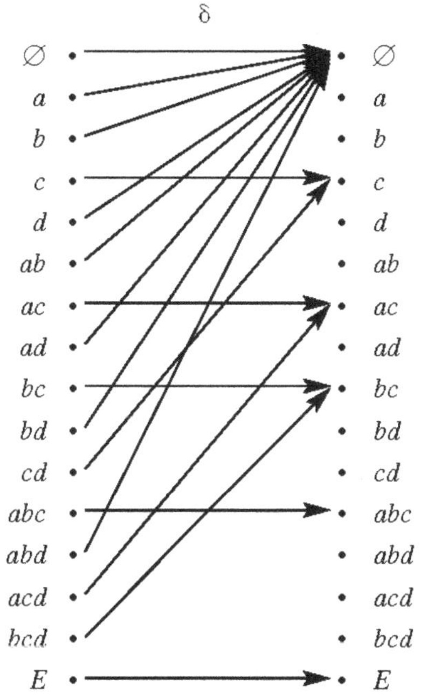

In the same way, given the set of closes:

$$\{\emptyset, \{d\}, \{a, d\}, \{b, d\}, \{a, b, d\}, E\}$$

the adherent application $\mathbf{\Gamma}$ is:

$\Gamma \emptyset = \emptyset$	$\Gamma \{a\} = \{a, d\}$	$\Gamma \{b\} = \{b, d\}$	$\Gamma \{c\} = E$
$\Gamma \{d\} = \{d\}$	$\Gamma \{a, b\} = \{a, b, d\}$	$\Gamma \{a, c\} = E$	$\Gamma \{a, d\} = \{a, d\}$
$\Gamma \{b, c\} = E$	$\Gamma \{b, d\} = \{b, d\}$	$\Gamma \{c, d\} = E$	$\Gamma \{a, b, c\} = E$
$\Gamma \{a, b, d\} = \{a, b, d\}$	$\Gamma \{a, c, d\} = E$	$\Gamma \{b, c, d\} = E$	$\Gamma E = E$

In continuation we present the corresponding graphic representation:

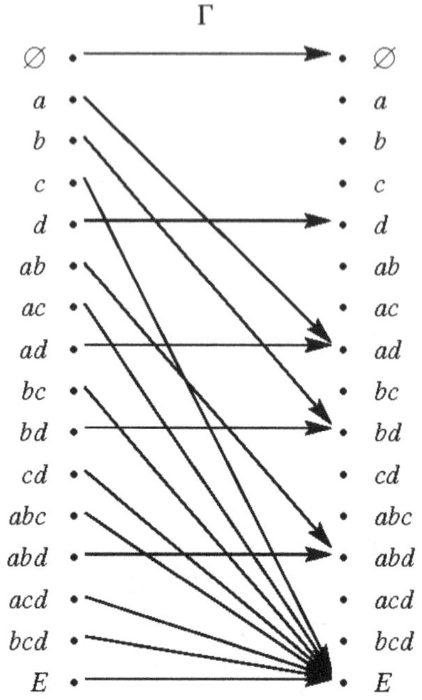

It is well-known that the compliments of the closes of this adherent application form the topology.

Let us confirm the fulfillment of the isotonia:

$$(A_j \subset A_k) \Rightarrow (\Gamma A_j \subset \Gamma A_k)$$

$$\begin{array}{ll}
\{a\} \subset \{a,b\} & \Gamma\,\{a\} = \{a,d\} \subset \Gamma\,\{a,b\} = \{a,b,d\} \\
\{a\} \subset \{a,c\} & \Gamma\,\{a\} = \{a,d\} \subset \Gamma\,\{a,c\} = E \\
\{a\} \subset \{a,d\} & \Gamma\,\{a\} = \{a,d\} \subset \Gamma\,\{a,d\} = \{a,d\} \\
\{a\} \subset \{a,b,c\} & \Gamma\,\{a\} = \{a,d\} \subset \Gamma\,\{a,b,c\} = \{a,d\} \\
\{a\} \subset \{a,b,d\} & \Gamma\,\{a\} = \{a,d\} \subset \Gamma\,\{a,b,d\} = \{a,b,d\} \\
\{a\} \subset \{a,c,d\} & \Gamma\,\{a\} = \{a,d\} \subset \Gamma\,\{a,c,d\} = E \\
\{a\} \subset E & \Gamma\,\{a\} = \{a,d\} \subset \Gamma\,E = E
\end{array}$$

$$\begin{array}{ll}
\{b\} \subset \{a, b\} & \Gamma\,\{b\} = \{b, d\} \subset \Gamma\,\{a, b\} = \{a, b, d\} \\
\{b\} \subset \{b, c\} & \Gamma\,\{b\} = \{b, d\} \subset \Gamma\,\{b, c\} = E \\
\{b\} \subset \{b, d\} & \Gamma\,\{b\} = \{b, d\} \subset \Gamma\,\{b, d\} = \{b, d\} \\
\{b\} \subset \{a, b, c\} & \Gamma\,\{b\} = \{b, d\} \subset \Gamma\,\{a, b, c\} = E \\
\{b\} \subset \{a, b, d\} & \Gamma\,\{b\} = \{b, d\} \subset \Gamma\,\{a, b, d\} = \{a, b, d\} \\
\{b\} \subset \{b, c, d\} & \Gamma\,\{b\} = \{b, d\} \subset \Gamma\,\{b, c, d\} = E \\
\{b\} \subset E & \Gamma\,\{b\} = \{b, d\} \subset \Gamma\,E = E
\end{array}$$

Given that $\mathbf{\Gamma}\,\{c\} = E$ for all A_j which include $\boldsymbol{c}$ should have E as an adherent application, as it is possible to confirm with a simple look at the corresponding figure representative to $\mathbf{\Gamma}$.

On the other hand, with $\mathbf{\Gamma}\,\{d\} = \{d\}$, all A_j which include $\boldsymbol{d}$ have an adherent application which fulfils $\mathbf{\Gamma}\,\{d\} \subset \mathbf{\Gamma}\,\{A_k\}$.

$$\begin{array}{ll}
\{a, b\} \subset \{a, b, c\} & \Gamma\,\{a, b\} = \{a, b, d\} \subset \Gamma\,\{a, b, c\} = E \\
\{a, b\} \subset \{a, b, d\} & \Gamma\,\{a, b\} = \{a, b, d\} \subset \Gamma\,\{a, b, d\} = \{a, b, d\} \\
\{a, b\} \subset E & \Gamma\,\{a, b\} = \{a, b, d\} \subset \Gamma\,E = E
\end{array}$$

With $\mathbf{\Gamma}\,\{a, c\} = E$, $\mathbf{\Gamma}\,\{a, b, c\}$, $\mathbf{\Gamma}\,\{a, c, d\}$ and $\mathbf{\Gamma}\,E$ should be, and are, all equal to E.

Here we will leave the development of the process, having taken into account that the isotonia may be easily confirmed visually in the explanatory graph of $\mathbf{\Gamma}A_j$.

In continuation we will move on to confirm the axiom:

$\forall\, A_j, A_k \in T(E)$:

$$\begin{aligned}
(A_j, A_k \in T(E)) &\Rightarrow (A_j \cap A_k \in T(E)) \\
&\Rightarrow (A_j \cup A_k \in T(E))
\end{aligned}$$

$$\begin{array}{lll}
\varnothing \cap \{c\} = \varnothing & \{c\} \cap \{a, c\} = \{c\} & \{a, c\} \cap \{a, b, c\} = \{a, c\} \\
\varnothing \cap \{a, c\} = \varnothing & \{c\} \cap \{b, c\} = \{c\} & \{a, c\} \cap E = \{a, c\} \\
\varnothing \cap \{b, c\} = \varnothing & \{c\} \cap \{a, b, c\} = \{c\} & \{b, c\} \cap \{a, b, c\} = \{b, c\} \\
\varnothing \cap \{a, b, c\} = \varnothing & \{c\} \cap E = \{c\} & \{b, c\} \cap E = \{b, c\} \\
\varnothing \cap E = \varnothing & \{a, c\} \cap \{b, c\} = \{c\} & \{a, b, c\} \cap E = \{a, b, c\}
\end{array}$$

$$\begin{array}{lll}
\varnothing \cup \{c\} = \{c\} & \{c\} \cup \{a, c\} = \{a, c\} & \{a, c\} \cup \{a, b, c\} = \{a, b, c\} \\
\varnothing \cup \{a, c\} = \{a, c\} & \{c\} \cup \{b, c\} = \{a, c\} & \{a, c\} \cup E = E \\
\varnothing \cup \{b, c\} = \{b, c\} & \{c\} \cup \{a, b, c\} = \{a, c\} & \{b, c\} \cup \{a, b, c\} = \{a, b, c\} \\
\varnothing \cup \{a, b, c\} = \{a, b, c\} & \{c\} \cup E = E & \{b, c\} \cup E = E \\
\varnothing \cup E = E & \{a, c\} \cup \{b, c\} = \{a, b, c\} & \{a, b, c\} \cup E = E
\end{array}$$

in this way it is evident that:

$$T(E) = \{\emptyset, \{c\}, \{a, c\}, \{b, c\}, \{a, b, c\}, E\}$$

is a topology.

These six elements of $T(E)$ are the open-T subsets or opens of $P(E)$. As has been expounded, these opens allow the construction of the interior application δ. From δ taking all the compliments of all of the elements of $T(E)$ the adherent application **Γ** is constructed. These compliments are the closed-T subsets or closes of $P(E)$.

In continuation we indicate the topology of our example within a Boole lattice:

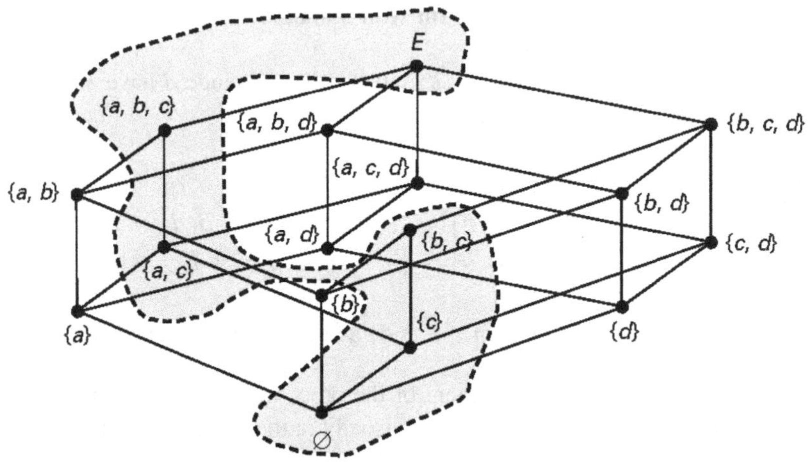

In a topology, an open may also be a close. The following example shows this:

$$E = \{a, b, c, d\}$$
$$T(E) = \{\emptyset, \{a\}, \{b, c, d\}, E\}$$

Considering the interior and adherent applications:

$$\delta\,\emptyset = \emptyset \qquad \delta\,\{a\} = \{a\} \qquad \delta\,\{b, c, d\} = \{b, c, d\} \qquad \delta\, E = E$$

all of the other elements of $P(E)$ give ∅.

$$\Gamma\,\emptyset = \emptyset \qquad \Gamma\,\{a\} = \{a\} \qquad \Gamma\,\{b, c, d\} = \{b, c, d\} \qquad \Gamma\, E = E$$

all of the other elements of $P(E)$ are equal to E.

It is easy to confirm that we are dealing with a topology, given that the extensivity and isotonia stand out and it also has:

$$\{a\} \cap \{b, c, d\} = \varnothing \in T(E)$$
$$\{a\} \cup \{b, c, d\} = E \in T(E)$$

As much $\{a\}$ as $\{b, c, d\}$ are opens and also closes. In effect, $\{b, c, d\} = \{a\}$. In this way $\{a\}$ and $\{b, c, d\}$ are complements of each other and are therefore both opens and closes at the same time.

We will now move onto an important topological notion: that of **neighbourhood**.

4 The Notion of Neighbourhood

Consider a topology $(E, T(E))$, defined by E and $T(E)$ (subset of opens).

$$(\mathrm{X} \subset \mathrm{E} \text{ is a neighboor of } \mathrm{A_j} \subset \mathrm{E}) \Rightarrow (\exists \mathrm{Y} \in \mathrm{T(E)} \,/\, \mathrm{A_j} \subset \mathrm{Y} \subset \mathrm{X})$$

We take an example of the previous lattice in which:

$$T(E) = \{\varnothing, \{c\}, \{a, c\}, \{b, c\}, \{a, b, c\}, E\}$$

We will confirm that $X = \{a, b, c\}$ is a neighbourhood of $A_j = \{a\}$.

An open $Y = \{a, c\}$ exists in such a way that $\{a, b, c\} \supset \{a, c\}$ and in this way $\{a, b, c\}$ is a neighbourhood of $\{a\}$.

In effect, $A_j \subset Y \subset X$ must be satisfied.

In continuation we give the relationship of all the neighbourhoods of the elements of $P(E)$ **distinct to those of $T(E)$**.

neighbourhood $\{a\} = \{ \{a, c\}, \{a, c, d\}, \{a, b, c\}, \{a, b, c, d\}\}$
neighbourhood $\{b\} = \{ \{\mathrm{b}, c\}, \{\mathrm{b}, c, d\}, \{a, b, c\}, \{a, b, c, d\}\}$
neighbourhood $\{d\} = \{\{a, b, c, d\}\}$
neighbourhood $\{a, b\} = \{\{a, b, c\}, \{a, b, c, d\}\}$
neighbourhood $\{a, d\} = \{\{a, b, c, d\}\}$
neighbourhood $\{b, d\} = \{\{a, b, c, d\}\}$
neighbourhood $\{c, d\} = \{\{a, b, c, d\}\}$
neighbourhood $\{a, b, d\} = \{\{a, b, c, d\}\}$
neighbourhood $\{a, c, d\} = \{\{a, b, c, d\}\}$
neighbourhood $\{b, c, d\} = \{\{a, b, c, d\}\}$

and for those of T(E):

neighbourhood$\{c\}$=$\{\{c\},\{a,c\},\{b,c\},\{c,d\},\{a,b,c\},\{a,c,d\},\{b,c,d\},\{a,b,c,d\}\}$
neighbourhood $\{a, c\} = \{ \{a, c\}, \{a, b, c\}, \{a, c, d\}, \{a, b, c, d\}\}$
neighbourhood $\{b, c\} = \{ \{b, c\}, \{a, b, c\}, \{b, c, d\}, \{a, b, c, d\}\}$
neighbourhood $\{a, b, c\} = \{\{a, b, c\}, \{a, b, c, d\}\}$
neighbourhood $E = \{ E \}$

As can be observed, if $Y = A_j$, Y is a neighbourhood of Y, if $Y \neq A_j$, Y is not a neighbourhood of Y.

In the following figures we show the neighbourhoods of each of the elements of $P(E)$ in a Boole lattice.

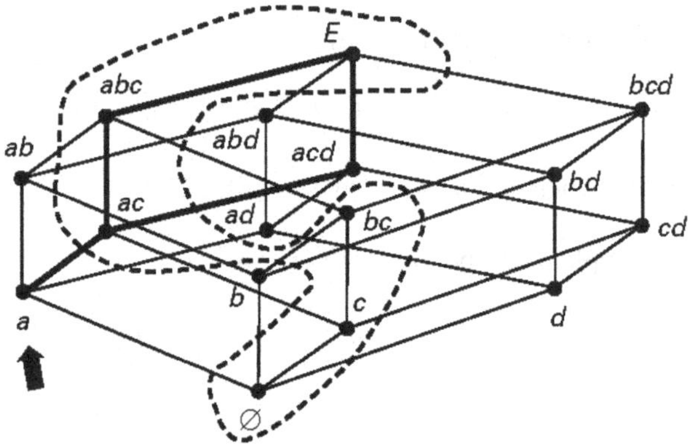

neighbourhood of ***a*** for the topology

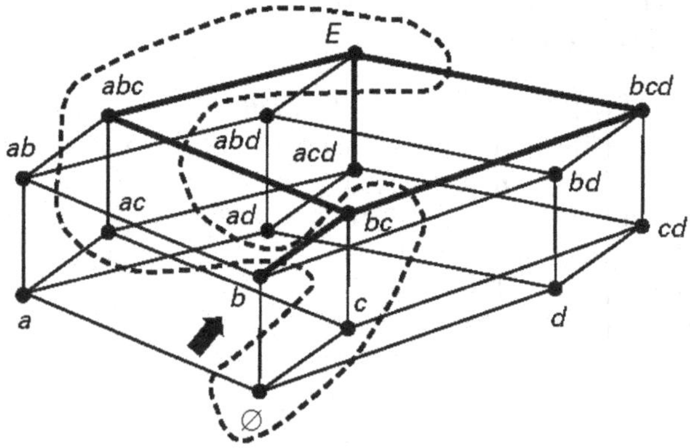

neighbourhood of ***b*** for the topology

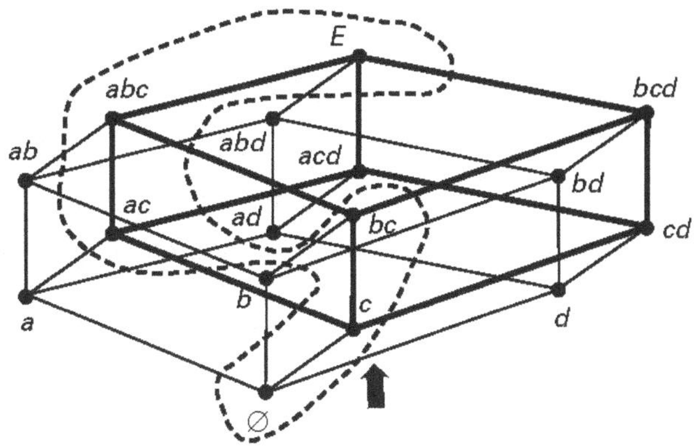

neighbourhood of ***c*** for the topology

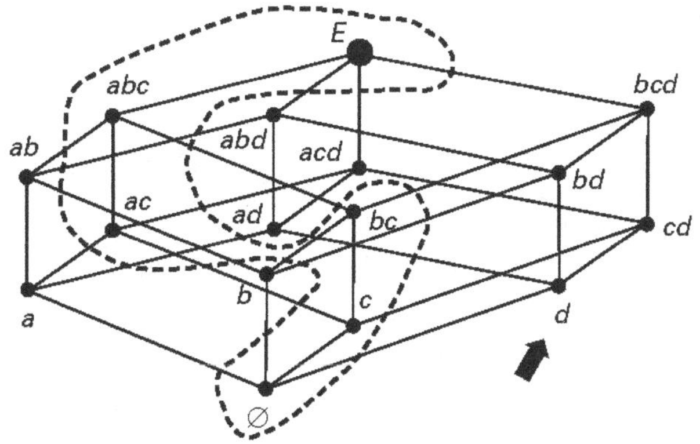

neighbourhood of ***d*** for the topology

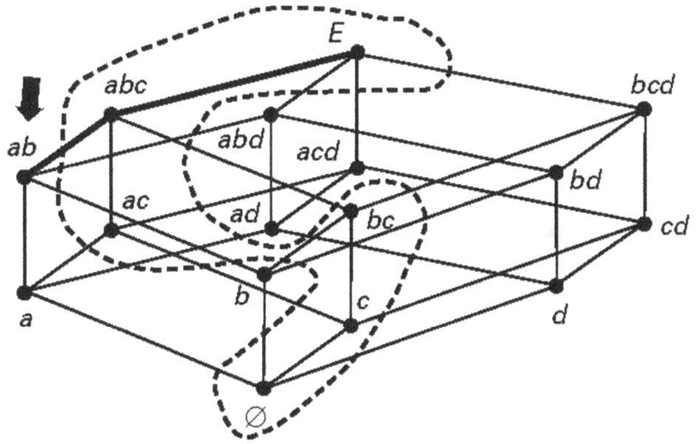

neighbourhood of ***ab*** for the topology

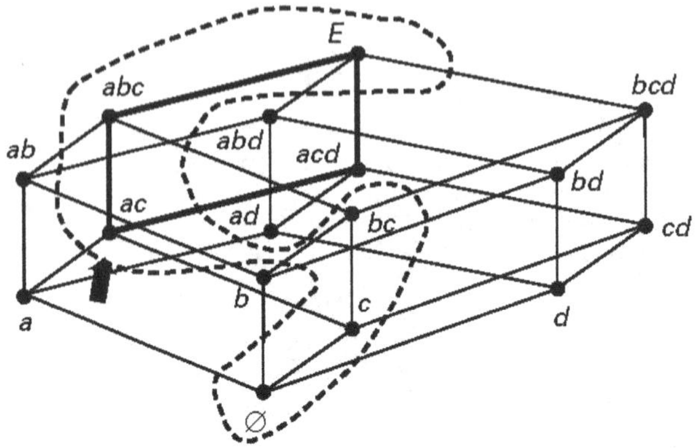

neighbourhood of ***ac*** for the topology

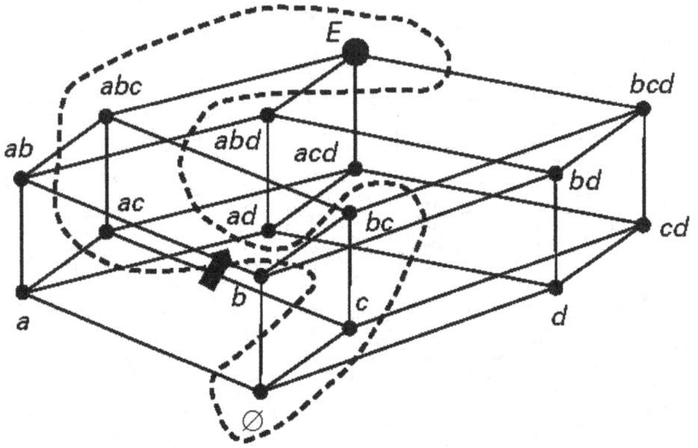

neighbourhood of ***ad*** for the topology

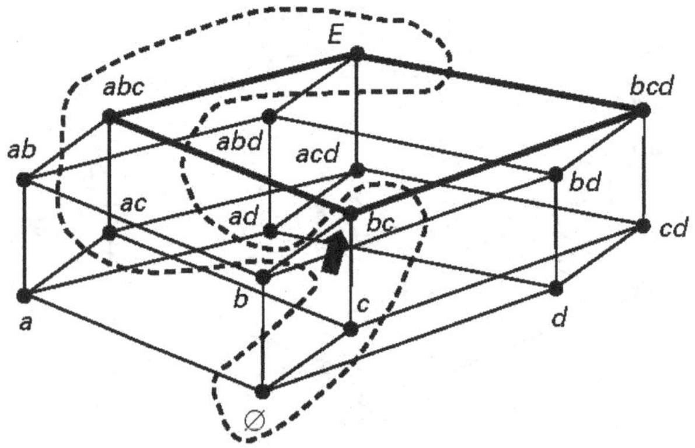

neighbourhood of ***bc*** for the topology

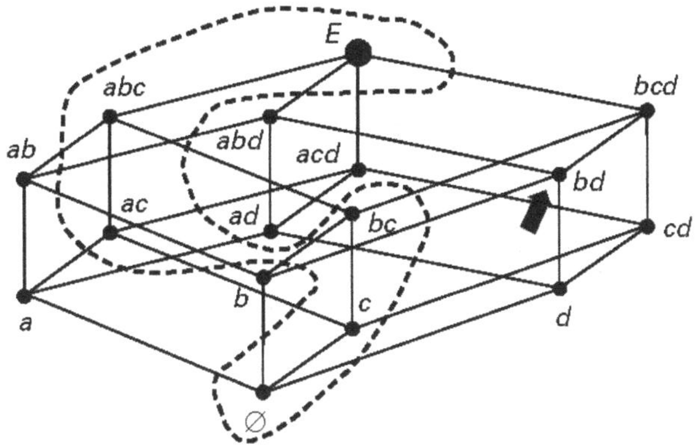

neighbourhood of ***bd*** for the topology

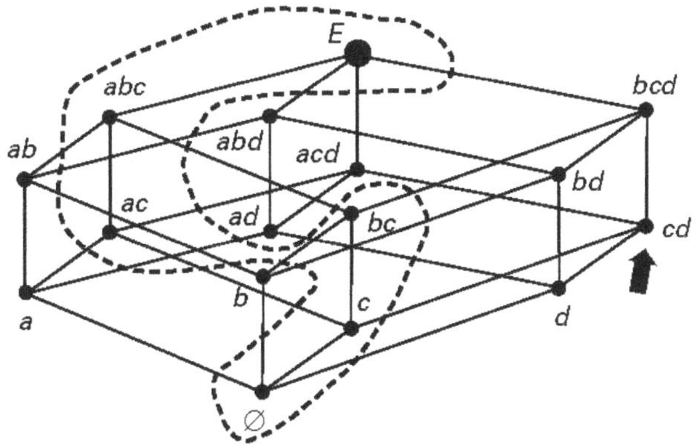

neighbourhood of ***cd*** for the topology

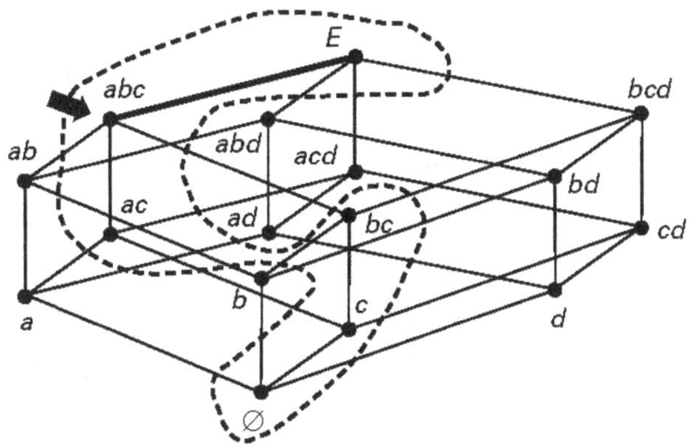

neighbourhood of ***abc*** for the topology

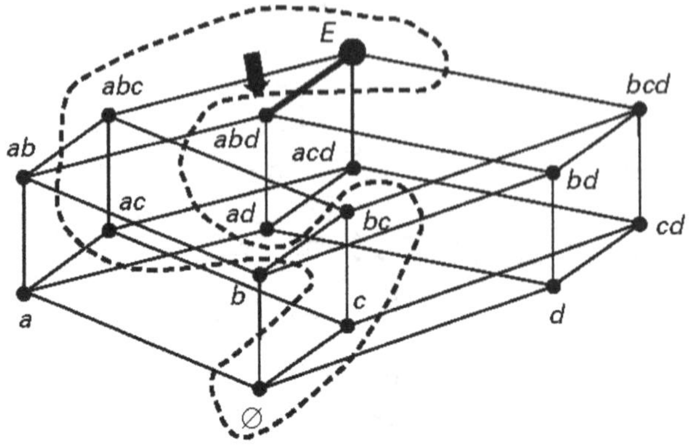

neighbourhood of ***abd*** for the topology

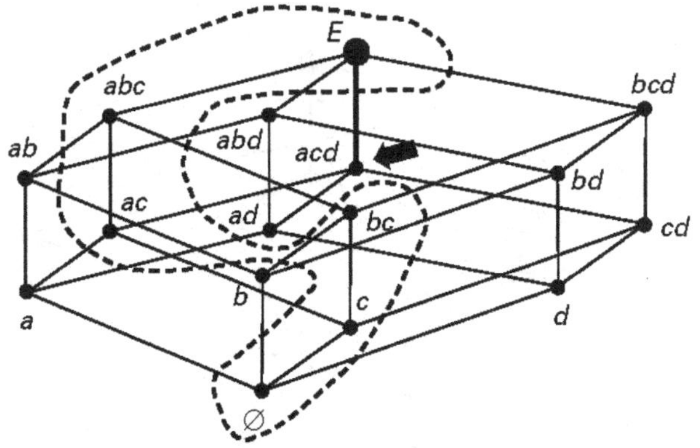

neighbourhood of ***acd*** for the topology

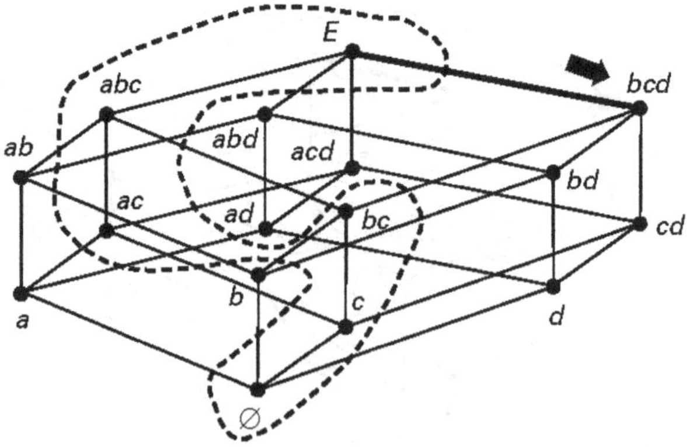

neighbourhood of ***bcd*** for the topology

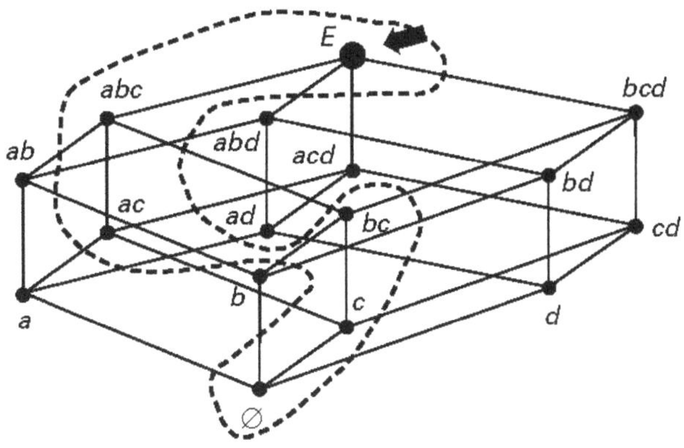

neighbourhood of ***E*** for the topology

As can be seen, which is verified by construction, the neighbourhood of $A_j \in P(E)$ forms a simple Boole sublattice of $P(E)$ (we remind ourselves that a Boole sublattice always contains E). A neighbourhood of may equally be an open or a close.

5 Topology, Open, Closed and Neighbourhoods

We will develop another example with the same referential, in which we have chosen as an example a topology so that the set of opens **coincides** with the set of closes. Let:

$$E = \{a, b, c, d\}$$

We assume that the opens of the topology are available.

$$A_b = \{ \varnothing, \{a\}, \{b\}, \{\mathrm{a, b}\}, \{c, d\}, \{a, c, d\}, \{b, c, d\}, E\}$$

Let us see if it fulfils the axioms:

$$\begin{aligned}(A_j \in T(E), A_k \in T(E)) &\Rightarrow (A_j \cap A_k \in T(E)) \\ &\Rightarrow (A_j \cup A_k \in T(E))\end{aligned}$$

Let us see the first:

$\{a\} \cap \{b\} = \varnothing$	$\{b\} \cap \{a, b\} = \{b\}$	$\{a, b\} \cap \{a, c, d\} = \{a\}$	$\{a, c, d\} \cap \{b, c, d\} = \{c, d\}$
$\{a\} \cap \{a, b\} = \{a\}$	$\{b\} \cap \{c, d\} = \varnothing$	$\{a, b\} \cap \{b, c, d\} = \{b\}$	$\{a, c, d\} \cap E = \{a, c, d\}$
$\{a\} \cap \{c, d\} = \varnothing$	$\{b\} \cap \{a, c, d\} = \varnothing$	$\{a, b\} \cap E = \{a, b\}$	$\{b, c, d\} \cap E = \{b, c, d\}$
$\{a\} \cap \{a, c, d\} = \{a\}$	$\{b\} \cap \{b, c, d\} = \{b\}$	$\{c, d\} \cap \{a, c, d\} = \{c, d\}$	
$\{a\} \cap \{b, c, d\} = \varnothing$	$\{b\} \cap E = \{b\}$	$\{c, d\} \cap \{b, c, d\} = \{c, d\}$	
$\{a\} \cap E = \{a\}$	$\{a, b\} \cap \{c, d\} = \varnothing$	$\{c, d\} \cap E = \{c, d\}$	

We move on to the second:

$\{a\} \cup \{b\} = \{a, b\}$	$\{b\} \cup \{a, b\} = \{a, b\}$	$\{a, b\} \cup \{a, c, d\} = E$	$\{a, c, d\} \cup \{b, c, d\} = E$
$\{a\} \cup \{a, b\} = \{a, b\}$	$\{b\} \cup \{c, d\} = \{b, c, d\}$	$\{a, b\} \cup \{b, c, d\} = E$	$\{a, c, d\} \cup E = E$
$\{a\} \cup \{c, d\} = \{a, c, d\}$	$\{b\} \cup \{a, c, d\} = E$	$\{a, b\} \cup E = E$	$\{b, c, d\} \cup E = E$
$\{a\} \cup \{a, c, d\} = \{a, c, d\}$	$\{b\} \cup \{b, c, d\} = \{b, c, d\}$	$\{c, d\} \cup \{a, c, d\} = \{a, c, d\}$	
$\{a\} \cup \{b, c, d\} = E$	$\{b\} \cup E = E$	$\{c, d\} \cup \{b, c, d\} = \{b, c, d\}$	
$\{a\} \cup E = E$	$\{a, b\} \cup \{c, d\} = E$	$\{c, d\} \cup E = E$	

It therefore evidently results that we are dealing with a topology.

We immediately establish the interior application corresponding to this topology. As we have stated, the interior of $A_j \in P(E)$ is the largest open contained in A_j. This rule is applied and the following is found:

$\delta \varnothing = \varnothing$	$\delta \{a\} = \{a\}$	$\delta \{b\} = \{b\}$	$\delta \{c\} = \varnothing$
$\delta \{d\} = \varnothing$	$\delta \{a, b\} = \{a, b\}$	$\delta \{a, c\} = \{a\}$	$\delta \{a, d\} = \{a\}$
$\delta \{b, c\} = \{b\}$	$\delta \{b, d\} = \{b\}$	$\delta \{c, d\} = \{c, d\}$	$\delta \{a, b, c\} = \{a, b\}$
$\delta \{a, b, d\} = \{a, b\}$	$\delta \{a, c, d\} = \{a, c, d\}$	$\delta \{b, c, d\} = \{b, c, d\}$	$\delta E = E$

The following figure shows the interior application of the topology $T(E)$.

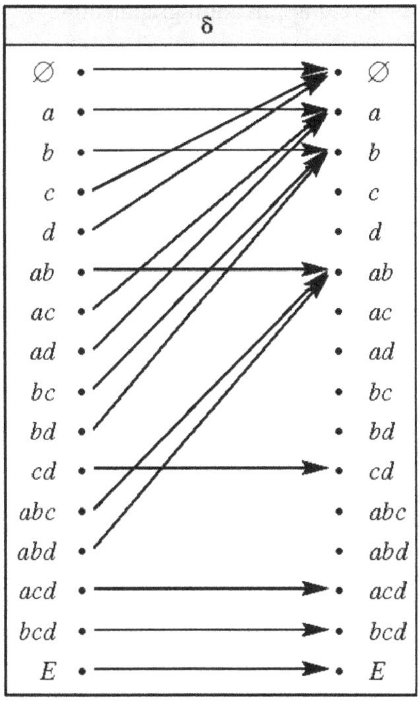

We remind ourselves that the adherence of $A_j \in P(E)$ is the smallest close which contains A_j. And therefore:

$\Gamma\{\varnothing\} = \{\varnothing\}$ $\Gamma\{a\} = \{a\}$ $\Gamma\{b\} = \{b\}$ $\Gamma\{c\} = \{c, d\}$ $\Gamma\{d\} = \{c, d\}$
$\Gamma\{a, b\} = \{a, b\}$ $\Gamma\{a, c\} = \{a, c, d\}$ $\Gamma\{a, d\} = \{a, c, d\}$ $\Gamma\{b, c\} = \{b, c, d\}$
$\Gamma\{b, d\} = \{b, c, d\}$ $\Gamma\{c, d\} = \{c, d\}$ $\Gamma\{a, b, c\} = E$ $\Gamma\{a, b, d\} = E$
$\Gamma\{a, c, d\} = \{a, c, d\}$ $\Gamma\{b, c, d\} = \{b, c, d\}$ $\Gamma E = E$

We now show this adherent application graphically:

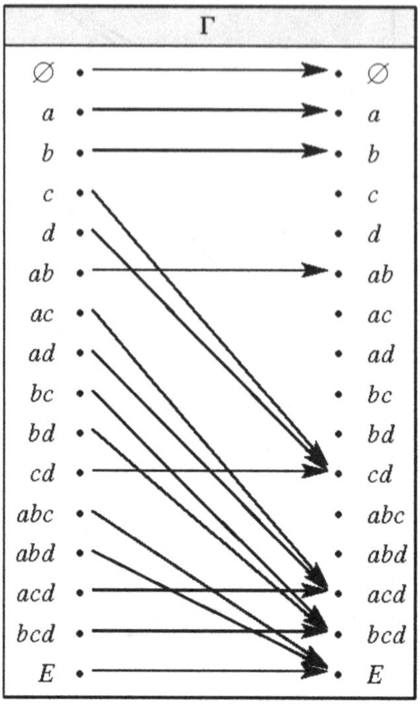

The closes are:

$$C_e = \{\ \emptyset, \{a\}, \{b\}, \{\mathrm{a, b}\}, \{c, d\}, \{a, c, d\}, \{b, c, d\}, E\}$$

It is easily confirmed that the set of closes coincides with the set of opens. In the following we will describe the neighbourhoods for each element of the "power set".

We have:

neighbourhood of $\emptyset = \emptyset$
neighbourhood of$\{a\} = \{\{\mathrm{a}\},\{a, b\},\{a, c\},\{a, d\},\{a, b, c\},\{a, b, d\},\{a, c, d\},E\}$
neighbourhood of$\{b\} = \{\{b\},\{a, b\},\{b, c\},\{b, d\},\{a, b, c\},\{a, b, d\},\{b, c, d\},E\}$
neighbourhood of $\{c\}$=$\{\{c, d\},\{a, c, d\},\{b, c, d\},E\}$
neighbourhood of $\{d\} = \{\{c, d\}, \{a, c, d\}'\{b, c, d\}, E\}$
neighbourhood of $\{a, b\} = \{\{a, b\}, \{a, b, c\}, \{a, b, d\}, E\}$
neighbourhood of $\{a, c\} = \{\{a, c, d\}, E\}$
neighbourhood of $\{a, d\} = \{\{a, c, d\}, E\}$
neighbourhood of $\{b, c\} = \{\{b, c, d\}, E\}$
neighbourhood of $\{b, d\} = \{\{b, c, d\}, E\}$
neighbourhood of $\{c, d\} = \{\{c, d\}, \{a, c, d\}, \{b, c, d\}, E\}$
neighbourhood of $\{a, b, c\} = E$

neighbourhood of $\{a, b, d\} = E$
neighbourhood of $\{a, c, d\} = \{\{a, c, d\}, E\}$
neighbourhood of $\{b, c, d\} = \{\{b, c, d\}, E\}$
neighbourhood of $E = E$

It is evident that in the hypothetical case of having a set of neighbourhoods it is possible to find the interior application and the adherent application. For this it is enough to take from all of the family of neighbourhoods those in which the intersection of all of the neighbours is equal to the generating element. The set of all of these forms the topology and from this the interior application and the adherent application are obtained.

In this way, for example:

$$\text{neighbourhood of } \{c\}=\{\{c, d\},\{a, c, d\},\{b, c, d\},E\}$$

with:

$$\{c, d\} \cap \{a, c, d\} \cap \{b, c, d\} \cap E = \{c, d\} \neq \{c\}$$

not forming a part of the topology $\{c, d\}$.

On the other hand:

$$\text{neighbourhood of } \{a, b\} = \{\{a, b\}, \{a, b, c\}, \{a, b, d\}, E\}$$

as:

$$\{a, b\} \cap \{a, b, c\} \cap \{a, b, d\} \cap E = \{a, b\}$$

in this case $\{a, b\}$ does form a part of the topology.

In this way a sequential and symmetrical diagram can be constructed which we show in the following figure:

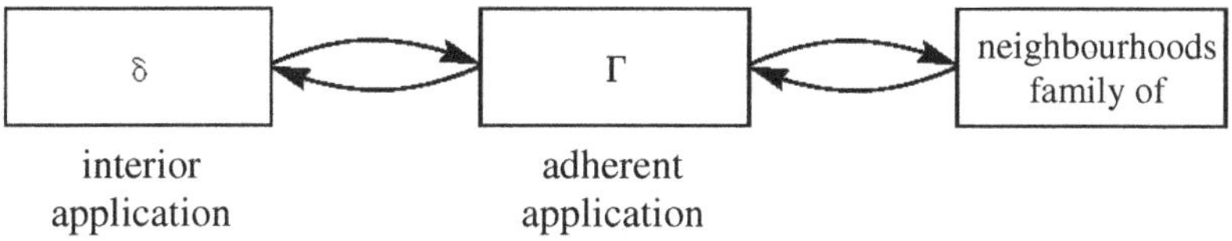

Finally, it is possible to ask the question of if only a single interior application is obtained from a family of neighbourhoods. It is evidently this way. The same occurs for δ and **Γ**. As a rule the knowledge of one of the three gives the other two.

6 Topological Continuity, Open Functions and Closed Functions

To end this first block of the second part of our work we will centre our attention on two concepts which result of special interest for the application of uncertain topology in the area of economics and management. We are refering to **topological continuity** and **homeomorphism**. We will begin with the notion of continuity.

Given two topological spaces $(E_1, T_1(E_1))$ and $(E_2, T_2(E_2))$, it is said that a function Γ of E_1 in E_2 is **continuous with respect to T_1 and T_2, T_1 - T_2 continuous** or simply **continuous**, if the reciprocal image $\Gamma^{-1}(A_j)$ of all of the open T_2 subset B_k of E_2 is an open T_1 subset of E_1, which is to say:

$$B_k \in T_2 \;\Rightarrow\; \Gamma^{-1}(B_k) \in T_1$$

We now move on to a specific supposition. The following sets are assumed as given:

$$E_1 = \{a, b, c, d\}$$
$$E_2 = \{\alpha, \beta, \gamma, \delta\}$$

and the following topologies:

$$T_1(E_1) = \{\varnothing, \{a\}, \{a, b\}, \{a, b, c\}, E_1\}$$
$$T_2(E_2) = \{\varnothing, \{\alpha\}, \{\beta\}, \{\alpha, \beta\}, \{\beta, \gamma, \delta\}, E_2\}$$

We will consider the following functional application:

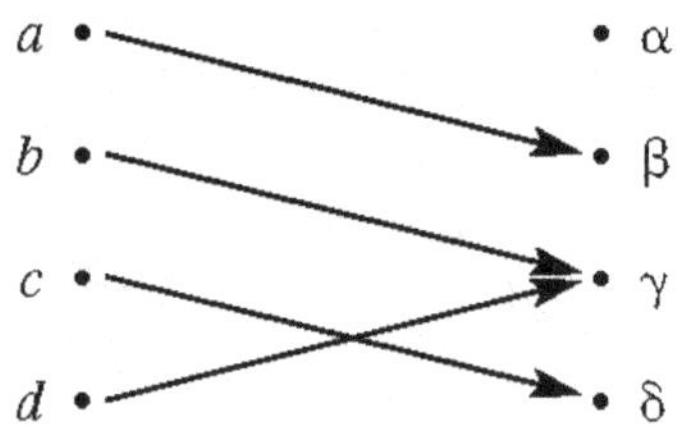

Let us confirm if the established function is continuous. For this we will investigate the elements of the reciprocal image of each of the elements of the topology $T_2(E_2)$, seeing if they form a part of the topology $T_1(E_1)$.

$$\Gamma^{-1}\{\alpha\} = \varnothing$$
$$\Gamma^{-1}\{\beta\} = \{a\}$$
$$\Gamma^{-1}\{\alpha, \beta\} = \{a\}$$
$$\Gamma^{-1}\{\beta, \gamma, \delta\} = E_1$$

As, ∅, $\{a\}$, E_1 in effect form part of the topology $T_1(E_1)$ we say that the cited function **is continuous**.

We now move onto another functional application such as:

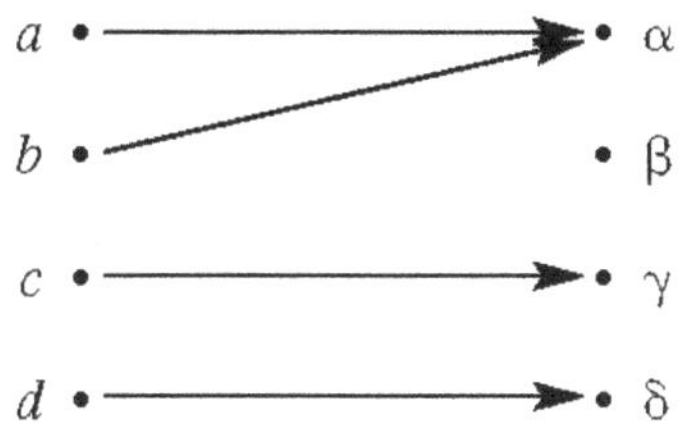

The inverse or reciprocal image of each of the elements of $T_2(E_2)$ is:

$$\Gamma^{-1}\{\alpha\} = \{a, b\}$$
$$\Gamma^{-1}\{\beta\} = \varnothing$$
$$\Gamma^{-1}\{\alpha, \beta\} = \{a, b\}$$
$$\Gamma^{-1}\{\beta, \gamma, \delta\} = \{c, d\}$$

It can be seen that the inverse image of { β, γ, δ } is $\{c, d\}$ which does not form a part of the subsets of opens of E_1 and, therefore, does not belong to the topology $T_1(E_1)$. This functional application is therefore **not** continuous.

The supposition of **topological continuity** can also be expressed from the notion of closed. Here a function **Γ** of E_1 in E_2 is continuous if the inverse image or reciprocal of all of the subset of closes of E_2 is a subset of closes of E_1.

We are speaking of the **continuity of an element of *E***.

Given a topological space $(E, T(E))$, it is said that an element x of E is an **adherence element** of a subset $A_j \subset E$ if:

$$x \subset A_j \qquad A_j \in T(E)$$

$\Gamma(x) \subset A_k \in T(E)$:

$$\Gamma^{-1} A_k \in T(E)$$

It is also said that x is an accumulation element[2] of A_j.

Let us return to the topology:

$$T_2 = \{\varnothing, \{\alpha\}, \{\beta\}, \{\alpha, \beta\}, \{\beta, \gamma, \delta\}, E\}$$

[2] An element or accumulation point of a subset A_j of E fulfils the property of which **all open set** which contains x contains some element (point) of A_j distinct to x.

and we will consider the following functional application:

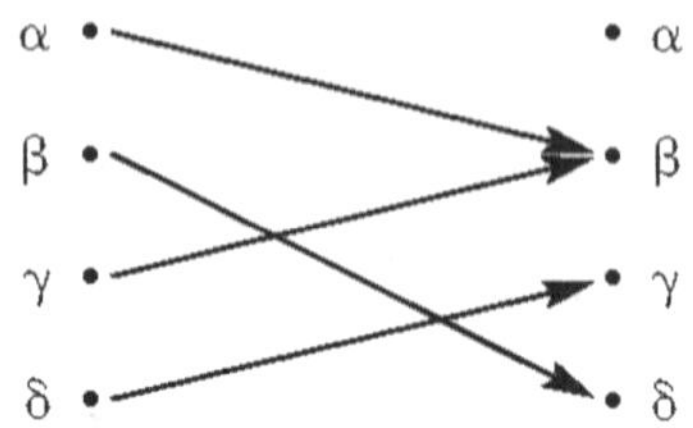

Let us see if this function is continuous in the element δ.

The only open subsets which contain **Γ** (δ) = γ are {β, γ, δ} and *E*.

On confirming that:

$$\Gamma^{-1}(\{\beta, \gamma, \delta\}) = E$$

$$\Gamma^{-1} E = E$$

gives *E* which forms part of *T*(*E*) as a result, it can be confirmed that **Γ is continuous in δ**. In this way it is verified that the inverse image of every open subset which contains **Γ** (δ) is a subset which contains δ.

The same does not occur with the point γ. It is easy to confirm that the function **Γ** is not continuous in this element or point. In effect:

The subsets of opens which contain **Γ**(δ) = β are {β}, {α, β}, {β, γ, δ} and *E*.

We have:

$$\Gamma^{-1}(\{\beta\}) = \{\alpha, \gamma\}$$

$$\Gamma^{-1}(\{\alpha, \beta\}) = \{\alpha, \gamma\}$$

$$\Gamma^{-1}(\{\beta, \gamma, \delta\}) = E$$

$$\Gamma^{-1} E = E$$

But as no open subset which contains {α, γ} and which at the same time is contained in {α, γ} exists, the function **Γ** established **is not continuous** in γ.

On the other hand, we have seen that **continuous functions** have the property that every **open subset** has an **open reciprocal image** and any **closed subset** has a **closed reciprocal image**.

It is therefore said that a **function is open in $E_1 \rightarrow E_2$** if every open subset has an open as an image and that a **function is closed in $E_1 \rightarrow E_2$** when the image of every closed subset is closed.

We will move on to some examples beginning from some functional applications already used. The following, as we have already seen, was continuous:

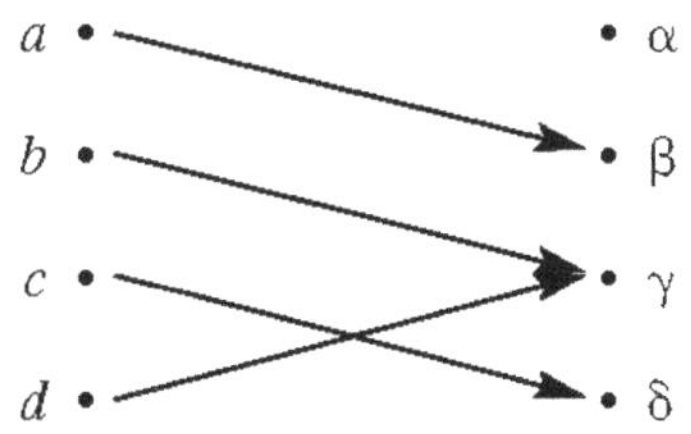

Given that we have established as subsets of opens:

$$A_b^{(1)} = \{\varnothing, \{a\}, \{a, b\}, \{a, b, c\}, E_1\}$$
$$A_b^{(2)} = \{\varnothing, \{\alpha\}, \{\beta\}, \{\alpha, \beta\}, \{\beta, \gamma, \delta\}, E_2\}$$

the subsets of closes are:

$$C_e^{(1)} = \{\varnothing, \{d\}, \{c, d\}, \{b, c, d\}, E_1\}$$
$$C_e^{(2)} = \{\varnothing, \{\alpha\}, \{\gamma, \delta\}, \{\alpha, \gamma, \delta\}, \{\beta, \gamma, \delta\}, E_2\}$$

Let us see if the image of every closed subset is a close:

$$\Gamma^{-1}\{\alpha\} = \varnothing$$
$$\Gamma^{-1}\{\gamma, \delta\} = \{b, c, d\}$$
$$\Gamma^{-1}\{\alpha, \gamma, \delta\} = \{b, c, d\}$$
$$\Gamma^{-1}\{\beta, \gamma, \delta\} = E_1$$
$$\Gamma^{-1} E_2 = E_1$$

In this case we are dealing with an open and closed function.

In continuation we will see if the function which we initially studied which was not continuous upon not being open is, in exchange, closed. Let us remind ourselves of the functional application:

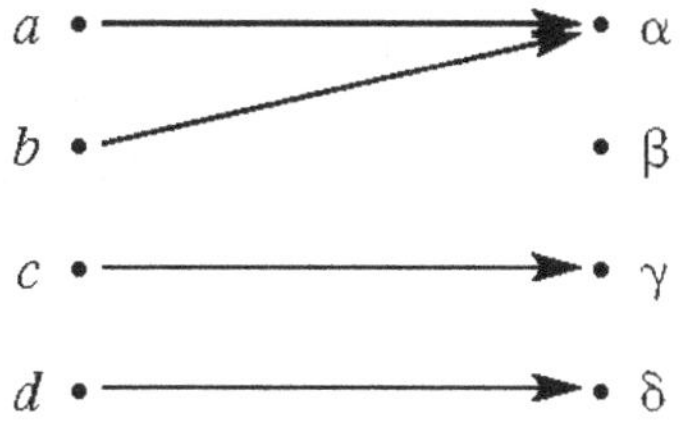

We obtain the images of all of the closed subsets:

$$\Gamma^{-1}\{\alpha\} = \{a, b\}$$
$$\Gamma^{-1}\{\gamma, \delta\} = \{c, d\}$$
$$\Gamma^{-1}\{\alpha, \gamma, \delta\} = E_1$$
$$\Gamma^{-1}\{\beta, \gamma, \delta\} = \{c, d\}$$
$$\Gamma^{-1} E_2 = E_1$$

In this way we find ourselves before a function which is **neither open nor closed**.

We summarise by saying that a topological space $(E, T(E))$ is, ultimately, a set E and a special class $T(E)$ of subsets of E which fulfil certain axioms. We will also say that between two topological spaces $(E_1, T_1(E_1))$ and $(E_2, T_2(E_2))$ there may exist an elevated number of functional applications. Among these we have paid attention to **continuous functions**, **open functions** and **closed functions** for how these functions preserve certain properties of the structure of the topological spaces $(E_1, T_1(E_1))$ and $(E_2, T_2(E_2))$.

7 Topological Homeomorphism

If we have a **bijective function**[3] $E_1 \rightarrow E_2$, then this functional application also induces a bijective function $P(E_1) \rightarrow Q(E_2)$, which is to say of the "power set" of E_1 in the "power set" of E_2. So, if this induced function also defines a univocal correspondence between the open subsets of E_1 and those of E_2, which is to say bijectively applies $T_1(E_1)$ in $T_2(E_2)$, then the topological spaces $(E_1, T_1(E_1))$ and $(E_2, T_2(E_2))$ **are identical from a topological point of view**. From here we reach the notion of **homeomorphism**.

It is said that two topological spaces $(E_1, T_1(E_1))$ and $(E_2, T_2(E_2))$ are **homeomorphic** or **topologically equivalent** when a bijective function $E_1 \rightarrow E_2$, $\mathbf{\Gamma}$ exists in a way that as much $\mathbf{\Gamma}$ as $\mathbf{\Gamma}^{-1}$ are continuous. The function $\mathbf{\Gamma}$ is given the denomination of **homeomorphism**.

With the aim of adequately affirming this important element of topology we are going to turn to a simple but significant example which, we hope, will shed some light on the concept of homeomorphism. For this we will begin with a functional application **created in such a way** that it fulfils the demanded requisites for it to be **bijective**. We represent it in the following way:

[3] As is well-known, a function is **injective** if distinct elements of E_1 have distinct images. A function is **surjective** if all $Q_k \in E_2$ is image of a $P_1 \in E_1$. A function is injective and surjective is known by the name **bijective**.

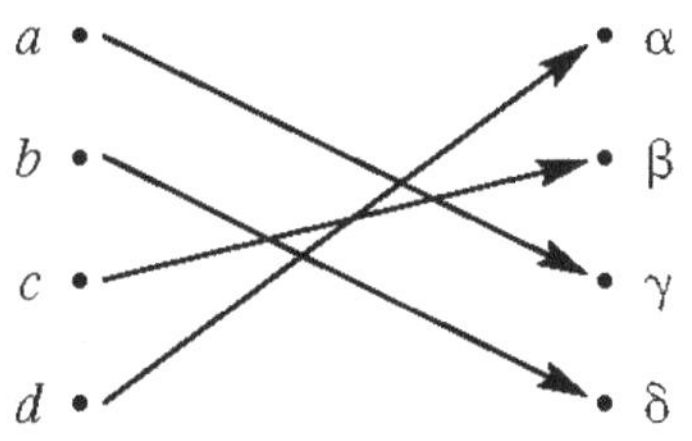

We assume the existence of the following topologies:

$$T_1(E_1) = \{\varnothing, \{a\}, \{b\}, \{a, b\}, \{a, b, c\}, E_1\}$$
$$T_2(E_2) = \{\varnothing, \{\alpha\}, \{\delta\}, \{\gamma, \delta\}, \{\beta, \gamma, \delta\}, E_2\}$$

Given that the application $E_1 \rightarrow E_2$ is bijective we will confirm that it **induces** a bijective family $P(E_1) \rightarrow Q(E_2)$. For this we will establish all of the connections between the elements of the "power set" of E_1 and those of the "power set" of E_2. The result is the following:

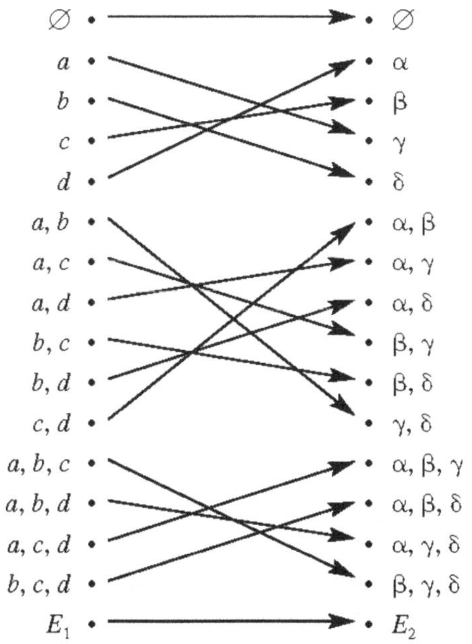

As it could not be any other way, the application $P(E_1\ ) \rightarrow Q(E_2)$ is also bijective.

At this time it will be necessary to confirm if this **induced function** defines a bijective application of $T_1(E_1)$ in $T_2\,(E_2)$. Let us see:

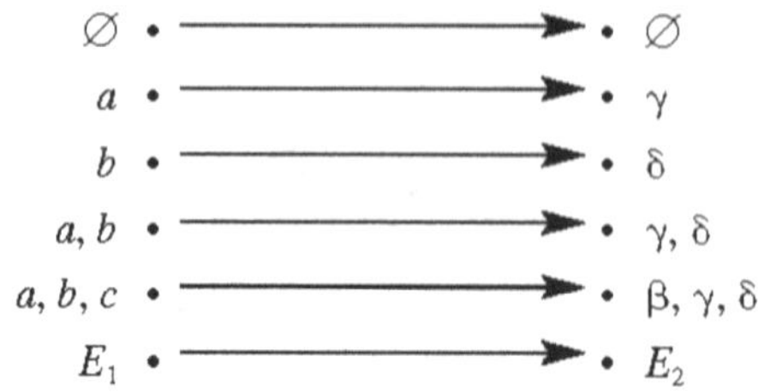

In effect, it is shown that (E_1, $T_1(E_1)$), (E_2, $T_2(E_2)$) are **homeomorphs** by the bijective function Γ ($E_1 \rightarrow E_2$).

We will now present the application in the corresponding representative lattices of the two topologies.

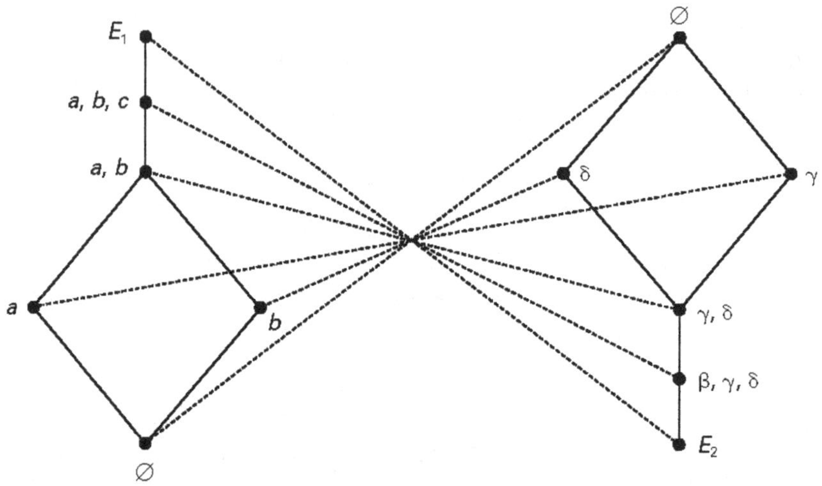

Throughout this expository block we have principally dealt with two aspects of deterministic topological space. In the first of these we have centred our attention in the study of the **self structure of topology**, its conceptualization, its characteristics, its most significant properties, always looking at the possibilities of transition to **uncertain topological spaces**. In the second, we have moved **into the comparison between two or more topologies**, in its most significant relationships. We have reached **homeomorphism**, emphasizing lattice representation, in which it did not result as difficult to discern the importance which it will acquire for the visualisation of **affinities** with the help of **Galois lattices**.

The concept of **combinatory topology** is rich in nuances and susceptible to support interesting developments. However, we consider it prudent to stop ourselves here, having taken into account that, with that which has been expounded, we have the essential elements available to deal with the problem of **uncertain topology**.

Chapter 4
Uncertain Topological Spaces

1 Some Notions of Uncertain Topology

We will now begin with one of the basic themes of our work. We will start by saying that the axiomatic of combinatorial topology continues to be valid when moving on to the area of uncertainty, obviously with the necessary adaptations.

An uncertain topology of a set E, such as we understand it, can be expressed in two distinct aspects: as a non-fuzzy subset of fuzzy subsets of L^E, or as a non-fuzzy subset of elements of the "power set" of a referential E of fuzzy subsets. But as occurs in an uncertain pretopology, in an uncertain topology a finite number of fuzzy subsets of E must be taken into account from one perspective and form another a finite number of elements of the "power set" $P(E)$ of E. In both $\varnothing$ and E are found. An uncertain topology $T(E)$ is usually defined from the following axioms:

1. $\varnothing \in T(E)$
2. $E \in T(E)$
3. $(\underset{\sim}{A}_j, \underset{\sim}{A}_k \in T(E)) \;\Rightarrow\; (\underset{\sim}{A}_j \cap \underset{\sim}{A}_k \in T(E))$
4. $(\underset{\sim}{A}_j, \underset{\sim}{A}_k \in T(E)) \;\Rightarrow\; (\underset{\sim}{A}_j \cup \underset{\sim}{A}_k \in T(E))$

$\underset{\sim}{A}_j, \underset{\sim}{A}_k$ are fuzzy subsets in one of the focuses, in the other elements of the "power set" of a referential of fuzzy subsets.

These axioms may be expressed saying:

- $\varnothing$ and E are open.
- The intersection of a finite number of opens is an open.
- The union of a finite (or infinite in its case) number of opens is an open.

The existence of these axioms means that the following axioms are also fulfilled:

5. $\Gamma \varnothing = \varnothing$

J. Gil Aluja & A.M. Gil Lafuente: Towards an Advanced Modelling, STUDFUZZ 276, pp. 107–183.
springerlink.com

6. $\forall \underset{\sim}{A}_j \in T(E)$:

$$\underset{\sim}{A}_j \subset \Gamma \underset{\sim}{A}_j$$

7. $\forall \underset{\sim}{A}_j, \underset{\sim}{A}_k \in T(E)$:

$$(\underset{\sim}{A}_j \subset \underset{\sim}{A}_k) \Rightarrow (\Gamma \underset{\sim}{A}_j \subset \Gamma \underset{\sim}{A}_k)$$

These last three axioms lead us to the definition of an **uncertain topology** as an **isotone uncertain pretopology** which also has the axioms three and four. Due to this, also in uncertainty, it may be defined as much by the adherent application **Γ** as by the interior application δ, but we have taken the custom of designating it by the subset $T(E) \in P(E)$ of its "opens".

Once again, to summarize here, it is possible to express an uncertain topology in *E*:

1. By the subset $T(E) \in P(E)$ composed of the **opens**
2. By the functional application δ, called the **interior application**. The interior application of $\underset{\sim}{A}_j \in P(E)$ is the largest open within $\underset{\sim}{A}_j$.
3. By the functional application **Γ**, called the **adherent application**. The adherence of $\underset{\sim}{A}_j \in P(E)$ is the smallest close within $\underset{\sim}{A}_j$.

The pair $(E, T(E))$ is called **uncertain topological space**[1]. The elements $T(E)$ are the **open-Γ** s or **opens** of the uncertain topology. The same occurs in an ordinary topology, the fuzzy subsets or groups of fuzzy subsets such as:

$$\Gamma \underset{\sim}{A}_j = \overline{\delta \overline{\underset{\sim}{A}_j}}$$

form the adherent application, when $\delta \underset{\sim}{A}_j$ forms the interior application (always dealing with functional applications). If $\underset{\sim}{A}_j$ is an open of δ, then the compliment of $\underset{\sim}{A}_j$, which is to say $\underset{\sim}{A}_j$, will be a close of **Γ**. δ is also designated as the **opening** and **Γ** as the **close**.

In the same way as in **pretopology**, it is now also convenient to perform a **distinction in the way of "fuzzifying"** the concept of topology.

A) In this aspect it fits to consider a **first way** which allows two alternatives:

1st A finite family K(*E*) of fuzzy subsets of a referential *E* formed by the referentials of the fuzzy subsets and taking into account the functional applications δ

[1] The same capital letter T is used as in topology, with no accent below, thinking that this would not create confusion. It may also be written as T, but we have thought that it does not seem necessary for the correct identification of the concept.

and Γ may be considered, as done in the concept of uncertain pretopology. The opens of δ should satisfy the four specified axioms.

2nd All of the fuzzy subsets of L^E are considered, and at first a topology $T(E)$ **is defined** for the opens that it contains, which should fulfil the four established axioms.

B) The **second way**, as we proposed for pretopologies, consists of establishing some fuzzy subsets as elements of the referential E, in such a way that the elements of the "power set" are fuzzy subsets or sets of fuzzy subsets.

In the following we will expound each of these conceptions, beginning with the first way.

We will perform the exposition by means of a simple example as this alternative will be abandoned.

Beginning with the referential:

$$E = \{a, b, c\}$$

and the finite subset of fuzzy subsets of E, $K(E)$, previously used in the study of **isotone uncertain pretopology**.

$$K(E) = \left\{ \begin{array}{ccccc}
\varnothing & \underset{\sim}{A}_1 & \overline{\underset{\sim}{A}_1} & \underset{\sim}{A}_2 & \overline{\underset{\sim}{A}_2} \\
\begin{array}{|c|c|c|} \hline a & b & c \\ \hline 0 & 0 & 0 \\ \hline \end{array} & \begin{array}{|c|c|c|} \hline a & b & c \\ \hline 0 & 0 & .1 \\ \hline \end{array} & \begin{array}{|c|c|c|} \hline a & b & c \\ \hline 1 & 1 & .9 \\ \hline \end{array} & \begin{array}{|c|c|c|} \hline a & b & c \\ \hline 1 & .2 & .8 \\ \hline \end{array} & \begin{array}{|c|c|c|} \hline a & b & c \\ \hline 0 & .8 & .2 \\ \hline \end{array}
\end{array} \right.,$$

$$\begin{array}{ccccc}
\underset{\sim}{A}_3 & \overline{\underset{\sim}{A}_3} & \underset{\sim}{A}_4 & \overline{\underset{\sim}{A}_4} & \underset{\sim}{A}_5 \\
\begin{array}{|c|c|c|} \hline a & b & c \\ \hline .1 & .8 & .2 \\ \hline \end{array} & \begin{array}{|c|c|c|} \hline a & b & c \\ \hline .9 & .2 & .8 \\ \hline \end{array} & \begin{array}{|c|c|c|} \hline a & b & c \\ \hline .6 & .1 & .8 \\ \hline \end{array} & \begin{array}{|c|c|c|} \hline a & b & c \\ \hline .4 & .9 & .2 \\ \hline \end{array} & \begin{array}{|c|c|c|} \hline a & b & c \\ \hline 1 & .3 & .8 \\ \hline \end{array}
\end{array},$$

$$\left. \begin{array}{cccc}
\overline{\underset{\sim}{A}_5} & \underset{\sim}{A}_6 & \overline{\underset{\sim}{A}_6} & E \\
\begin{array}{|c|c|c|} \hline a & b & c \\ \hline 0 & .7 & .2 \\ \hline \end{array} & \begin{array}{|c|c|c|} \hline a & b & c \\ \hline .5 & 1 & 1 \\ \hline \end{array} & \begin{array}{|c|c|c|} \hline a & b & c \\ \hline .5 & 0 & 0 \\ \hline \end{array} & \begin{array}{|c|c|c|} \hline a & b & c \\ \hline 1 & 1 & 1 \\ \hline \end{array}
\end{array} \right\}$$

From the interior application δ, already established upon the treatment of isotone uncertain pretopology, the subset of opens is found, which gives an uncertain topology as it fulfils the previously presented axioms 3 and 4. We will reproduce this previously presented topology as a subset of opens of the pretopology which can be expressed in the following way:

$$T(E) = \{\varnothing, \overline{\underset{\sim}{A}}_2, \underset{\sim}{A}_3, \overline{\underset{\sim}{A}}_5, \underset{\sim}{A}_6, E\}$$

or:

$$T(E) = \left\{ \begin{array}{|c|c|c|} \multicolumn{1}{c}{a} & \multicolumn{1}{c}{b} & \multicolumn{1}{c}{c} \\ \hline 0 & .8 & .2 \\ \hline \end{array}\,, \begin{array}{|c|c|c|} \multicolumn{1}{c}{a} & \multicolumn{1}{c}{b} & \multicolumn{1}{c}{c} \\ \hline .1 & .8 & .2 \\ \hline \end{array}\,, \begin{array}{|c|c|c|} \multicolumn{1}{c}{a} & \multicolumn{1}{c}{b} & \multicolumn{1}{c}{c} \\ \hline 0 & .7 & .2 \\ \hline \end{array}\,, \begin{array}{|c|c|c|} \multicolumn{1}{c}{a} & \multicolumn{1}{c}{b} & \multicolumn{1}{c}{c} \\ \hline .5 & 1 & 1 \\ \hline \end{array}\,, E \right\}$$

In continuation we confirm the fulfillment of the axioms 3 and 4:

$$\begin{array}{|c|c|c|} \multicolumn{1}{c}{a} & \multicolumn{1}{c}{b} & \multicolumn{1}{c}{c} \\ \hline 0 & .8 & .2 \\ \hline \end{array} \cap \begin{array}{|c|c|c|} \multicolumn{1}{c}{a} & \multicolumn{1}{c}{b} & \multicolumn{1}{c}{c} \\ \hline .1 & .8 & .2 \\ \hline \end{array} = \begin{array}{|c|c|c|} \multicolumn{1}{c}{a} & \multicolumn{1}{c}{b} & \multicolumn{1}{c}{c} \\ \hline 0 & .8 & .2 \\ \hline \end{array}$$

$$\begin{array}{|c|c|c|} \multicolumn{1}{c}{a} & \multicolumn{1}{c}{b} & \multicolumn{1}{c}{c} \\ \hline 0 & .8 & .2 \\ \hline \end{array} \cap \begin{array}{|c|c|c|} \multicolumn{1}{c}{a} & \multicolumn{1}{c}{b} & \multicolumn{1}{c}{c} \\ \hline 0 & .7 & .2 \\ \hline \end{array} = \begin{array}{|c|c|c|} \multicolumn{1}{c}{a} & \multicolumn{1}{c}{b} & \multicolumn{1}{c}{c} \\ \hline 0 & .7 & .2 \\ \hline \end{array}$$

$$\begin{array}{|c|c|c|} \multicolumn{1}{c}{a} & \multicolumn{1}{c}{b} & \multicolumn{1}{c}{c} \\ \hline 0 & .8 & .2 \\ \hline \end{array} \cap \begin{array}{|c|c|c|} \multicolumn{1}{c}{a} & \multicolumn{1}{c}{b} & \multicolumn{1}{c}{c} \\ \hline .5 & 1 & 1 \\ \hline \end{array} = \begin{array}{|c|c|c|} \multicolumn{1}{c}{a} & \multicolumn{1}{c}{b} & \multicolumn{1}{c}{c} \\ \hline 0 & .8 & .2 \\ \hline \end{array}$$

$$\begin{array}{|c|c|c|} \multicolumn{1}{c}{a} & \multicolumn{1}{c}{b} & \multicolumn{1}{c}{c} \\ \hline 0 & .8 & .2 \\ \hline \end{array} \cap \begin{array}{|c|c|c|} \multicolumn{1}{c}{a} & \multicolumn{1}{c}{b} & \multicolumn{1}{c}{c} \\ \hline 1 & 1 & 1 \\ \hline \end{array} = \begin{array}{|c|c|c|} \multicolumn{1}{c}{a} & \multicolumn{1}{c}{b} & \multicolumn{1}{c}{c} \\ \hline 0 & .8 & .2 \\ \hline \end{array}$$

$$\begin{array}{|c|c|c|} \multicolumn{1}{c}{a} & \multicolumn{1}{c}{b} & \multicolumn{1}{c}{c} \\ \hline .1 & .8 & .2 \\ \hline \end{array} \cap \begin{array}{|c|c|c|} \multicolumn{1}{c}{a} & \multicolumn{1}{c}{b} & \multicolumn{1}{c}{c} \\ \hline 0 & .7 & .2 \\ \hline \end{array} = \begin{array}{|c|c|c|} \multicolumn{1}{c}{a} & \multicolumn{1}{c}{b} & \multicolumn{1}{c}{c} \\ \hline 0 & .7 & .2 \\ \hline \end{array}$$

$$\begin{array}{|c|c|c|} \multicolumn{1}{c}{a} & \multicolumn{1}{c}{b} & \multicolumn{1}{c}{c} \\ \hline .1 & .8 & .2 \\ \hline \end{array} \cap \begin{array}{|c|c|c|} \multicolumn{1}{c}{a} & \multicolumn{1}{c}{b} & \multicolumn{1}{c}{c} \\ \hline .5 & 1 & 1 \\ \hline \end{array} = \begin{array}{|c|c|c|} \multicolumn{1}{c}{a} & \multicolumn{1}{c}{b} & \multicolumn{1}{c}{c} \\ \hline .1 & .8 & .2 \\ \hline \end{array}$$

$$\begin{array}{|c|c|c|} \multicolumn{1}{c}{a} & \multicolumn{1}{c}{b} & \multicolumn{1}{c}{c} \\ \hline .1 & .8 & .2 \\ \hline \end{array} \cap \begin{array}{|c|c|c|} \multicolumn{1}{c}{a} & \multicolumn{1}{c}{b} & \multicolumn{1}{c}{c} \\ \hline 1 & 1 & 1 \\ \hline \end{array} = \begin{array}{|c|c|c|} \multicolumn{1}{c}{a} & \multicolumn{1}{c}{b} & \multicolumn{1}{c}{c} \\ \hline .1 & .8 & .2 \\ \hline \end{array}$$

$$\begin{array}{|c|c|c|} \multicolumn{1}{c}{a} & \multicolumn{1}{c}{b} & \multicolumn{1}{c}{c} \\ \hline 0 & .7 & .2 \\ \hline \end{array} \cap \begin{array}{|c|c|c|} \multicolumn{1}{c}{a} & \multicolumn{1}{c}{b} & \multicolumn{1}{c}{c} \\ \hline .5 & 1 & 1 \\ \hline \end{array} = \begin{array}{|c|c|c|} \multicolumn{1}{c}{a} & \multicolumn{1}{c}{b} & \multicolumn{1}{c}{c} \\ \hline 0 & .7 & .2 \\ \hline \end{array}$$

$$\begin{array}{|c|c|c|} \multicolumn{1}{c}{a} & \multicolumn{1}{c}{b} & \multicolumn{1}{c}{c} \\ \hline 0 & .7 & .2 \\ \hline \end{array} \cap \begin{array}{|c|c|c|} \multicolumn{1}{c}{a} & \multicolumn{1}{c}{b} & \multicolumn{1}{c}{c} \\ \hline 1 & 1 & 1 \\ \hline \end{array} = \begin{array}{|c|c|c|} \multicolumn{1}{c}{a} & \multicolumn{1}{c}{b} & \multicolumn{1}{c}{c} \\ \hline 0 & .7 & .2 \\ \hline \end{array}$$

$$\begin{array}{|c|c|c|} \multicolumn{1}{c}{a} & \multicolumn{1}{c}{b} & \multicolumn{1}{c}{c} \\ \hline .5 & 1 & 1 \\ \hline \end{array} \cap \begin{array}{|c|c|c|} \multicolumn{1}{c}{a} & \multicolumn{1}{c}{b} & \multicolumn{1}{c}{c} \\ \hline 1 & 1 & 1 \\ \hline \end{array} = \begin{array}{|c|c|c|} \multicolumn{1}{c}{a} & \multicolumn{1}{c}{b} & \multicolumn{1}{c}{c} \\ \hline .5 & 1 & 1 \\ \hline \end{array}$$

The third axiom has been confirmed, let us move on to the fourth:

$$\begin{array}{|c|c|c|}\multicolumn{1}{c}{a}&\multicolumn{1}{c}{b}&\multicolumn{1}{c}{c}\\\hline 0&.8&.2\\\hline\end{array}\cup\begin{array}{|c|c|c|}\multicolumn{1}{c}{a}&\multicolumn{1}{c}{b}&\multicolumn{1}{c}{c}\\\hline .1&.8&.2\\\hline\end{array}=\begin{array}{|c|c|c|}\multicolumn{1}{c}{a}&\multicolumn{1}{c}{b}&\multicolumn{1}{c}{c}\\\hline .1&.8&.2\\\hline\end{array}$$

$$\begin{array}{|c|c|c|}\multicolumn{1}{c}{a}&\multicolumn{1}{c}{b}&\multicolumn{1}{c}{c}\\\hline 0&.8&.2\\\hline\end{array}\cup\begin{array}{|c|c|c|}\multicolumn{1}{c}{a}&\multicolumn{1}{c}{b}&\multicolumn{1}{c}{c}\\\hline 0&.7&.2\\\hline\end{array}=\begin{array}{|c|c|c|}\multicolumn{1}{c}{a}&\multicolumn{1}{c}{b}&\multicolumn{1}{c}{c}\\\hline 0&.8&.2\\\hline\end{array}$$

$$\begin{array}{|c|c|c|}\multicolumn{1}{c}{a}&\multicolumn{1}{c}{b}&\multicolumn{1}{c}{c}\\\hline 0&.8&.2\\\hline\end{array}\cup\begin{array}{|c|c|c|}\multicolumn{1}{c}{a}&\multicolumn{1}{c}{b}&\multicolumn{1}{c}{c}\\\hline .5&1&1\\\hline\end{array}=\begin{array}{|c|c|c|}\multicolumn{1}{c}{a}&\multicolumn{1}{c}{b}&\multicolumn{1}{c}{c}\\\hline .5&1&1\\\hline\end{array}$$

$$\begin{array}{|c|c|c|}\multicolumn{1}{c}{a}&\multicolumn{1}{c}{b}&\multicolumn{1}{c}{c}\\\hline 0&.8&.2\\\hline\end{array}\cup\begin{array}{|c|c|c|}\multicolumn{1}{c}{a}&\multicolumn{1}{c}{b}&\multicolumn{1}{c}{c}\\\hline 1&1&1\\\hline\end{array}=\begin{array}{|c|c|c|}\multicolumn{1}{c}{a}&\multicolumn{1}{c}{b}&\multicolumn{1}{c}{c}\\\hline 1&1&1\\\hline\end{array}$$

$$\begin{array}{|c|c|c|}\multicolumn{1}{c}{a}&\multicolumn{1}{c}{b}&\multicolumn{1}{c}{c}\\\hline .1&.8&.2\\\hline\end{array}\cup\begin{array}{|c|c|c|}\multicolumn{1}{c}{a}&\multicolumn{1}{c}{b}&\multicolumn{1}{c}{c}\\\hline 0&.7&.2\\\hline\end{array}=\begin{array}{|c|c|c|}\multicolumn{1}{c}{a}&\multicolumn{1}{c}{b}&\multicolumn{1}{c}{c}\\\hline .1&.8&.2\\\hline\end{array}$$

$$\begin{array}{|c|c|c|}\multicolumn{1}{c}{a}&\multicolumn{1}{c}{b}&\multicolumn{1}{c}{c}\\\hline .1&.8&.2\\\hline\end{array}\cup\begin{array}{|c|c|c|}\multicolumn{1}{c}{a}&\multicolumn{1}{c}{b}&\multicolumn{1}{c}{c}\\\hline .5&1&1\\\hline\end{array}=\begin{array}{|c|c|c|}\multicolumn{1}{c}{a}&\multicolumn{1}{c}{b}&\multicolumn{1}{c}{c}\\\hline .5&1&1\\\hline\end{array}$$

$$\begin{array}{|c|c|c|}\multicolumn{1}{c}{a}&\multicolumn{1}{c}{b}&\multicolumn{1}{c}{c}\\\hline .1&.8&.2\\\hline\end{array}\cup\begin{array}{|c|c|c|}\multicolumn{1}{c}{a}&\multicolumn{1}{c}{b}&\multicolumn{1}{c}{c}\\\hline 1&1&1\\\hline\end{array}=\begin{array}{|c|c|c|}\multicolumn{1}{c}{a}&\multicolumn{1}{c}{b}&\multicolumn{1}{c}{c}\\\hline 1&1&1\\\hline\end{array}$$

$$\begin{array}{|c|c|c|}\multicolumn{1}{c}{a}&\multicolumn{1}{c}{b}&\multicolumn{1}{c}{c}\\\hline 0&.7&.2\\\hline\end{array}\cup\begin{array}{|c|c|c|}\multicolumn{1}{c}{a}&\multicolumn{1}{c}{b}&\multicolumn{1}{c}{c}\\\hline .5&1&1\\\hline\end{array}=\begin{array}{|c|c|c|}\multicolumn{1}{c}{a}&\multicolumn{1}{c}{b}&\multicolumn{1}{c}{c}\\\hline .5&1&1\\\hline\end{array}$$

$$\begin{array}{|c|c|c|}\multicolumn{1}{c}{a}&\multicolumn{1}{c}{b}&\multicolumn{1}{c}{c}\\\hline 0&.7&.2\\\hline\end{array}\cup\begin{array}{|c|c|c|}\multicolumn{1}{c}{a}&\multicolumn{1}{c}{b}&\multicolumn{1}{c}{c}\\\hline 1&1&1\\\hline\end{array}=\begin{array}{|c|c|c|}\multicolumn{1}{c}{a}&\multicolumn{1}{c}{b}&\multicolumn{1}{c}{c}\\\hline 1&1&1\\\hline\end{array}$$

$$\begin{array}{|c|c|c|}\multicolumn{1}{c}{a}&\multicolumn{1}{c}{b}&\multicolumn{1}{c}{c}\\\hline .5&1&1\\\hline\end{array}\cup\begin{array}{|c|c|c|}\multicolumn{1}{c}{a}&\multicolumn{1}{c}{b}&\multicolumn{1}{c}{c}\\\hline 1&1&1\\\hline\end{array}=\begin{array}{|c|c|c|}\multicolumn{1}{c}{a}&\multicolumn{1}{c}{b}&\multicolumn{1}{c}{c}\\\hline 1&1&1\\\hline\end{array}$$

In this way it has been confirmed that the third and fourth axioms are fulfilled. In effect:

$$
\begin{array}{lllll}
\overline{\underset{\sim}{A}}_2 \cap \underset{\sim}{A}_3 = \overline{\underset{\sim}{A}}_2 & \overline{\underset{\sim}{A}}_2 \cap \overline{\underset{\sim}{A}}_5 = \overline{\underset{\sim}{A}}_5 & \overline{\underset{\sim}{A}}_2 \cap \underset{\sim}{A}_6 = \overline{\underset{\sim}{A}}_2 & \overline{\underset{\sim}{A}}_2 \cap E = \overline{\underset{\sim}{A}}_2 & \underset{\sim}{A}_3 \cap \overline{\underset{\sim}{A}}_5 = \overline{\underset{\sim}{A}}_5 \\
\underset{\sim}{A}_3 \cap \underset{\sim}{A}_6 = \underset{\sim}{A}_3 & \underset{\sim}{A}_3 \cap E = \underset{\sim}{A}_3 & \overline{\underset{\sim}{A}}_5 \cap \underset{\sim}{A}_6 = \overline{\underset{\sim}{A}}_5 & \overline{\underset{\sim}{A}}_5 \cap E = \overline{\underset{\sim}{A}}_5 & \underset{\sim}{A}_6 \cap E = \underset{\sim}{A}_6 \\
\overline{\underset{\sim}{A}}_2 \cup \underset{\sim}{A}_3 = \underset{\sim}{A}_3 & \overline{\underset{\sim}{A}}_2 \cup \overline{\underset{\sim}{A}}_5 = \overline{\underset{\sim}{A}}_2 & \overline{\underset{\sim}{A}}_2 \cup \underset{\sim}{A}_6 = \underset{\sim}{A}_6 & \overline{\underset{\sim}{A}}_2 \cup E = E & \underset{\sim}{A}_3 \cup \overline{\underset{\sim}{A}}_5 = \underset{\sim}{A}_3 \\
\underset{\sim}{A}_3 \cup \underset{\sim}{A}_6 = \overline{\underset{\sim}{A}}_2 & \underset{\sim}{A}_3 \cup E = E & \overline{\underset{\sim}{A}}_5 \cup \underset{\sim}{A}_6 = \underset{\sim}{A}_6 & \overline{\underset{\sim}{A}}_5 \cup E = E & \underset{\sim}{A}_6 \cup E = E
\end{array}
$$

In this way it can be said that *T*(*E*) is an uncertain topology of *K*(*E*).

We will not continue more in this conception of uncertain topology, which we consider to be of little use in economics and management applications. We will move on to the second, which may result as more fertile for our purposes.

We have already shown that the fuzzy subsets of a referential *E* give way to a set of parts of L^E which is not finite, even if *E* is, as a consequence of the fact that $L = [0.1]$ this is so. On the contrary, a finite subset $T(E) \subset L^E$ can be established, formed, **a priori**, of opens and which therefore will be considered a topology. **It will also be assumed** that all of the elements of L^E, in infinite quantity, form a functional application **Γ** which is an **isotone pretopology**. We have indicated on various occasions that in this work we limit ourselves to finite sets, but when topological uncertainty follows this path, **except in the first limited conception** previously expounded, this is not possible. All valuation of an element *E* belongs to [0.1] and not {0.1}. Perhaps in this form uncertain topology may result as useful. However, we repeat that *T*(*E*), which is to say the subset of opens, is finite.

Therefore, *T*(*E*) is the **subset of opens** $\underset{\sim}{A}_j \in T(E)$ **from which one begins**, directly or not, **as a previous estimation**. The set of closes will be formed considering for all $\underset{\sim}{A}_j \in T(E)$ its compliment $\underset{\sim}{A}_j$.

Let us move on to an example:

$$E = \{a, b, c\}$$

$$
T(E) = \left\{ \varnothing ,\;
\begin{array}{ccc} a & b & c \\ \hline \multicolumn{1}{|c|}{.1} & \multicolumn{1}{c|}{0} & \multicolumn{1}{c|}{.3} \\ \hline \end{array} ,\;
\begin{array}{ccc} a & b & c \\ \hline \multicolumn{1}{|c|}{.2} & \multicolumn{1}{c|}{.2} & \multicolumn{1}{c|}{.6} \\ \hline \end{array} ,\;
\begin{array}{ccc} a & b & c \\ \hline \multicolumn{1}{|c|}{.4} & \multicolumn{1}{c|}{.2} & \multicolumn{1}{c|}{.6} \\ \hline \end{array} ,\;
\begin{array}{ccc} a & b & c \\ \hline \multicolumn{1}{|c|}{.2} & \multicolumn{1}{c|}{.5} & \multicolumn{1}{c|}{.7} \\ \hline \end{array} ,
\right.
$$

$$
\begin{array}{ccc} a & b & c \\ \hline \multicolumn{1}{|c|}{.6} & \multicolumn{1}{c|}{.2} & \multicolumn{1}{c|}{.6} \\ \hline \end{array} ,\;
\begin{array}{ccc} a & b & c \\ \hline \multicolumn{1}{|c|}{.4} & \multicolumn{1}{c|}{.5} & \multicolumn{1}{c|}{.7} \\ \hline \end{array} ,\;
\begin{array}{ccc} a & b & c \\ \hline \multicolumn{1}{|c|}{.2} & \multicolumn{1}{c|}{.8} & \multicolumn{1}{c|}{1} \\ \hline \end{array} ,\;
\begin{array}{ccc} a & b & c \\ \hline \multicolumn{1}{|c|}{.6} & \multicolumn{1}{c|}{.5} & \multicolumn{1}{c|}{.7} \\ \hline \end{array} ,
$$

$$
\left.
\begin{array}{ccc} a & b & c \\ \hline \multicolumn{1}{|c|}{.4} & \multicolumn{1}{c|}{.8} & \multicolumn{1}{c|}{1} \\ \hline \end{array} ,\;
\begin{array}{ccc} a & b & c \\ \hline \multicolumn{1}{|c|}{.6} & \multicolumn{1}{c|}{.8} & \multicolumn{1}{c|}{1} \\ \hline \end{array} ,\; E \right\}
$$

The closes of this topology are:

$$C_e = \left\{ E \,,\; \begin{array}{ccc} a & b & c \\ \hline \multicolumn{1}{|c|}{.9} & \multicolumn{1}{c|}{1} & \multicolumn{1}{c|}{.7} \\ \hline \end{array} \,,\; \begin{array}{ccc} a & b & c \\ \hline \multicolumn{1}{|c|}{.8} & \multicolumn{1}{c|}{.8} & \multicolumn{1}{c|}{.4} \\ \hline \end{array} \,,\; \begin{array}{ccc} a & b & c \\ \hline \multicolumn{1}{|c|}{.6} & \multicolumn{1}{c|}{.8} & \multicolumn{1}{c|}{.4} \\ \hline \end{array} \,,\; \begin{array}{ccc} a & b & c \\ \hline \multicolumn{1}{|c|}{.8} & \multicolumn{1}{c|}{.5} & \multicolumn{1}{c|}{.3} \\ \hline \end{array} \,, \right.$$

$$\begin{array}{ccc} a & b & c \\ \hline \multicolumn{1}{|c|}{.4} & \multicolumn{1}{c|}{.8} & \multicolumn{1}{c|}{.4} \\ \hline \end{array} \,,\; \begin{array}{ccc} a & b & c \\ \hline \multicolumn{1}{|c|}{.6} & \multicolumn{1}{c|}{.5} & \multicolumn{1}{c|}{.3} \\ \hline \end{array} \,,\; \begin{array}{ccc} a & b & c \\ \hline \multicolumn{1}{|c|}{.8} & \multicolumn{1}{c|}{.2} & \multicolumn{1}{c|}{0} \\ \hline \end{array} \,,\; \begin{array}{ccc} a & b & c \\ \hline \multicolumn{1}{|c|}{.4} & \multicolumn{1}{c|}{.5} & \multicolumn{1}{c|}{.3} \\ \hline \end{array} \,,$$

$$\left. \begin{array}{ccc} a & b & c \\ \hline \multicolumn{1}{|c|}{.6} & \multicolumn{1}{c|}{.2} & \multicolumn{1}{c|}{0} \\ \hline \end{array} \,,\; \begin{array}{ccc} a & b & c \\ \hline \multicolumn{1}{|c|}{.4} & \multicolumn{1}{c|}{.2} & \multicolumn{1}{c|}{0} \\ \hline \end{array} \,,\; \varnothing \right\}$$

The only fuzzy subsets which are at the same time open and closes are $\varnothing$ and E. For other elements distinct to $\varnothing$ and E to be both open and closes at the same time it would be necessary that the topology contains $\bar{\underset{\sim}{A}}$ if it contains $\underset{\sim}{A}$.

We will leave this description here to move on to the second proposed focus.

The path which we will follow in this **second focus** is completely distinct to the two alternatives proposed in the first and has been adopted by us to expound uncertain pretopology. As we have previously shown, it deals with establishing a referential E of fuzzy subsets in such a way that:

$$E = \{\underset{\sim}{P}_1, \underset{\sim}{P}_2, \ldots, \underset{\sim}{P}_n\}$$

in order to then choose a subset $T(E)$ of the "power set" which fulfils the axioms of the topology.

With the aim of elucidating in the best way that which we have just stated we propose the development of an example. For this we will begin from a referential E of fuzzy subsets:

$$E = \{\underset{\sim}{P}_1, \underset{\sim}{P}_2, \underset{\sim}{P}_3, \underset{\sim}{P}_4\}$$

in which:

$$\underset{\sim}{P}_1 = \begin{array}{ccc} a & b & c \\ \hline \multicolumn{1}{|c|}{.7} & \multicolumn{1}{c|}{.2} & \multicolumn{1}{c|}{.9} \\ \hline \end{array} \qquad \underset{\sim}{P}_2 = \begin{array}{ccc} a & b & c \\ \hline \multicolumn{1}{|c|}{.3} & \multicolumn{1}{c|}{.9} & \multicolumn{1}{c|}{.7} \\ \hline \end{array} \qquad \underset{\sim}{P}_3 = \begin{array}{ccc} a & b & c \\ \hline \multicolumn{1}{|c|}{.8} & \multicolumn{1}{c|}{.2} & \multicolumn{1}{c|}{.4} \\ \hline \end{array} \qquad \underset{\sim}{P}_4 = \begin{array}{ccc} a & b & c \\ \hline \multicolumn{1}{|c|}{.2} & \multicolumn{1}{c|}{.9} & \multicolumn{1}{c|}{.8} \\ \hline \end{array}$$

We will choose a subset of the "power set" such as:

$$T(E) = \{\varnothing, \{\underset{\sim}{P}_1\}, \{\underset{\sim}{P}_3\}, \{\underset{\sim}{P}_2, \underset{\sim}{P}_4\}, \{\underset{\sim}{P}_1, \underset{\sim}{P}_3\}, \{\underset{\sim}{P}_1, \underset{\sim}{P}_2, \underset{\sim}{P}_4\}, \{\underset{\sim}{P}_2, \underset{\sim}{P}_3, \underset{\sim}{P}_4\}, E\}$$

We now ask ourselves, could this subset be a topology? We achieve the answer by checking if it fulfils the established axioms.

It is immediately clear that it fulfils the first two. Let us see if it also fulfils the third and the fourth.

Let us see the third:

$\forall \underset{\sim}{A}_j, \underset{\sim}{A}_k \in T(E)$:

$$\underset{\sim}{A}_j \cap \underset{\sim}{A}_k \in T(E)$$

We have:

$$\begin{aligned}
\{\underset{\sim}{P}_1\} \cap \{\underset{\sim}{P}_3\} &= \varnothing \\
\{\underset{\sim}{P}_1\} \cap \{\underset{\sim}{P}_1, \underset{\sim}{P}_3\} &= \{\underset{\sim}{P}_1\} \\
\{\underset{\sim}{P}_1\} \cap \{\underset{\sim}{P}_2, \underset{\sim}{P}_4\} &= \varnothing \\
\{\underset{\sim}{P}_1\} \cap \{\underset{\sim}{P}_1, \underset{\sim}{P}_2, \underset{\sim}{P}_4\} &= \{\underset{\sim}{P}_1\}
\end{aligned}$$

$$\begin{aligned}
\{\underset{\sim}{P}_1\} \cap \{\underset{\sim}{P}_2, \underset{\sim}{P}_3, \underset{\sim}{P}_4\} &= \varnothing \\
\{\underset{\sim}{P}_1\} \cap E &= \{\underset{\sim}{P}_1\}
\end{aligned}$$

$$\begin{aligned}
\{\underset{\sim}{P}_3\} \cap \{\underset{\sim}{P}_1, \underset{\sim}{P}_3\} &= \{\underset{\sim}{P}_3\} \\
\{\underset{\sim}{P}_3\} \cap \{\underset{\sim}{P}_2, \underset{\sim}{P}_4\} &= \varnothing \\
\{\underset{\sim}{P}_3\} \cap \{\underset{\sim}{P}_1, \underset{\sim}{P}_2, \underset{\sim}{P}_4\} &= \varnothing \\
\{\underset{\sim}{P}_3\} \cap \{\underset{\sim}{P}_2, \underset{\sim}{P}_3, \underset{\sim}{P}_4\} &= \{\underset{\sim}{P}_3\} \\
\{\underset{\sim}{P}_3\} \cap E &= \{\underset{\sim}{P}_3\}
\end{aligned}$$

$$\begin{aligned}
\{\underset{\sim}{P}_1, \underset{\sim}{P}_3\} \cap \{\underset{\sim}{P}_2, \underset{\sim}{P}_4\} &= \varnothing \\
\{\underset{\sim}{P}_1, \underset{\sim}{P}_3\} \cap \{\underset{\sim}{P}_1, \underset{\sim}{P}_2, \underset{\sim}{P}_4\} &= \{\underset{\sim}{P}_1\} \\
\{\underset{\sim}{P}_1, \underset{\sim}{P}_3\} \cap \{\underset{\sim}{P}_2, \underset{\sim}{P}_3, \underset{\sim}{P}_4\} &= \{\underset{\sim}{P}_3\} \\
\{\underset{\sim}{P}_1, \underset{\sim}{P}_3\} \cap E &= \{\underset{\sim}{P}_1, \underset{\sim}{P}_3\}
\end{aligned}$$

$$\begin{aligned}
\{\underset{\sim}{P}_2, \underset{\sim}{P}_4\} \cap \{\underset{\sim}{P}_1, \underset{\sim}{P}_2, \underset{\sim}{P}_4\} &= \{\underset{\sim}{P}_2, \underset{\sim}{P}_4\} \\
\{\underset{\sim}{P}_2, \underset{\sim}{P}_4\} \cap \{\underset{\sim}{P}_2, \underset{\sim}{P}_3, \underset{\sim}{P}_4\} &= \{\underset{\sim}{P}_2, \underset{\sim}{P}_4\} \\
\{\underset{\sim}{P}_2, \underset{\sim}{P}_4\} \cap E &= \{\underset{\sim}{P}_2, \underset{\sim}{P}_4\}
\end{aligned}$$

$$\begin{aligned}
\{\underset{\sim}{P}_1, \underset{\sim}{P}_2, \underset{\sim}{P}_4\} \cap \{\underset{\sim}{P}_2, \underset{\sim}{P}_3, \underset{\sim}{P}_4\} &= \{\underset{\sim}{P}_2, \underset{\sim}{P}_4\} \\
\{\underset{\sim}{P}_1, \underset{\sim}{P}_2, \underset{\sim}{P}_4\} \cap E &= \{\underset{\sim}{P}_1, \underset{\sim}{P}_2, \underset{\sim}{P}_4\}
\end{aligned}$$

$$\{\underset{\sim}{P}_2, \underset{\sim}{P}_3, \underset{\sim}{P}_4\} \cap E = \{\underset{\sim}{P}_2, \underset{\sim}{P}_3, \underset{\sim}{P}_4\}$$

There is no doubt that it has fulfilled the demands of this axiom.
Refering to the fourth axiom:

$\forall \underset{\sim}{A}_j, \underset{\sim}{A}_k \in T(E)$:

$$\underset{\sim}{A}_j \cup \underset{\sim}{A}_k \in T(E)$$

we perform the confirming process in the following way:

$$\begin{aligned}
\{\underset{\sim}{P}_1\} \cup \{\underset{\sim}{P}_3\} &= \{\underset{\sim}{P}_1, \underset{\sim}{P}_3\}\\
\{\underset{\sim}{P}_1\} \cup \{\underset{\sim}{P}_1, \underset{\sim}{P}_3\} &= \{\underset{\sim}{P}_1, \underset{\sim}{P}_3\}\\
\{\underset{\sim}{P}_1\} \cup \{\underset{\sim}{P}_2, \underset{\sim}{P}_4\} &= \{\underset{\sim}{P}_1, \underset{\sim}{P}_2, \underset{\sim}{P}_4\}\\
\{\underset{\sim}{P}_1\} \cup \{\underset{\sim}{P}_1, \underset{\sim}{P}_2, \underset{\sim}{P}_4\} &= \{\underset{\sim}{P}_1, \underset{\sim}{P}_2, \underset{\sim}{P}_4\}\\
\{\underset{\sim}{P}_1\} \cup \{\underset{\sim}{P}_2, \underset{\sim}{P}_3, \underset{\sim}{P}_4\} &= E\\
\{\underset{\sim}{P}_1\} \cup E &= E
\end{aligned}$$

$$\begin{aligned}
\{\underset{\sim}{P}_3\} \cup \{\underset{\sim}{P}_1, \underset{\sim}{P}_3\} &= \{\underset{\sim}{P}_1, \underset{\sim}{P}_3\}\\
\{\underset{\sim}{P}_3\} \cup \{\underset{\sim}{P}_2, \underset{\sim}{P}_4\} &= \{\underset{\sim}{P}_2, \underset{\sim}{P}_3, \underset{\sim}{P}_4\}
\end{aligned}$$

$$\begin{aligned}
\{\underset{\sim}{P}_3\} \cup \{\underset{\sim}{P}_1, \underset{\sim}{P}_2, \underset{\sim}{P}_4\} &= E\\
\{\underset{\sim}{P}_3\} \cup \{\underset{\sim}{P}_2, \underset{\sim}{P}_3, \underset{\sim}{P}_4\} &= \{\underset{\sim}{P}_2, \underset{\sim}{P}_3, \underset{\sim}{P}_4\}\\
\{\underset{\sim}{P}_3\} \cup E &= E
\end{aligned}$$

$$\begin{aligned}
\{\underset{\sim}{P}_1, \underset{\sim}{P}_3\} \cup \{\underset{\sim}{P}_2, \underset{\sim}{P}_4\} &= E\\
\{\underset{\sim}{P}_1, \underset{\sim}{P}_3\} \cup \{\underset{\sim}{P}_1, \underset{\sim}{P}_2, \underset{\sim}{P}_4\} &= E\\
\{\underset{\sim}{P}_1, \underset{\sim}{P}_3\} \cup \{\underset{\sim}{P}_2, \underset{\sim}{P}_3, \underset{\sim}{P}_4\} &= E\\
\{\underset{\sim}{P}_1, \underset{\sim}{P}_3\} \cup E &= E
\end{aligned}$$

$$\begin{aligned}
\{\underset{\sim}{P}_2, \underset{\sim}{P}_4\} \cup \{\underset{\sim}{P}_1, \underset{\sim}{P}_2, \underset{\sim}{P}_4\} &= \{\underset{\sim}{P}_1, \underset{\sim}{P}_2, \underset{\sim}{P}_4\}\\
\{\underset{\sim}{P}_2, \underset{\sim}{P}_4\} \cup \{\underset{\sim}{P}_1, \underset{\sim}{P}_3, \underset{\sim}{P}_4\} &= E\\
\{\underset{\sim}{P}_2, \underset{\sim}{P}_4\} \cup E &= E
\end{aligned}$$

$$\begin{aligned}
\{\underset{\sim}{P}_1, \underset{\sim}{P}_2, \underset{\sim}{P}_4\} \cup \{\underset{\sim}{P}_2, \underset{\sim}{P}_3, \underset{\sim}{P}_4\} &= E\\
\{\underset{\sim}{P}_1, \underset{\sim}{P}_2, \underset{\sim}{P}_4\} \cup E &= E
\end{aligned}$$

$$\{\underset{\sim}{P}_2, \underset{\sim}{P}_3, \underset{\sim}{P}_4\} \cup E = E$$

It also fulfils this last axiom.

Therefore the subset $T(E)$ is a topology.

This topology, defined by $(E, T(E))$, forms a Boole lattice, as can be seen in the following figure:

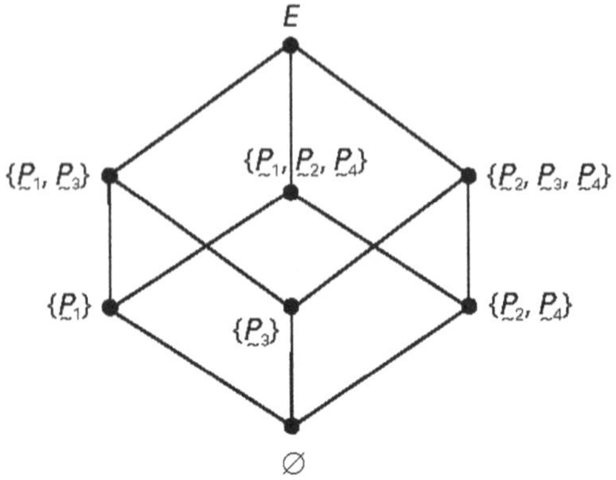

But, as we have stated, **a topology** may also be defined through **the pair** ($\boldsymbol{E}$, $\boldsymbol{\Gamma}$) or, in the same way, **by the pair** ($\boldsymbol{E}$, $\boldsymbol{\delta}$). To show this we will obtain the interior, δ, and adherent, $\boldsymbol{\Gamma}$, applications.

As is well-known, to find the **interior application**, for each of the elements of the "power set" $P(E)$ its corresponding application is established, choosing the **largest open** contained in the element of the "power set".

In our case we have:

$$\delta\varnothing = \varnothing \qquad \delta\{\underset{\sim}{P}_1\} = \{\mathrm{P}_1\} \qquad \delta\{\underset{\sim}{P}_2\} = \varnothing \qquad \delta\{\underset{\sim}{P}_3\} = \{\underset{\sim}{P}_3\}$$

$$\delta\{\underset{\sim}{P}_4\} = \varnothing \qquad \delta\{\underset{\sim}{P}_1, \underset{\sim}{P}_2\} = \{\underset{\sim}{P}_1\} \qquad \delta\{\underset{\sim}{P}_1, \underset{\sim}{P}_3\} = \{\underset{\sim}{P}_1, \underset{\sim}{P}_3\}$$

$$\delta\{\underset{\sim}{P}_1, \underset{\sim}{P}_4\} = \{\underset{\sim}{P}_1\} \qquad \delta\{\underset{\sim}{P}_2, \underset{\sim}{P}_3\} = \{\underset{\sim}{P}_3\} \qquad \delta\{\underset{\sim}{P}_2, \underset{\sim}{P}_4\} = \{\underset{\sim}{P}_2, \underset{\sim}{P}_4\}$$

$$\delta\{\underset{\sim}{P}_3, \underset{\sim}{P}_4\} = \{\underset{\sim}{P}_3\} \qquad \delta\{\underset{\sim}{P}_1, \underset{\sim}{P}_2, \underset{\sim}{P}_3\} = \{\underset{\sim}{P}_1, \underset{\sim}{P}_3\} \qquad \delta\{\underset{\sim}{P}_1, \underset{\sim}{P}_2, \underset{\sim}{P}_4\} = \{\underset{\sim}{P}_1, \underset{\sim}{P}_2, \underset{\sim}{P}_4\}$$

$$\delta\{\underset{\sim}{P}_1, \underset{\sim}{P}_3, \underset{\sim}{P}_4\} = \{\underset{\sim}{P}_1, \underset{\sim}{P}_3\} \qquad \delta\{\underset{\sim}{P}_2, \underset{\sim}{P}_3, \underset{\sim}{P}_4\} = \{\underset{\sim}{P}_2, \underset{\sim}{P}_3, \underset{\sim}{P}_4\} \qquad \delta E = E$$

In continuation we present the graph which represents the interior application.

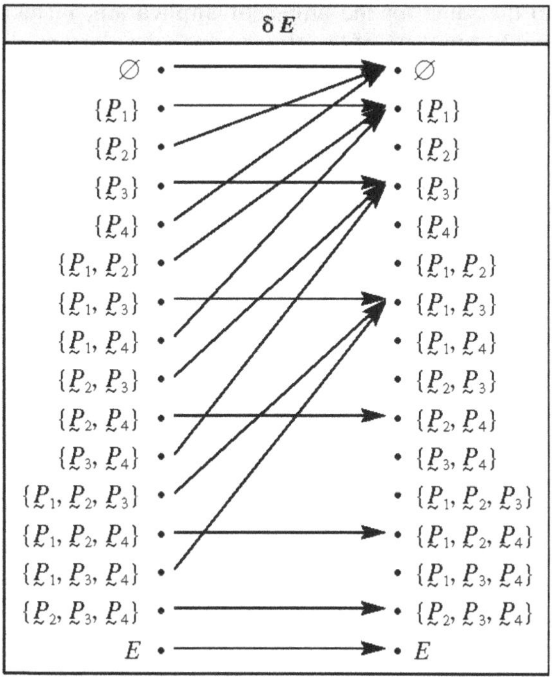

We have seen that the formative elements of topology have the configuration of a Boole lattice. So, these elements are registered as a sub-lattice of the Boole lattice corresponding to the "power set" as can be seen in the following:

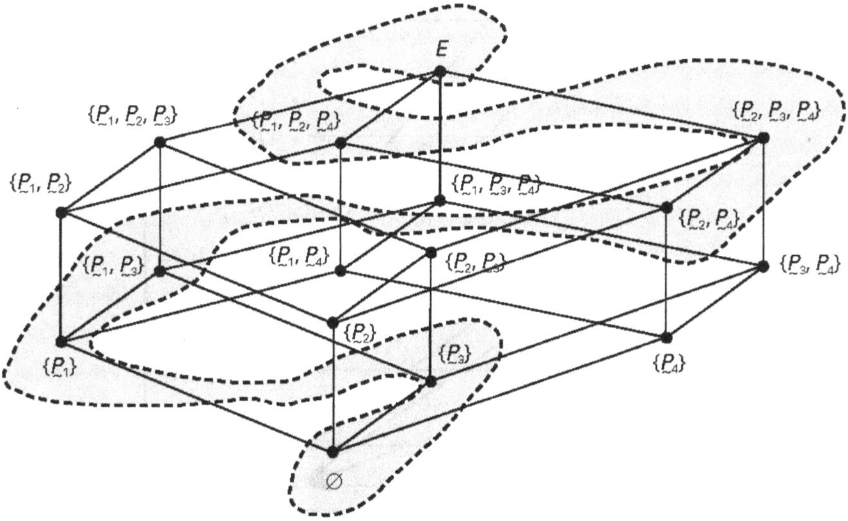

We could do the same for the **adherent application, Γ**, taking into account that, in this case, the **smallest** of the closes which the element of the "power set" contains should be chosen as **adherent**.

To find the closes it is enough to obtain the compliments of each of the opens. They would be:

$$C_e = \{\varnothing, \{P_1\}, \{P_3\}, \{P_1, P_3\}, \{P_2, P_4\}, \{P_1, P_2, P_4\}, \{P_2, P_3, P_4\}, E\}$$

The circumstance has been seen that, exceptionally, in this case specifically the opens coincide with the closes.

We will move on to expound the adherent application:

$\Gamma\varnothing = \varnothing$ $\quad\Gamma\{P_1\} = \{P_1\}$ $\quad\Gamma\{P_2\} = \{P_2, P_4\}$ $\quad\Gamma\{P_3\} = \{P_3\}$
$\Gamma\{P_4\} = \{P_2, P_4\}$ $\quad\Gamma\{P_1, P_2\} = \{P_1, P_2, P_4\}$ $\quad\Gamma\{P_1, P_3\} = \{P_1, P_3\}$,
$\Gamma\{P_1, P_4\} = \{P_1, P_2, P_4\}$ $\quad\Gamma\{P_2, P_3\} = \{P_2, P_3, P_4\}$ $\quad\Gamma\{P_2, P_4\} = \{P_2, P_4\}$,
$\Gamma\{P_3, P_4\} = \{P_2, P_3, P_4\}$ $\quad\Gamma\{P_1, P_2, P_3\} = E$ $\quad\Gamma\{P_1, P_2, P_4\} = \{P_1, P_2, P_4\}$,
$\Gamma\{P_1, P_3, P_4\} = E$ $\quad\Gamma\{P_2, P_3, P_4\} = \{P_2, P_3, P_4\}$ $\quad\Gamma E = E$

we will represent it by means of the following graph:

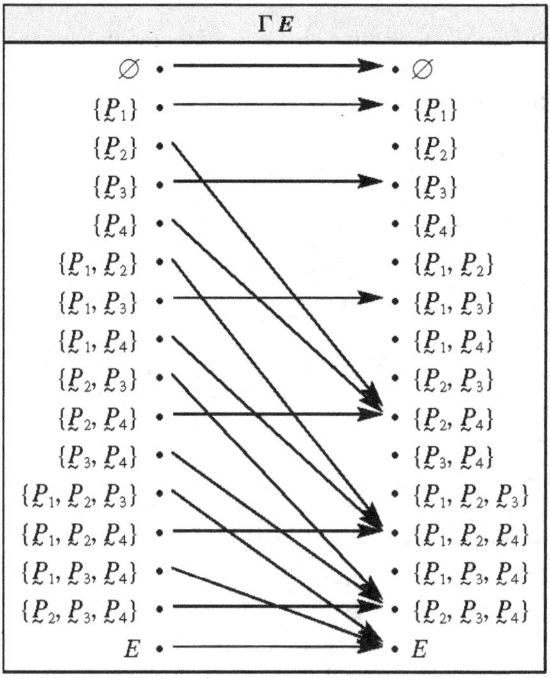

The configuration of the closes (which in this case, we remind ourselves, coincides with the opens) gives, evidently in this case, the same Boole lattice and therefore exactly the same occurs as with the opens in the corresponding lattice to the "power set".

We now move on to confirm that our proposal also constitutes an uncertain topology in the form (E, **Γ**). For this we will confirm, one by one, each axiom in which **Γ** intervenes. We will begin with the following:

$\forall \underset{\sim}{A}_j \in P(E)$:

$$\underset{\sim}{A}_j \subset \Gamma \underset{\sim}{A}_j$$

We do not believe the detail of each of the elements necessary as a simple look at the previous graph clearly shows the fulfillment of this axiom.

The final axiom gives the isotonia, which can be formalized without distinction by:

$\forall \underset{\sim}{A}_j, \underset{\sim}{A}_k \in P(E)$:

$$(\underset{\sim}{A}_j \subset A_k) \Rightarrow (\Gamma \underset{\sim}{A}_j \subset \Gamma \underset{\sim}{A}_k)$$

or by:

$\forall \underset{\sim}{A}_j, \underset{\sim}{A}_k \in P(E)$:

$$(\underset{\sim}{A}_j \subset A_k) \Rightarrow (\delta \underset{\sim}{A}_j \subset \delta \underset{\sim}{A}_k)$$

We arbitrarily choose the first:

$$
\begin{array}{lcl}
\{\underset{\sim}{P}_1\} \subset \{\underset{\sim}{P}_1, \underset{\sim}{P}_2\} & \Rightarrow & \{\underset{\sim}{P}_1\} \subset \{\underset{\sim}{P}_1, \underset{\sim}{P}_2, \underset{\sim}{P}_4\} \\
\{\underset{\sim}{P}_1\} \subset \{\underset{\sim}{P}_1, \underset{\sim}{P}_3\} & \Rightarrow & \{\underset{\sim}{P}_1\} \subset \{\underset{\sim}{P}_1, \underset{\sim}{P}_3\} \\
\{\underset{\sim}{P}_1\} \subset \{\underset{\sim}{P}_1, \underset{\sim}{P}_4\} & \Rightarrow & \{\underset{\sim}{P}_1\} \subset \{\underset{\sim}{P}_1, \underset{\sim}{P}_2, \underset{\sim}{P}_4\} \\
\{\underset{\sim}{P}_1\} \subset \{\underset{\sim}{P}_1, \underset{\sim}{P}_2, \underset{\sim}{P}_3\} & \Rightarrow & \{\underset{\sim}{P}_1\} \subset E \\
\{\underset{\sim}{P}_1\} \subset \{\underset{\sim}{P}_1, \underset{\sim}{P}_2, \underset{\sim}{P}_4\} & \Rightarrow & \{\underset{\sim}{P}_1\} \subset \{\underset{\sim}{P}_1, \underset{\sim}{P}_2, \underset{\sim}{P}_4\} \\
\{\underset{\sim}{P}_1\} \subset \{\underset{\sim}{P}_1, \underset{\sim}{P}_3, \underset{\sim}{P}_4\} & \Rightarrow & \{\underset{\sim}{P}_1\} \subset E \\
\{\underset{\sim}{P}_1\} \subset E & \Rightarrow & \{\underset{\sim}{P}_1\} \subset E
\end{array}
$$

$$
\begin{array}{ll}
\{\underset{\sim}{P}_2\} \subset \{\underset{\sim}{P}_1, \underset{\sim}{P}_2\} & \Rightarrow \{\underset{\sim}{P}_2, \underset{\sim}{P}_4\} \subset \{\underset{\sim}{P}_1, \underset{\sim}{P}_2, \underset{\sim}{P}_4\} \\
\{\underset{\sim}{P}_2\} \subset \{\underset{\sim}{P}_2, \underset{\sim}{P}_3\} & \Rightarrow \{\underset{\sim}{P}_2, \underset{\sim}{P}_4\} \subset \{\underset{\sim}{P}_2, \underset{\sim}{P}_3, \underset{\sim}{P}_4\} \\
\{\underset{\sim}{P}_2\} \subset \{\underset{\sim}{P}_2, \underset{\sim}{P}_4\} & \Rightarrow \{\underset{\sim}{P}_2, \underset{\sim}{P}_4\} \subset \{\underset{\sim}{P}_2, \underset{\sim}{P}_4\} \\
\{\underset{\sim}{P}_2\} \subset \{\underset{\sim}{P}_1, \underset{\sim}{P}_2, \underset{\sim}{P}_3\} & \Rightarrow \{\underset{\sim}{P}_2, \underset{\sim}{P}_4\} \subset E \\
\{\underset{\sim}{P}_2\} \subset \{\underset{\sim}{P}_1, \underset{\sim}{P}_2, \underset{\sim}{P}_4\} & \Rightarrow \{\underset{\sim}{P}_2, \underset{\sim}{P}_4\} \subset \{\underset{\sim}{P}_1, \underset{\sim}{P}_2, \underset{\sim}{P}_4\} \\
\{\underset{\sim}{P}_2\} \subset \{\underset{\sim}{P}_2, \underset{\sim}{P}_3, \underset{\sim}{P}_4\} & \Rightarrow \{\underset{\sim}{P}_2, \underset{\sim}{P}_4\} \subset \{\underset{\sim}{P}_2, \underset{\sim}{P}_3, \underset{\sim}{P}_4\} \\
\{\underset{\sim}{P}_2\} \subset E & \Rightarrow \{\underset{\sim}{P}_2, \underset{\sim}{P}_4\} \subset E
\end{array}
$$

$$
\begin{array}{ll}
\{\underset{\sim}{P}_3\} \subset \{\underset{\sim}{P}_1, \underset{\sim}{P}_3\} & \Rightarrow \{\underset{\sim}{P}_3\} \subset \{\underset{\sim}{P}_1, \underset{\sim}{P}_3\} \\
\{\underset{\sim}{P}_3\} \subset \{\underset{\sim}{P}_2, \underset{\sim}{P}_3\} & \Rightarrow \{\underset{\sim}{P}_3\} \subset \{\underset{\sim}{P}_2, \underset{\sim}{P}_3, \underset{\sim}{P}_4\} \\
\{\underset{\sim}{P}_3\} \subset \{\underset{\sim}{P}_3, \underset{\sim}{P}_4\} & \Rightarrow \{\underset{\sim}{P}_3\} \subset \{\underset{\sim}{P}_2, \underset{\sim}{P}_3, \underset{\sim}{P}_4\} \\
\{\underset{\sim}{P}_3\} \subset \{\underset{\sim}{P}_1, \underset{\sim}{P}_2, \underset{\sim}{P}_3\} & \Rightarrow \{\underset{\sim}{P}_3\} \subset E \\
\{\underset{\sim}{P}_3\} \subset \{\underset{\sim}{P}_1, \underset{\sim}{P}_3, \underset{\sim}{P}_4\} & \Rightarrow \{\underset{\sim}{P}_3\} \subset E \\
\{\underset{\sim}{P}_3\} \subset \{\underset{\sim}{P}_2, \underset{\sim}{P}_3, \underset{\sim}{P}_4\} & \Rightarrow \{\underset{\sim}{P}_3\} \subset \{\underset{\sim}{P}_2, \underset{\sim}{P}_3, \underset{\sim}{P}_4\} \\
\{\underset{\sim}{P}_3\} \subset E & \Rightarrow \{\underset{\sim}{P}_3\} \subset E
\end{array}
$$

....

$$
\begin{array}{ll}
\{\underset{\sim}{P}_1, \underset{\sim}{P}_3\} \subset \{\underset{\sim}{P}_1, \underset{\sim}{P}_2, \underset{\sim}{P}_3\} & \Rightarrow \{\underset{\sim}{P}_1, \underset{\sim}{P}_3\} \subset E \\
\{\underset{\sim}{P}_1, \underset{\sim}{P}_3\} \subset \{\underset{\sim}{P}_1, \underset{\sim}{P}_3, \underset{\sim}{P}_4\} & \Rightarrow \{\underset{\sim}{P}_1, \underset{\sim}{P}_3\} \subset E \\
\{\underset{\sim}{P}_1, \underset{\sim}{P}_3\} \subset E & \Rightarrow \{\underset{\sim}{P}_1, \underset{\sim}{P}_3\} \subset E
\end{array}
$$

$$
\begin{array}{ll}
\{\underset{\sim}{P}_1, \underset{\sim}{P}_4\} \subset \{\underset{\sim}{P}_1, \underset{\sim}{P}_2, \underset{\sim}{P}_4\} & \Rightarrow \{\underset{\sim}{P}_1, \underset{\sim}{P}_4\} \subset \{\underset{\sim}{P}_1, \underset{\sim}{P}_2, \underset{\sim}{P}_4\} \\
\{\underset{\sim}{P}_1, \underset{\sim}{P}_4\} \subset \{\underset{\sim}{P}_1, \underset{\sim}{P}_3, \underset{\sim}{P}_4\} & \Rightarrow \{\underset{\sim}{P}_1, \underset{\sim}{P}_4\} \subset E \\
\{\underset{\sim}{P}_1, \underset{\sim}{P}_4\} \subset E & \Rightarrow \{\underset{\sim}{P}_1, \underset{\sim}{P}_4\} \subset E
\end{array}
$$

....

$$\{\underset{\sim}{P}_1, \underset{\sim}{P}_2, \underset{\sim}{P}_4\} \subset E \Rightarrow \{\underset{\sim}{P}_1, \underset{\sim}{P}_2, \underset{\sim}{P}_4\} \subset E$$

....

$$\{\underset{\sim}{P}_2, \underset{\sim}{P}_3, \underset{\sim}{P}_4\} \subset E \Rightarrow \{\underset{\sim}{P}_2, \underset{\sim}{P}_3, \underset{\sim}{P}_4\} \subset E$$

In this way the isotonia is tangible and, therefore, there is the existence of an **uncertain topological space** also through the adherent application in the sense which we have given this term.

2 The Notion of Neighbourhood in an Uncertain Topological Space

The concept of uncertain topological space, in such a way that it has been expounded, allows the development of a wide range of elements which may result as useful for a greater comprehension and efficient treatment of complex phenomena of economics and management. One of these elements takes the name of "neighbourhood".

The notion of **neighbourhood** as a subset of **neighbours** of an element of the "power set" is not new in the studies of ordinary topology.

Here we propose the presentation of this interesting element, **first** by taking **a referential of fuzzy subsets** as a starting point. In this way each neighbour could well be a fuzzy subset or an ordinary subset of fuzzy subsets and, in this way, each neighbourhood will be enriched by groupings which will be, at the same time, elements of the "power set". **Later** this element will be considered when we start from **a referential** formed **by the referentials of the fuzzy subsets**[2].

All that has gone before allows us to define the concept of neighbour, through the following expression:

$$(\underset{\sim}{X} \in P(E) \text{ is a neighbour of } \underset{\sim}{A}_j \in P(E))$$
$$\Rightarrow (\exists\, \underset{\sim}{Y} \in T(E) \,/\, \underset{\sim}{A}_j \subset \underset{\sim}{Y} \subset \underset{\sim}{X})$$

The set of neighbours of $\underset{\sim}{A}_j$ is named neighbourhood of $\underset{\sim}{A}_j$.

In a colloquial way it may be said that a neighbourhood of an element or of a point is formed by "all sets which **contain an open** which, at the same time, contain this point".

To illustrate the concept of neighbourhood we will consider the same referential used up to now:

$$E = \{\underset{\sim}{P}_1, \underset{\sim}{P}_2, \underset{\sim}{P}_3, \underset{\sim}{P}_4\}$$

In which, also here:

$\underset{\sim}{P}_1 =$

a	b	c
.7	.2	.9

$\underset{\sim}{P}_2 =$

a	b	c
.3	.9	.7

$\underset{\sim}{P}_3 =$

a	b	c
.8	.2	.4

$\underset{\sim}{P}_4 =$

a	b	c
.2	.9	.8

and the topology:

$$T(E) = \{\emptyset, \{\underset{\sim}{P}_1\}, \{\underset{\sim}{P}_3\}, \{\underset{\sim}{P}_2, \underset{\sim}{P}_4\}, \{\underset{\sim}{P}_1, \underset{\sim}{P}_3\}, \{\underset{\sim}{P}_1, \underset{\sim}{P}_2, \underset{\sim}{P}_4\}, \{\underset{\sim}{P}_2, \underset{\sim}{P}_3, \underset{\sim}{P}_4\}, E\}$$

[2] Kaufmann, A.: «La prétopologie ordinaire et la prétopologie floue». Note de travail 115. La Tronche, 1983.

We remind ourselves that in a Boole lattice one has:

$$\underset{\sim}{A}_j \cap \overline{\underset{\sim}{A}}_j = \varnothing$$
$$\underset{\sim}{A}_j \cup \overline{\underset{\sim}{A}}_j = E$$

which, in our case, performing

$$\underset{\sim}{A}_1 = \{\underset{\sim}{P}_1\} \qquad \underset{\sim}{A}_2 = \{\underset{\sim}{P}_3\} \qquad \underset{\sim}{A}_3 = \{\underset{\sim}{P}_2, \underset{\sim}{P}_4\}$$
$$\underset{\sim}{A}_4 = \overline{\underset{\sim}{A}}_1 = \{\underset{\sim}{P}_2, \underset{\sim}{P}_3, \underset{\sim}{P}_4\} \qquad \underset{\sim}{A}_5 = \overline{\underset{\sim}{A}}_2 = \{\underset{\sim}{P}_1, \underset{\sim}{P}_2, \underset{\sim}{P}_4\} \qquad \underset{\sim}{A}_6 = \overline{\underset{\sim}{A}}_3 = \{\underset{\sim}{P}_1, \underset{\sim}{P}_3\}$$

the fulfillment of the previous requisite is easily confirmed, always with the condition of considering the **fuzzy subsets** $\underset{\sim}{P}_1$, $\underset{\sim}{P}_2$, $\underset{\sim}{P}_3$, $\underset{\sim}{P}_4$ as elements with their **own identity**. An example which we believe to be representative would be the case in which these fuzzy subsets describe people or products (real or financial) and their referentials *a*, *b*, *c*, properties, characteristics or singularities[3].

We will now move on to present, as a guideline, the first neighbourhoods of *P*(*E*):

neighbourhood $\{\underset{\sim}{P}_1\} = \{\{\underset{\sim}{P}_1\}, \{\underset{\sim}{P}_1, \underset{\sim}{P}_2\}, \{\underset{\sim}{P}_1, \underset{\sim}{P}_3\}, \{\underset{\sim}{P}_1, \underset{\sim}{P}_4\}, \{\underset{\sim}{P}_1, \underset{\sim}{P}_2, \underset{\sim}{P}_3\}, \{\underset{\sim}{P}_1, \underset{\sim}{P}_2, \underset{\sim}{P}_4\}, \{\underset{\sim}{P}_1, \underset{\sim}{P}_3, \underset{\sim}{P}_4\}, E\}$

neighbourhood $\{\underset{\sim}{P}_2\} = \{\{\underset{\sim}{P}_2, \underset{\sim}{P}_4\}, \{\underset{\sim}{P}_1, \underset{\sim}{P}_2, \underset{\sim}{P}_4\}, \{\underset{\sim}{P}_2, \underset{\sim}{P}_3, \underset{\sim}{P}_4\}, E\}$

neighbourhood $\{\underset{\sim}{P}_3\} = \{\{\underset{\sim}{P}_3\}, \{\underset{\sim}{P}_1, \underset{\sim}{P}_3\}, \{\underset{\sim}{P}_2, \underset{\sim}{P}_3\}, \{\underset{\sim}{P}_3, \underset{\sim}{P}_4\}, \{\underset{\sim}{P}_1, \underset{\sim}{P}_2, \underset{\sim}{P}_3\}, \{\underset{\sim}{P}_1, \underset{\sim}{P}_3, \underset{\sim}{P}_4\}, \{\underset{\sim}{P}_2, \underset{\sim}{P}_3, \underset{\sim}{P}_4\}, E\}$

neighbourhood $\{\underset{\sim}{P}_4\} = \{\{\underset{\sim}{P}_2, \underset{\sim}{P}_4\}, \{\underset{\sim}{P}_1, \underset{\sim}{P}_2, \underset{\sim}{P}_4\}, \{\underset{\sim}{P}_2, \underset{\sim}{P}_3, \underset{\sim}{P}_4\}, E\}$

neighbourhood $\{\underset{\sim}{P}_1, \underset{\sim}{P}_2\} = \{\{\underset{\sim}{P}_1, \underset{\sim}{P}_2, \underset{\sim}{P}_4\}, E\}$

neighbourhood $\{\underset{\sim}{P}_1, \underset{\sim}{P}_3\} = \{\{\underset{\sim}{P}_1, \underset{\sim}{P}_3\}, \{\underset{\sim}{P}_1, \underset{\sim}{P}_2, \underset{\sim}{P}_3\}, \{\underset{\sim}{P}_1, \underset{\sim}{P}_3, \underset{\sim}{P}_4\}, E\}$

and successively in this way.

---

[3] The intersection or union in our example refers to, in its case, the people or the products but not at the level of their attributes (qualities, characteristics or singularities).

We represent these neighbourhoods in the Boole lattice of the "power set".

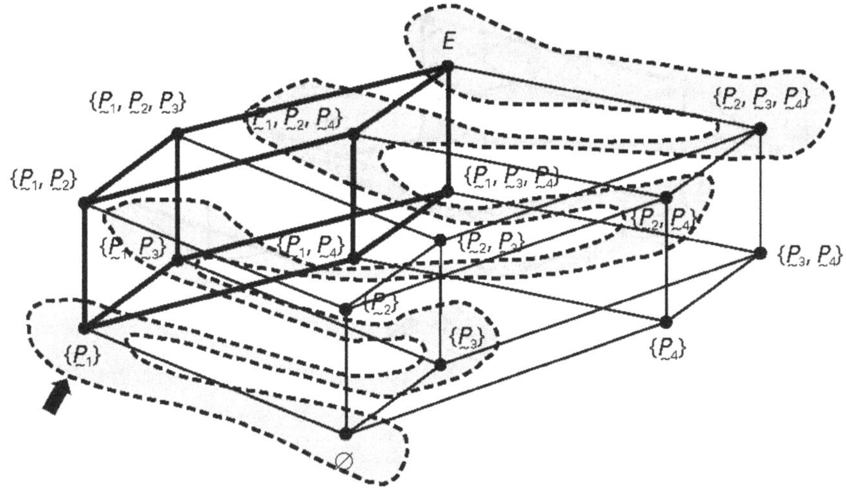

Neighbourhood $\{\underset{\sim}{P}_1\}$

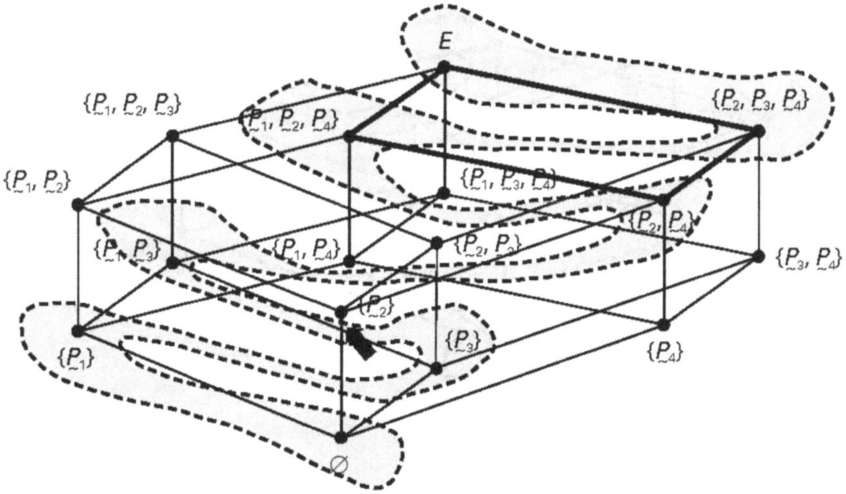

Neighbourhood $\{\underset{\sim}{P}_2\}$

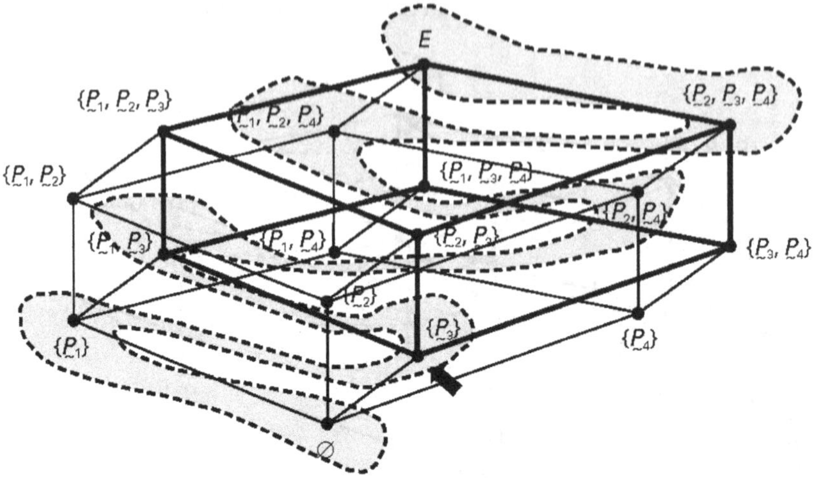

Neighbourhood $\{\underset{\sim}{P}_3\}$

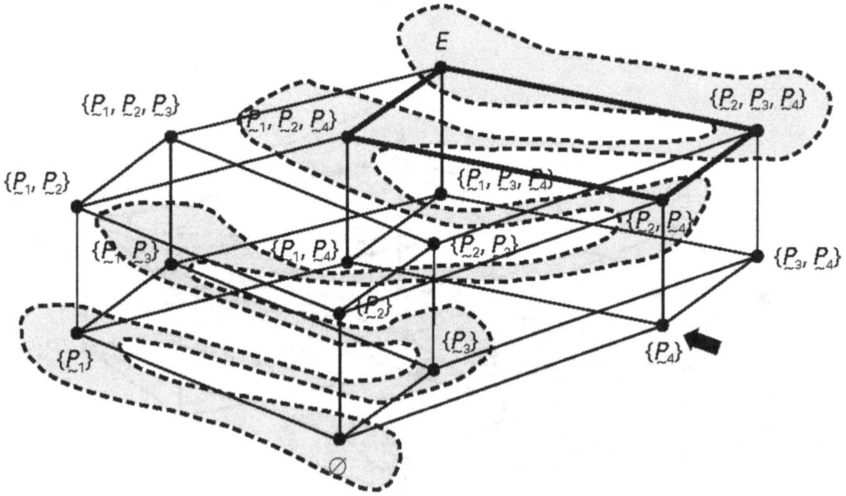

Neighbourhood $\{\underset{\sim}{P}_4\}$

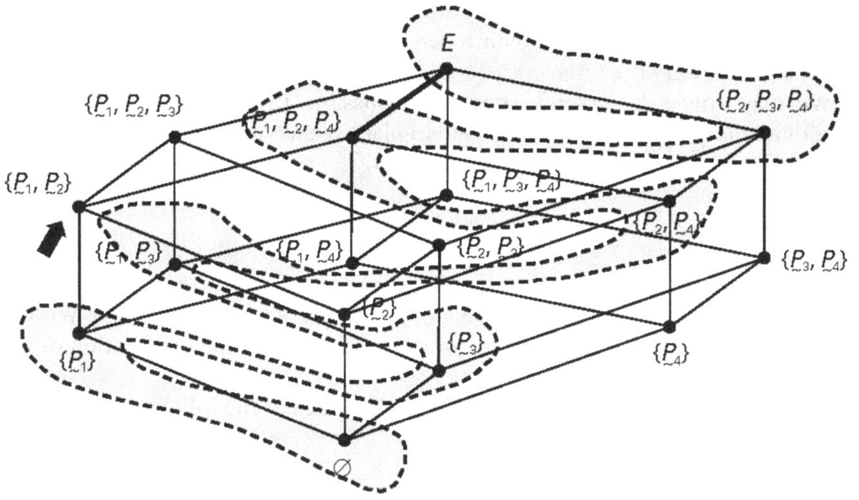

Neighbourhood $\{\underset{\sim}{P}_1, \underset{\sim}{P}_2\}$

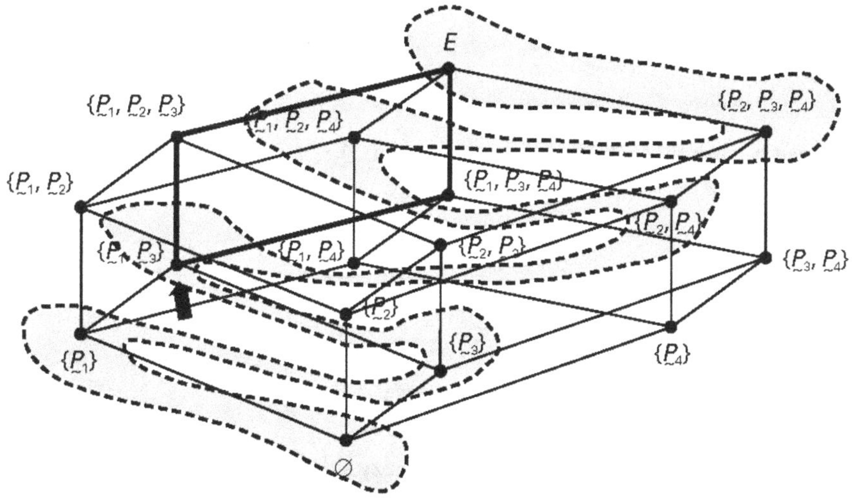

Neighbourhood $\{\underset{\sim}{P}_1, \underset{\sim}{P}_3\}$

We believe that the visual clarity of these figures exempts any comment.

We consider that with these specifications we will have been capable of illustrating the technical aspects of the interesting concept of neighbourhood. It may now be convenient to give a brief outline on its potential representation in the area of economics finance.

A product or group of financial products with certain attributes (qualities, characteristics, singularities) may in some way form a part of a "hard nucleus" (element of the topology) which constitutes the base of other more numerous groupings which are constructed around and complete the cited hard nucleus.

The possibilities for the formation of groups and, consequently, the separation of financial products into groups with a determined homogeneity will not have escaped the attention of our readers.

We may now ask ourselves **how it is possible to conceive the notion of neighbourhood** if we abandon this path to place ourselves in that where **the referential *E*** comes given by:

$$E = \{a, b, c, \dots s\}$$

where its elements are also elements of the referential of the fuzzy subsets. For later exposition we will choose the second of the presented alternatives. We will also continue with the same example which we develop in the explanation of this alternative.

In first place we choose a fuzzy subset of $P(E)$ such as the following:

$$\underset{\sim}{A}_j = \begin{array}{c|c|c} a & b & c \\ \hline .6 & .4 & .7 \end{array}$$

in which:

$$\underset{\sim}{A}_j \in P(E)$$

It can be seen that a $\underset{\sim}{Y} \in T(E)$ exists close to $\underset{\sim}{A}_j$:

$$\underset{\sim}{Y} = \begin{array}{c|c|c} a & b & c \\ \hline .6 & .5 & .7 \end{array}$$

for which it may be written:

$$\underset{\sim}{A}_j \subset \underset{\sim}{Y}$$

So, it may be established that all of the fuzzy subsets $\underset{\sim}{X} \subset L^E$ of the infinite isotone pretopology δ, for which:

$$\mu_{\underset{\sim}{x}}(a) \geq 0{,}6 \qquad \mu_{\underset{\sim}{x}}(b) \geq 0{,}5 \qquad \mu_{\underset{\sim}{x}}(c) \geq 0{,}7$$

form the neighbourhood of $\underset{\sim}{A}_j$, and this whether they belong to $T(E)$ or not.

The configuration of an uncertain topology from this perspective cannot now be represented by means of a Boolean lattice, but by means of a distributive lattice. In this way the uncertain topology which we have previously proposed:

$$T(E) = \left\{ \varnothing,\ \begin{array}{ccc} a & b & c \\ \hline .1 & 0 & .3 \end{array},\ \begin{array}{ccc} a & b & c \\ \hline .2 & .2 & .6 \end{array},\ \begin{array}{ccc} a & b & c \\ \hline .4 & .2 & .6 \end{array},\ \begin{array}{ccc} a & b & c \\ \hline .2 & .5 & .7 \end{array},\ \begin{array}{ccc} a & b & c \\ \hline .6 & .2 & .6 \end{array},\ \begin{array}{ccc} a & b & c \\ \hline .4 & .5 & .7 \end{array},\ \begin{array}{ccc} a & b & c \\ \hline .2 & .8 & 1 \end{array},\ \begin{array}{ccc} a & b & c \\ \hline .6 & .5 & .7 \end{array},\ \begin{array}{ccc} a & b & c \\ \hline .4 & .8 & 1 \end{array},\ \begin{array}{ccc} a & b & c \\ \hline .6 & .8 & 1 \end{array},\ E \right\}$$

can be represented by means of the following pseudo-complimentary distributive lattice:

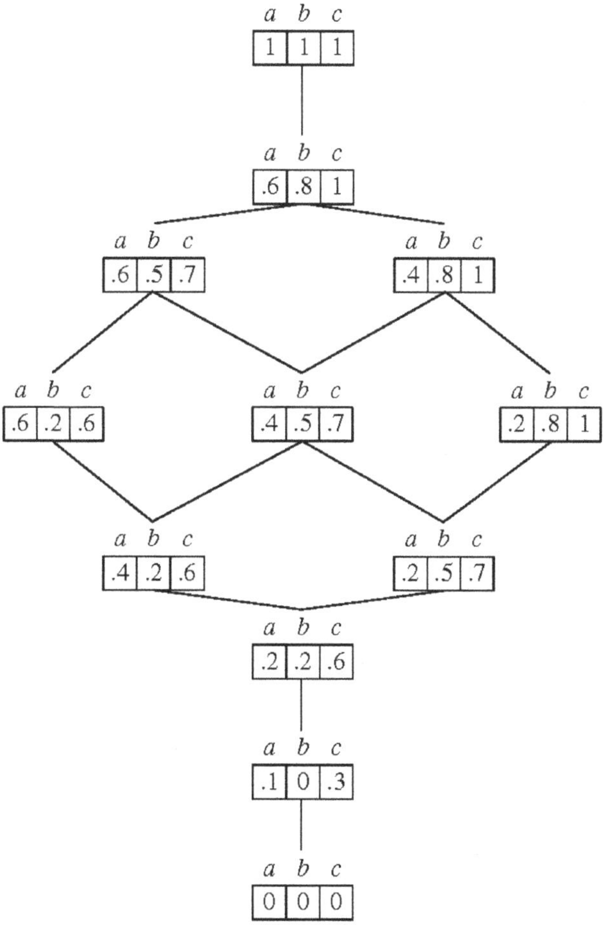

We once again remind ourselves that if one has in a Boole lattice:

$$A_j \cap \overline{A}_j = \varnothing$$
$$A_j \cup \overline{A}_j = E$$

the same does not occur in those distributive lattices whose elements are fuzzy subsets.

$$\underset{\sim}{A}_j \cap \overline{\underset{\sim}{A}}_j \neq \varnothing$$
$$\underset{\sim}{A}_j \cup \overline{\underset{\sim}{A}}_j \neq E$$

We will conclude by saying that the rejection or choice of one way or the other constitutes a fundamental decision for the correct usage of topological spaces in economics and management problems.

3 Properties of Uncertain Topologies

Before presenting some of the most interesting properties of uncertain topologies, we would like to state that in their enunciation it must be taken into account **which the option chosen for the configuration of the referential set *E* is**. For this we are refering to option ***A*** (referential of the referentials of the fuzzy subsets) and option ***B*** (referential of fuzzy subsets).

We will now move on to the properties:

1. A fuzzy subset or in its case an element of the power set $\underset{\sim}{X} \subset E$ is **open** in the fuzzy topology T if and only if for all of $\underset{\sim}{A}_j \subset \underset{\sim}{X}$, $\underset{\sim}{X}$ is a neighbour of $\underset{\sim}{A}_j$.

If $\underset{\sim}{X}$ is an open and includes $\underset{\sim}{A}_j$, it is evidently a neighbour of all $\underset{\sim}{A}_j$ included in $\underset{\sim}{X}$.

Reciprocally, if for all $\underset{\sim}{A}_j$ included in $\underset{\sim}{X}$, $\underset{\sim}{X}$ is a neighbour of $\underset{\sim}{A}_j$, then $\underset{\sim}{X} \ \subset \underset{\sim}{X}$.

It results in this way that $\underset{\sim}{X}$ is a neighbour of $\underset{\sim}{X}$ and therefore a $\underset{\sim}{Y}$ exists which is open, which is to say $\underset{\sim}{X} \subset \ \underset{\sim}{Y} \subset \underset{\sim}{X}$, where $\underset{\sim}{X} = \underset{\sim}{Y}$ and, as a consequence $\underset{\sim}{X}$ is an open.

2. The family of all of the neighbours of $\underset{\sim}{A}_j$ is called the neighbourhood of $\underset{\sim}{A}_j$.

If $S(\underset{\sim}{A}_j)$ is a neighbourhood of $\underset{\sim}{A}_j$, then the intersections of the elements of $S(\underset{\sim}{A}_j)$ belong to $S(\underset{\sim}{A}_j)$, and all fuzzy subset or element of the "power set" which contains an element of $S(\underset{\sim}{A}_j)$ belongs to $S(\underset{\sim}{A}_j)$.

We will move on to demonstrate this. Let $\underset{\sim}{X}_1$ and $\underset{\sim}{X}_2$ be two neighbours of the same element . Some $\underset{\sim}{X}'_1$ and $\underset{\sim}{X}'_2$ exist so that:

$$\underset{\sim}{A}_j \subset \underset{\sim}{X}'_1 \subset \underset{\sim}{X}_1$$
$$\underset{\sim}{A}_j \subset \underset{\sim}{X}'_2 \subset \underset{\sim}{X}_2$$

From which results:

$$\underset{\sim}{A}_j \subset (\underset{\sim}{X}'_1 \cap \underset{\sim}{X}'_2) \subset (\underset{\sim}{X}_1 \cap \underset{\sim}{X}_2)$$

Therefore $\underset{\sim}{X}'_1 \cap \underset{\sim}{X}'_2$ is an open by definition, and so $\underset{\sim}{X}'_1 \cap \underset{\sim}{X}'_2$ is also a neighbour of $\underset{\sim}{A}_j$.

On the other hand, if $\underset{\sim}{A}_k$ is a neighbour of $\underset{\sim}{A}_j$ and $\underset{\sim}{C}$ contains $\underset{\sim}{A}_k$, then an open $\underset{\sim}{Y}$ exists such as:

$$\underset{\sim}{A}_j \subset \underset{\sim}{Y} \subset \underset{\sim}{A}_k \subset \underset{\sim}{C}$$

from which $\underset{\sim}{C}$ results as a neighbour of $\underset{\sim}{A}_j$.

To continue with our proposal, it seems apt to remember certain notion which we have already expounded:

1. The **adherence or close** of $\underset{\sim}{A}_j$ is the intersection of all of the **closes** which contain $\underset{\sim}{A}_j$ and if $\underset{\sim}{A}_j$ is a closed then it is its own close. We will continue with our examples.

A) We will begin with that which we have taken as a base in order to develop the idea of topology from a referential of referentials. The subset of opens has been presented corresponding to the uncertain topology.

$$T(E) = \left\{ \emptyset, \begin{array}{ccc} a & b & c \\ \hline .1 & 0 & .3 \end{array}, \begin{array}{ccc} a & b & c \\ \hline .2 & .2 & .6 \end{array}, \begin{array}{ccc} a & b & c \\ \hline .4 & .2 & .6 \end{array}, \begin{array}{ccc} a & b & c \\ \hline .2 & .5 & .7 \end{array}, \begin{array}{ccc} a & b & c \\ \hline .6 & .2 & .6 \end{array}, \begin{array}{ccc} a & b & c \\ \hline .4 & .5 & .7 \end{array}, \begin{array}{ccc} a & b & c \\ \hline .2 & .8 & 1 \end{array}, \begin{array}{ccc} a & b & c \\ \hline .6 & .5 & .7 \end{array}, \begin{array}{ccc} a & b & c \\ \hline .4 & .8 & 1 \end{array}, \begin{array}{ccc} a & b & c \\ \hline .6 & .8 & 1 \end{array}, E \right\}$$

The closes of this topology are:

$$C_e = \left\{ E, \begin{array}{ccc} a & b & c \\ \hline .9 & 1 & .7 \end{array}, \begin{array}{ccc} a & b & c \\ \hline .8 & .8 & .4 \end{array}, \begin{array}{ccc} a & b & c \\ \hline .6 & .8 & .4 \end{array}, \begin{array}{ccc} a & b & c \\ \hline .8 & .5 & .3 \end{array}, \begin{array}{ccc} a & b & c \\ \hline .4 & .8 & .4 \end{array}, \begin{array}{ccc} a & b & c \\ \hline .6 & .5 & .3 \end{array}, \begin{array}{ccc} a & b & c \\ \hline .8 & .2 & 0 \end{array}, \begin{array}{ccc} a & b & c \\ \hline .4 & .5 & .3 \end{array}, \begin{array}{ccc} a & b & c \\ \hline .6 & .2 & 0 \end{array}, \begin{array}{ccc} a & b & c \\ \hline .4 & .2 & 0 \end{array}, \emptyset \right\}$$

We will now consider any element of L^E, such as:

$$\underset{\sim}{A}_j = \begin{array}{ccc} a & b & c \\ \hline \multicolumn{1}{|c|}{.5} & \multicolumn{1}{c|}{.5} & \multicolumn{1}{c|}{.3} \\ \hline \end{array}$$

We search for the subset of all of the closes which contain $\underset{\sim}{A}_j$. They will be:

$$C_e = \left\{ E \, , \begin{array}{ccc} a & b & c \\ \hline \multicolumn{1}{|c|}{.9} & \multicolumn{1}{c|}{1} & \multicolumn{1}{c|}{.7} \\ \hline \end{array} , \begin{array}{ccc} a & b & c \\ \hline \multicolumn{1}{|c|}{.8} & \multicolumn{1}{c|}{.8} & \multicolumn{1}{c|}{.4} \\ \hline \end{array} , \begin{array}{ccc} a & b & c \\ \hline \multicolumn{1}{|c|}{.6} & \multicolumn{1}{c|}{.8} & \multicolumn{1}{c|}{.4} \\ \hline \end{array} , \right.$$

$$\left. \begin{array}{ccc} a & b & c \\ \hline \multicolumn{1}{|c|}{.8} & \multicolumn{1}{c|}{.5} & \multicolumn{1}{c|}{.3} \\ \hline \end{array} , \begin{array}{ccc} a & b & c \\ \hline \multicolumn{1}{|c|}{.6} & \multicolumn{1}{c|}{.5} & \multicolumn{1}{c|}{.3} \\ \hline \end{array} \right\}$$

The intersection of all of these fuzzy subsets:

$$\begin{array}{ccc} a & b & c \\ \hline \multicolumn{1}{|c|}{.6} & \multicolumn{1}{c|}{.5} & \multicolumn{1}{c|}{.3} \\ \hline \end{array}$$

gives the close of $\underset{\sim}{A}_j$.

B) We will now begin from a referential of fuzzy subsets and we will continue with one of the previous examples. We have:

$$C_e = \{\emptyset, \{\underset{\sim}{P}_1\}, \{\underset{\sim}{P}_3\}, \{\underset{\sim}{P}_1, \underset{\sim}{P}_3\}, \{\underset{\sim}{P}_2, \underset{\sim}{P}_4\}, \{\underset{\sim}{P}_1, \underset{\sim}{P}_2, \underset{\sim}{P}_4\}, \{\underset{\sim}{P}_2, \underset{\sim}{P}_3, \underset{\sim}{P}_4\}, E\}$$

In this exceptional case the set of closes coincides with the set of opens.

Let us take an element of the "power set" such as:

$$\underset{\sim}{A}_j = \{\underset{\sim}{P}_2\}$$

The subset of closes which it contains is:

$$\{\{\underset{\sim}{P}_2, \underset{\sim}{P}_4\}, \{\underset{\sim}{P}_1, \underset{\sim}{P}_2, \underset{\sim}{P}_4\}, \{\underset{\sim}{P}_2, \underset{\sim}{P}_3, \underset{\sim}{P}_4\}, E\}$$

and its intersection:

$$\{\underset{\sim}{P}_2, \underset{\sim}{P}_4\}$$

which is its adherence or close.

If we had chosen a close, for example:

$$\underset{\sim}{A}_j = \{\underset{\sim}{P}_1, \underset{\sim}{P}_3\}$$

the subset of closes which it contains is:

$$\{\{\underset{\sim}{P}_1, \underset{\sim}{P}_3\}, E\}$$

whose intersection is the same:

$$\{\underset{\sim}{P}_1, \underset{\sim}{P}_3\}$$

and, therefore, its own close.

2. **The interior** or the opening of $\underset{\sim}{A}_j$ is the union of all of the opens contained in $\underset{\sim}{A}_j$. We will once again consider the same examples.

A) For the first of the ways, we will consider the following fuzzy subset:

$$\underset{\sim}{A}_j = \begin{array}{ccc} a & b & c \\ \hline \multicolumn{1}{|c|}{.7} & \multicolumn{1}{c|}{.5} & \multicolumn{1}{c|}{.8} \\ \hline \end{array}$$

If the previously presented distributive lattice is examined, it can be seen that the opens contained in $\underset{\sim}{A}_j$ are all those which form a subset in which the upper extreme in the lattice is:

$$\begin{array}{ccc} a & b & c \\ \hline \multicolumn{1}{|c|}{.6} & \multicolumn{1}{c|}{.5} & \multicolumn{1}{c|}{.7} \\ \hline \end{array}$$

which gives:

$$\left\{ \begin{array}{ccc} a & b & c \\ \hline \multicolumn{1}{|c|}{.6} & \multicolumn{1}{c|}{.5} & \multicolumn{1}{c|}{.7} \\ \hline \end{array}, \begin{array}{ccc} a & b & c \\ \hline \multicolumn{1}{|c|}{.6} & \multicolumn{1}{c|}{.2} & \multicolumn{1}{c|}{.6} \\ \hline \end{array}, \begin{array}{ccc} a & b & c \\ \hline \multicolumn{1}{|c|}{.4} & \multicolumn{1}{c|}{.5} & \multicolumn{1}{c|}{.7} \\ \hline \end{array}, \begin{array}{ccc} a & b & c \\ \hline \multicolumn{1}{|c|}{.4} & \multicolumn{1}{c|}{.2} & \multicolumn{1}{c|}{.6} \\ \hline \end{array}, \begin{array}{ccc} a & b & c \\ \hline \multicolumn{1}{|c|}{.2} & \multicolumn{1}{c|}{.5} & \multicolumn{1}{c|}{.7} \\ \hline \end{array}, \begin{array}{ccc} a & b & c \\ \hline \multicolumn{1}{|c|}{.2} & \multicolumn{1}{c|}{.2} & \multicolumn{1}{c|}{.6} \\ \hline \end{array}, \begin{array}{ccc} a & b & c \\ \hline \multicolumn{1}{|c|}{.1} & \multicolumn{1}{c|}{0} & \multicolumn{1}{c|}{.3} \\ \hline \end{array}, \varnothing \right\}$$

whose union gives place to the previously stated open:

$$\begin{array}{ccc} a & b & c \\ \hline \multicolumn{1}{|c|}{.6} & \multicolumn{1}{c|}{.5} & \multicolumn{1}{c|}{.7} \\ \hline \end{array}$$

which is the opening of $\underset{\sim}{A}_j$.

B) when the option falls on the **second of the ways**, we will take as an example:

$$\underset{\sim}{A}_j = \{\underset{\sim}{P}_1, \underset{\sim}{P}_3, \underset{\sim}{P}_4\}$$

the subset of opens contained in this is:

$$\{\{\underset{\sim}{P}_1\}, \{\underset{\sim}{P}_3\}, \{\underset{\sim}{P}_1, \underset{\sim}{P}_3\}\}$$

and its union:

$$\{\underset{\sim}{P}_1, \underset{\sim}{P}_3\}$$

which is its interior application or its interior.

When an open is taken, such as:

$$\underset{\sim}{A}_j = \{\underset{\sim}{P}_1, \underset{\sim}{P}_2, \underset{\sim}{P}_4\}$$

the subset of opens contained in this is:

$$\{\{\underset{\sim}{P}_1\}, \{\underset{\sim}{P}_2, \underset{\sim}{P}_4\}, \{\underset{\sim}{P}_1, \underset{\sim}{P}_2, \underset{\sim}{P}_4\}\}$$

whose union:

$$\{\underset{\sim}{P}_1, \underset{\sim}{P}_2, \underset{\sim}{P}_4\}$$

is its own interior application or interior.

Finally we will summarise the main aspects to take into account in relationship with **the close** and with t**he opening** in a fuzzy topology:

$\underset{\sim}{A}_j \underset{\sim}{\supset}$ opening of $\underset{\sim}{A}_j$	decrease
$(\underset{\sim}{A}_j \underset{\sim}{\supset} \underset{\sim}{A}_k) \Rightarrow$ (opening $\underset{\sim}{A}_j \underset{\sim}{\supset}$ opening $\underset{\sim}{A}_k$)	isotonia
opening (opening $\underset{\sim}{A}_j$) =opening $\underset{\sim}{A}_j$	idempotency
opening $E = E$	invariance of the referential
$\underset{\sim}{A}_j \underset{\sim}{\supset}$ close of $\underset{\sim}{A}_j$	growth
$(\underset{\sim}{A}_j \underset{\sim}{\supset} \underset{\sim}{A}_k) \Rightarrow$ (close $\underset{\sim}{A}_j \underset{\sim}{\supset}$ close $\underset{\sim}{A}_k$)	isotonia
close (close $\underset{\sim}{A}_j$) = close $\underset{\sim}{A}_j$	idempotency
close $\varnothing = \varnothing$	invariance of the empty

$$\overline{\text{opening } \underset{\sim}{A}_j} = \text{close } \overline{\underset{\sim}{A}_j}$$

$$\overline{\text{close } \underset{\sim}{A}_j} = \text{opening } \overline{\underset{\sim}{A}_j}$$

$$\text{exterior } \underset{\sim}{A}_j = \text{opening } \overline{\underset{\sim}{A}_j} = \overline{\text{close } \underset{\sim}{A}_j}$$

To end this epigraph we will review a classic concept, that of **border**, to present a new concept, that of "uncertain border". The fuzzy border of $\underset{\sim}{A}_j$ is the fuzzy subset $\underset{\sim}{A}_j^*$ such as:

$$\underset{\sim}{A}_j^* = \overline{\text{opening } \underset{\sim}{A}_j \cup \text{exterior } \underset{\sim}{A}_j} = \overline{\text{opening } \underset{\sim}{A}_j \cup \text{opening } \overline{\underset{\sim}{A}_j}}$$
$$= \overline{\text{opening } \underset{\sim}{A}_j \cup \overline{\text{close } \underset{\sim}{A}_j}}$$
$$= \overline{\text{opening } \underset{\sim}{A}_j} \cap \text{close } \underset{\sim}{A}_j$$
$$= \text{close } \overline{\underset{\sim}{A}_j} \cap \text{opening } \underset{\sim}{A}_j$$

in which we have used the theorem of De Morgan:

In a colloquial way it may be considered that the border $\underset{\sim}{A}_j^*$ of an element $\underset{\sim}{A}_j$ is the subset of those elements which belong to neither the opening nor the exterior of $\underset{\sim}{A}_j$. The correspondence with the classical concept of border is unequivocal.

From these conceptions of uncertain topology we have been able to redefine concepts as interesting as **neighbourhood**, of promising perspectives to formally represent an economics and management phenomenon in which nuance and subtlety acquire a special relevance.

4 Induction of Deterministic Topologies from Uncertain Topologies

Once again we will refer to the two focuses in which we have supported uncertain topology.

When the referentials of the fuzzy subsets $\underset{\sim}{A}_j$ are chosen as referential *E*, an uncertain topology induces a finite and infinite quantity of non-uncertain topologies through the α-cuts. The same does not occur when we begin from a referential *E* of fuzzy subsets.

A) Let us assume an uncertain topology *T*(*E*) from a referential *E* = {*a*, *b*, *c*, ...*r*} and for $T_\alpha^*(E)$ we designate the non-uncertain topology obtained in the following way:

$$(\mu_{\underset{\sim}{A}_j}(x) \le \alpha) \Rightarrow (\mu_{A_j^{(\alpha)}}(x) = 0)$$
$$(\mu_{\underset{\sim}{A}_j}(x) > \alpha) \Rightarrow (\mu_{A_j^{(\alpha)}}(x) = 1)$$

It is easily verified, always in the admitted referential, that if $\underset{\sim}{A}_j$ and $\underset{\sim}{A}_k$ are opens of *T*(*E*):

$$(\underset{\sim}{A}_j \cap \underset{\sim}{A}_k \in T(E)) \Rightarrow (A_j^{(\alpha)} \cap A_k^{(\alpha)} \in T^*(E))$$
$$(\underset{\sim}{A}_j \cup \underset{\sim}{A}_k \in T(E)) \Rightarrow (A_j^{(\alpha)} \cup A_k^{(\alpha)} \in T^*(E))$$

As *E* and ∅ belong to *T*(*E*) they also belong to $T_\alpha^*(E)$.

We will illustrate that which we have just expounded by using the same examples which we have used up to now. Once again, let:

$$T(E) = \Big\{ \varnothing,\ \overset{a\ \ b\ \ c}{\boxed{.1\,|\,0\,|\,.3}},\ \overset{a\ \ b\ \ c}{\boxed{.2\,|\,.2\,|\,.6}},\ \overset{a\ \ b\ \ c}{\boxed{.4\,|\,.2\,|\,.6}},\ \overset{a\ \ b\ \ c}{\boxed{.2\,|\,.5\,|\,.7}},\ \overset{a\ \ b\ \ c}{\boxed{.6\,|\,.2\,|\,.6}},\ \overset{a\ \ b\ \ c}{\boxed{.4\,|\,.5\,|\,.7}},\ \overset{a\ \ b\ \ c}{\boxed{.2\,|\,.8\,|\,1}},\ \overset{a\ \ b\ \ c}{\boxed{.6\,|\,.5\,|\,.7}},\ \overset{a\ \ b\ \ c}{\boxed{.4\,|\,.8\,|\,1}},\ \overset{a\ \ b\ \ c}{\boxed{.6\,|\,.8\,|\,1}},\ E \Big\}$$

If we choose $\alpha = 0.5$, we have:

$$T^{*}_{(0,5)}(E) = \Big\{ \varnothing,\ \overset{a\ \ b\ \ c}{\boxed{0\,|\,0\,|\,0}},\ \overset{a\ \ b\ \ c}{\boxed{0\,|\,0\,|\,1}},\ \overset{a\ \ b\ \ c}{\boxed{0\,|\,0\,|\,1}},\ \overset{a\ \ b\ \ c}{\boxed{0\,|\,0\,|\,1}},\ \overset{a\ \ b\ \ c}{\boxed{1\,|\,0\,|\,1}},\ \overset{a\ \ b\ \ c}{\boxed{0\,|\,0\,|\,1}},\ \overset{a\ \ b\ \ c}{\boxed{0\,|\,1\,|\,1}},\ \overset{a\ \ b\ \ c}{\boxed{1\,|\,0\,|\,1}},\ \overset{a\ \ b\ \ c}{\boxed{0\,|\,1\,|\,1}},\ \overset{a\ \ b\ \ c}{\boxed{1\,|\,1\,|\,1}},\ E \Big\}$$

$$= \Big\{ \varnothing,\ \overset{a\ \ b\ \ c}{\boxed{0\,|\,0\,|\,1}},\ \overset{a\ \ b\ \ c}{\boxed{1\,|\,0\,|\,1}},\ \overset{a\ \ b\ \ c}{\boxed{0\,|\,1\,|\,1}},\ E \Big\}$$

$$= \{\varnothing, \{c\}, \{a, c\}, \{b, c\}, E\}$$

The closes are:

$$C_{e}^{(0,5)}(E) = \Big\{ E,\ \overset{a\ \ b\ \ c}{\boxed{1\,|\,1\,|\,1}},\ \overset{a\ \ b\ \ c}{\boxed{1\,|\,1\,|\,0}},\ \overset{a\ \ b\ \ c}{\boxed{1\,|\,1\,|\,0}},\ \overset{a\ \ b\ \ c}{\boxed{1\,|\,1\,|\,0}},\ \overset{a\ \ b\ \ c}{\boxed{0\,|\,1\,|\,0}},\ \overset{a\ \ b\ \ c}{\boxed{1\,|\,1\,|\,0}},\ \overset{a\ \ b\ \ c}{\boxed{1\,|\,0\,|\,0}},\ \overset{a\ \ b\ \ c}{\boxed{0\,|\,1\,|\,0}},\ \overset{a\ \ b\ \ c}{\boxed{1\,|\,0\,|\,0}},\ \overset{a\ \ b\ \ c}{\boxed{0\,|\,0\,|\,0}},\ \varnothing \Big\}$$

$$= \Big\{ \varnothing,\ \overset{a\ \ b\ \ c}{\boxed{1\,|\,0\,|\,0}},\ \overset{a\ \ b\ \ c}{\boxed{0\,|\,1\,|\,0}},\ \overset{a\ \ b\ \ c}{\boxed{1\,|\,1\,|\,0}},\ E \Big\}$$

$$= \{\varnothing, \{a\}, \{b\}, \{a, b\}, E\}$$

B) In the area of the second focus, the opens which formed the fuzzy topology were:

$$T(E) = \{\emptyset, \{\underset{\sim}{P}_1\}, \{\underset{\sim}{P}_3\}, \{\underset{\sim}{P}_1, \underset{\sim}{P}_3\}, \{\underset{\sim}{P}_2, \underset{\sim}{P}_4\}, \{\underset{\sim}{P}_1, \underset{\sim}{P}_2, \underset{\sim}{P}_4\}, \{\underset{\sim}{P}_2, \underset{\sim}{P}_3, \underset{\sim}{P}_4\}, E\}$$

and making the fuzzy subsets explicit one has:

$$\begin{aligned} T(E) = \{\emptyset, & \left\{\begin{array}{|c|c|c|} \multicolumn{1}{c}{a} & \multicolumn{1}{c}{b} & \multicolumn{1}{c}{c} \\ \hline .7 & .2 & .9 \\ \hline \end{array}\right\}, \left\{\begin{array}{|c|c|c|} \multicolumn{1}{c}{a} & \multicolumn{1}{c}{b} & \multicolumn{1}{c}{c} \\ \hline .8 & .2 & .4 \\ \hline \end{array}\right\}, \left\{\begin{array}{|c|c|c|} \multicolumn{1}{c}{a} & \multicolumn{1}{c}{b} & \multicolumn{1}{c}{c} \\ \hline .7 & .2 & .9 \\ \hline \end{array}, \begin{array}{|c|c|c|} \multicolumn{1}{c}{a} & \multicolumn{1}{c}{b} & \multicolumn{1}{c}{c} \\ \hline .8 & .2 & .4 \\ \hline \end{array}\right\}, \\ & \left\{\begin{array}{|c|c|c|} \multicolumn{1}{c}{a} & \multicolumn{1}{c}{b} & \multicolumn{1}{c}{c} \\ \hline .3 & .9 & .7 \\ \hline \end{array}, \begin{array}{|c|c|c|} \multicolumn{1}{c}{a} & \multicolumn{1}{c}{b} & \multicolumn{1}{c}{c} \\ \hline .2 & .9 & .8 \\ \hline \end{array}\right\}, \left\{\begin{array}{|c|c|c|} \multicolumn{1}{c}{a} & \multicolumn{1}{c}{b} & \multicolumn{1}{c}{c} \\ \hline .7 & .2 & .9 \\ \hline \end{array}, \begin{array}{|c|c|c|} \multicolumn{1}{c}{a} & \multicolumn{1}{c}{b} & \multicolumn{1}{c}{c} \\ \hline .3 & .9 & .7 \\ \hline \end{array}, \begin{array}{|c|c|c|} \multicolumn{1}{c}{a} & \multicolumn{1}{c}{b} & \multicolumn{1}{c}{c} \\ \hline .2 & .9 & .8 \\ \hline \end{array}\right\}, \\ & \left\{\begin{array}{|c|c|c|} \multicolumn{1}{c}{a} & \multicolumn{1}{c}{b} & \multicolumn{1}{c}{c} \\ \hline .3 & .9 & .7 \\ \hline \end{array}, \begin{array}{|c|c|c|} \multicolumn{1}{c}{a} & \multicolumn{1}{c}{b} & \multicolumn{1}{c}{c} \\ \hline .8 & .2 & .4 \\ \hline \end{array}, \begin{array}{|c|c|c|} \multicolumn{1}{c}{a} & \multicolumn{1}{c}{b} & \multicolumn{1}{c}{c} \\ \hline .2 & .9 & .8 \\ \hline \end{array}\right\}, E\} \end{aligned}$$

Upon making $\alpha = 0.5$, we have:

$$\begin{aligned} T^*_{0,5}(E) = \{\emptyset, & \left\{\begin{array}{|c|c|c|} \multicolumn{1}{c}{a} & \multicolumn{1}{c}{b} & \multicolumn{1}{c}{c} \\ \hline 1 & 0 & 1 \\ \hline \end{array}\right\}, \left\{\begin{array}{|c|c|c|} \multicolumn{1}{c}{a} & \multicolumn{1}{c}{b} & \multicolumn{1}{c}{c} \\ \hline 1 & 0 & 0 \\ \hline \end{array}\right\}, \left\{\begin{array}{|c|c|c|} \multicolumn{1}{c}{a} & \multicolumn{1}{c}{b} & \multicolumn{1}{c}{c} \\ \hline 1 & 0 & 1 \\ \hline \end{array}, \begin{array}{|c|c|c|} \multicolumn{1}{c}{a} & \multicolumn{1}{c}{b} & \multicolumn{1}{c}{c} \\ \hline 1 & 0 & 0 \\ \hline \end{array}\right\}, \\ & \left\{\begin{array}{|c|c|c|} \multicolumn{1}{c}{a} & \multicolumn{1}{c}{b} & \multicolumn{1}{c}{c} \\ \hline 0 & 1 & 1 \\ \hline \end{array}, \begin{array}{|c|c|c|} \multicolumn{1}{c}{a} & \multicolumn{1}{c}{b} & \multicolumn{1}{c}{c} \\ \hline 0 & 1 & 1 \\ \hline \end{array}\right\}, \left\{\begin{array}{|c|c|c|} \multicolumn{1}{c}{a} & \multicolumn{1}{c}{b} & \multicolumn{1}{c}{c} \\ \hline 1 & 0 & 1 \\ \hline \end{array}, \begin{array}{|c|c|c|} \multicolumn{1}{c}{a} & \multicolumn{1}{c}{b} & \multicolumn{1}{c}{c} \\ \hline 0 & 1 & 1 \\ \hline \end{array}, \begin{array}{|c|c|c|} \multicolumn{1}{c}{a} & \multicolumn{1}{c}{b} & \multicolumn{1}{c}{c} \\ \hline 0 & 1 & 1 \\ \hline \end{array}\right\}, \\ & \left\{\begin{array}{|c|c|c|} \multicolumn{1}{c}{a} & \multicolumn{1}{c}{b} & \multicolumn{1}{c}{c} \\ \hline 0 & 1 & 1 \\ \hline \end{array}, \begin{array}{|c|c|c|} \multicolumn{1}{c}{a} & \multicolumn{1}{c}{b} & \multicolumn{1}{c}{c} \\ \hline 1 & 0 & 0 \\ \hline \end{array}, \begin{array}{|c|c|c|} \multicolumn{1}{c}{a} & \multicolumn{1}{c}{b} & \multicolumn{1}{c}{c} \\ \hline 0 & 1 & 1 \\ \hline \end{array}\right\}, E\} \end{aligned}$$

A fact which will result as important can be seen here: upon establishing the cut ($\alpha = 0.5$ in our example) two fuzzy subsets, specifically:

$$\begin{array}{|c|c|c|} \multicolumn{1}{c}{a} & \multicolumn{1}{c}{b} & \multicolumn{1}{c}{c} \\ \hline .3 & .9 & .7 \\ \hline \end{array}, \begin{array}{|c|c|c|} \multicolumn{1}{c}{a} & \multicolumn{1}{c}{b} & \multicolumn{1}{c}{c} \\ \hline .2 & .9 & .8 \\ \hline \end{array}$$

are converted into the same boolean subset:

$$\begin{array}{|c|c|c|} \multicolumn{1}{c}{a} & \multicolumn{1}{c}{b} & \multicolumn{1}{c}{c} \\ \hline 0 & 1 & 1 \\ \hline \end{array}$$

which changes the content of those groups of boolean subsets in which they appear together. This makes the opens of our case become:

$$T^*_{0,5}(E) = \{\varnothing,\ \{\overset{a\ b\ c}{\boxed{1|0|1}}\},\ \{\overset{a\ b\ c}{\boxed{1|0|0}}\},\ \{\overset{a\ b\ c}{\boxed{1|0|1}},\ \overset{a\ b\ c}{\boxed{1|0|0}}\},$$

$$\{\overset{a\ b\ c}{\boxed{0|1|1}}\},\ \{\overset{a\ b\ c}{\boxed{1|0|1}},\ \overset{a\ b\ c}{\boxed{0|1|1}}\},\ \{\overset{a\ b\ c}{\boxed{0|1|1}},\ \overset{a\ b\ c}{\boxed{1|0|0}}\},\ E\}$$

$$= \left\{\varnothing,\ \{a,c\},\ \{a\},\ \{\{a,c\},\{a\}\},\ \{b,c\},\ \{\{a,c\},\{b,c\}\},\ \{\{b,c\},\{a\}\},\ E\right\}$$

A quick verification shows that it **does not fulfil** the axiomatic of the topology. Therefore, here appears a deficiency of this second way: **it is not possible to accept** that an uncertain topology generates ordinary topologies through α-cuts. For this reason we will leave this second path for the moment to centre our attention exclusively on the first.

A) We therefore return to the first focus. And in this we will propose the performing of a study of the **topology** of the referential $E = \{a, b, c\}$, $T(E)$, taken repeatedly, and its corresponding opens for all of the α-cuts, from $\alpha = 0$ up to $\alpha = 1$. We resort to the undecimal system:

$\alpha = 0$

$$T^*_0(E) = \left\{\varnothing,\ \overset{a\ b\ c}{\boxed{1|0|1}},\ E\right\} = \{\varnothing,\ \{a,c\},\ E\}$$

$\alpha = 0,1$

$$T^*_{0,1}(E) = \left\{\varnothing,\ \overset{a\ b\ c}{\boxed{0|0|1}},\ E\right\} = \{\varnothing,\ \{c\},\ E\}$$

$\alpha = 0,2$

$$T^*_{0,2}(E) = \left\{\varnothing,\ \overset{a\ b\ c}{\boxed{0|0|1}},\ \overset{a\ b\ c}{\boxed{1|0|1}},\ \overset{a\ b\ c}{\boxed{0|1|1}},\ E\right\}$$

$$= \{\varnothing,\ \{c\},\ \{a,c\},\ \{b,c\},\ E\}$$

$\alpha = 0{,}3$

$$T^*_{0,3}(E) = \left\{ \varnothing\ ,\ \begin{array}{|c|c|c|}\multicolumn{1}{c}{a} & \multicolumn{1}{c}{b} & \multicolumn{1}{c}{c} \\ \hline 0 & 0 & 1 \\ \hline \end{array}\ ,\ \begin{array}{|c|c|c|}\multicolumn{1}{c}{a} & \multicolumn{1}{c}{b} & \multicolumn{1}{c}{c} \\ \hline 1 & 0 & 1 \\ \hline \end{array}\ ,\ \begin{array}{|c|c|c|}\multicolumn{1}{c}{a} & \multicolumn{1}{c}{b} & \multicolumn{1}{c}{c} \\ \hline 0 & 1 & 1 \\ \hline \end{array}\ ,\ E \right\}$$

$$= \{\varnothing, \{c\}, \{a, c\}, \{b, c\}, E\}$$

$\alpha = 0{,}4$

$$T^*_{0,4}(E) = \left\{ \varnothing\ ,\ \begin{array}{|c|c|c|}\multicolumn{1}{c}{a} & \multicolumn{1}{c}{b} & \multicolumn{1}{c}{c} \\ \hline 0 & 0 & 1 \\ \hline \end{array}\ ,\ \begin{array}{|c|c|c|}\multicolumn{1}{c}{a} & \multicolumn{1}{c}{b} & \multicolumn{1}{c}{c} \\ \hline 0 & 1 & 1 \\ \hline \end{array}\ ,\ \begin{array}{|c|c|c|}\multicolumn{1}{c}{a} & \multicolumn{1}{c}{b} & \multicolumn{1}{c}{c} \\ \hline 1 & 0 & 1 \\ \hline \end{array}\ ,\ E \right\}$$

$$= \{\varnothing, \{c\}, \{b, c\}, \{a, c\}, E\}$$

$\alpha = 0{,}5$

$$T^*_{0,5}(E) = \left\{ \varnothing\ ,\ \begin{array}{|c|c|c|}\multicolumn{1}{c}{a} & \multicolumn{1}{c}{b} & \multicolumn{1}{c}{c} \\ \hline 0 & 0 & 1 \\ \hline \end{array}\ ,\ \begin{array}{|c|c|c|}\multicolumn{1}{c}{a} & \multicolumn{1}{c}{b} & \multicolumn{1}{c}{c} \\ \hline 1 & 0 & 1 \\ \hline \end{array}\ ,\ \begin{array}{|c|c|c|}\multicolumn{1}{c}{a} & \multicolumn{1}{c}{b} & \multicolumn{1}{c}{c} \\ \hline 0 & 1 & 1 \\ \hline \end{array}\ ,\ E \right\}$$

$$= \{\varnothing, \{c\}, \{a, c\}, \{b, c\}, E\}$$

$\alpha = 0{,}6$

$$T^*_{0,6}(E) = \left\{ \varnothing\ ,\ \begin{array}{|c|c|c|}\multicolumn{1}{c}{a} & \multicolumn{1}{c}{b} & \multicolumn{1}{c}{c} \\ \hline 0 & 0 & 1 \\ \hline \end{array}\ ,\ \begin{array}{|c|c|c|}\multicolumn{1}{c}{a} & \multicolumn{1}{c}{b} & \multicolumn{1}{c}{c} \\ \hline 0 & 1 & 1 \\ \hline \end{array}\ ,\ E \right\} = \{\varnothing, \{c\}, \{b, c\}, E\}$$

$\alpha = 0{,}7$

$$T^*_{0,7}(E) = \left\{ \varnothing\ ,\ \begin{array}{|c|c|c|}\multicolumn{1}{c}{a} & \multicolumn{1}{c}{b} & \multicolumn{1}{c}{c} \\ \hline 0 & 1 & 1 \\ \hline \end{array}\ ,\ E \right\} = \{\varnothing, \{b, c\}, E\}$$

$\alpha = 0{,}8$

$$T^*_{0,8}(E) = \left\{ \varnothing\ ,\ \begin{array}{|c|c|c|}\multicolumn{1}{c}{a} & \multicolumn{1}{c}{b} & \multicolumn{1}{c}{c} \\ \hline 0 & 0 & 1 \\ \hline \end{array}\ ,\ E \right\} = \{\varnothing, \{c\}, E\}$$

$\alpha = 0{,}9$

$$T^*_{0,9}(E) = \left\{ \varnothing,\ \begin{array}{ccc} a & b & c \\ \hline \multicolumn{1}{|c|}{0} & \multicolumn{1}{c|}{0} & \multicolumn{1}{c|}{1} \\ \hline \end{array},\ E \right\} = \{\varnothing, \{c\}, E\}$$

$\alpha = 1$

$$T^*_1(E) = \{\varnothing, E\}$$

The closes obtained from $T(E)$ do not give the same closes as $T^*_\alpha(E)$. But, on the other hand, all of $T^*_\alpha(E)$, $\forall \propto \subset [0.1]$, give the topologies.

We may now ask ourselves if this is also fulfilled when a definition of α–cuts distinct to that before is accepted, for example:

$$(\mu_{\underset{\sim}{A_j}}(x) < \alpha) \Rightarrow (\mu_{A_j^{(\alpha)}}(x) = 0)$$
$$(\mu_{\underset{\sim}{A_j}}(x) \geq \alpha) \Rightarrow (\mu_{A_j^{(\alpha)}}(x) = 1)$$

In this case it results:

$\alpha = 0$

$$T^*_0(E) = \{\varnothing, E\}$$

$\alpha = 0{,}1$

$$T^*_{0,1}(E) = \left\{ \varnothing,\ \begin{array}{ccc} a & b & c \\ \hline \multicolumn{1}{|c|}{1} & \multicolumn{1}{c|}{0} & \multicolumn{1}{c|}{1} \\ \hline \end{array},\ E \right\} = \{\varnothing, \{a, c\}, E\}$$

$\alpha = 0{,}2$

$$T^*_{0,2}(E) = \left\{ \varnothing,\ \begin{array}{ccc} a & b & c \\ \hline \multicolumn{1}{|c|}{0} & \multicolumn{1}{c|}{0} & \multicolumn{1}{c|}{1} \\ \hline \end{array},\ E \right\} = \{\varnothing, \{c\}, E\}$$

$\alpha = 0{,}3$

$$T^*_{0,3}(E) = \left\{ \varnothing,\ \begin{array}{ccc} a & b & c \\ \hline \multicolumn{1}{|c|}{0} & \multicolumn{1}{c|}{0} & \multicolumn{1}{c|}{1} \\ \hline \end{array},\ \begin{array}{ccc} a & b & c \\ \hline \multicolumn{1}{|c|}{1} & \multicolumn{1}{c|}{0} & \multicolumn{1}{c|}{1} \\ \hline \end{array},\ \begin{array}{ccc} a & b & c \\ \hline \multicolumn{1}{|c|}{0} & \multicolumn{1}{c|}{1} & \multicolumn{1}{c|}{1} \\ \hline \end{array},\ E \right\}$$

$$= \{\varnothing, \{c\}, \{a, c\}, \{b, c\}, E\}$$

$\alpha = 0{,}4$

$$T^*_{0,4}(E) = \left\{ \varnothing, \begin{array}{|c|c|c|} \multicolumn{1}{c}{a} & \multicolumn{1}{c}{b} & \multicolumn{1}{c}{c} \\ \hline 0 & 0 & 1 \\ \hline \end{array}, \begin{array}{|c|c|c|} \multicolumn{1}{c}{a} & \multicolumn{1}{c}{b} & \multicolumn{1}{c}{c} \\ \hline 1 & 0 & 1 \\ \hline \end{array}, \begin{array}{|c|c|c|} \multicolumn{1}{c}{a} & \multicolumn{1}{c}{b} & \multicolumn{1}{c}{c} \\ \hline 0 & 1 & 1 \\ \hline \end{array}, E \right\}$$

$$= \{\varnothing, \{c\}, \{a, c\}, \{b, c\}, E\}$$

$\alpha = 0{,}5$

$$T^*_{0,5}(E) = \left\{ \varnothing, \begin{array}{|c|c|c|} \multicolumn{1}{c}{a} & \multicolumn{1}{c}{b} & \multicolumn{1}{c}{c} \\ \hline 0 & 0 & 1 \\ \hline \end{array}, \begin{array}{|c|c|c|} \multicolumn{1}{c}{a} & \multicolumn{1}{c}{b} & \multicolumn{1}{c}{c} \\ \hline 0 & 1 & 1 \\ \hline \end{array}, \begin{array}{|c|c|c|} \multicolumn{1}{c}{a} & \multicolumn{1}{c}{b} & \multicolumn{1}{c}{c} \\ \hline 1 & 0 & 1 \\ \hline \end{array}, E \right\}$$

$$= \{\varnothing, \{c\}, \{a, c\}, \{b, c\}, E\}$$

$\alpha = 0{,}6$

$$T^*_{0,6}(E) = \left\{ \varnothing, \begin{array}{|c|c|c|} \multicolumn{1}{c}{a} & \multicolumn{1}{c}{b} & \multicolumn{1}{c}{c} \\ \hline 0 & 0 & 1 \\ \hline \end{array}, \begin{array}{|c|c|c|} \multicolumn{1}{c}{a} & \multicolumn{1}{c}{b} & \multicolumn{1}{c}{c} \\ \hline 1 & 0 & 1 \\ \hline \end{array}, \begin{array}{|c|c|c|} \multicolumn{1}{c}{a} & \multicolumn{1}{c}{b} & \multicolumn{1}{c}{c} \\ \hline 0 & 1 & 1 \\ \hline \end{array}, E \right\}$$

$$= \{\varnothing, \{c\}, \{a, c\}, \{b, c\}, E\}$$

$\alpha = 0{,}7$

$$T^*_{0,7}(E) = \left\{ \varnothing, \begin{array}{|c|c|c|} \multicolumn{1}{c}{a} & \multicolumn{1}{c}{b} & \multicolumn{1}{c}{c} \\ \hline 0 & 0 & 1 \\ \hline \end{array}, \begin{array}{|c|c|c|} \multicolumn{1}{c}{a} & \multicolumn{1}{c}{b} & \multicolumn{1}{c}{c} \\ \hline 0 & 1 & 1 \\ \hline \end{array}, E \right\} = \{\varnothing, \{c\}, \{b, c\}, E\}$$

$\alpha = 0{,}8$

$$T^*_{0,8}(E) = \left\{ \varnothing, \begin{array}{|c|c|c|} \multicolumn{1}{c}{a} & \multicolumn{1}{c}{b} & \multicolumn{1}{c}{c} \\ \hline 0 & 1 & 1 \\ \hline \end{array}, E \right\} = \{\varnothing, \{b, c\}, E\}$$

$\alpha = 0{,}9$

$$T^*_{0,9}(E) = \left\{ \varnothing, \begin{array}{|c|c|c|} \multicolumn{1}{c}{a} & \multicolumn{1}{c}{b} & \multicolumn{1}{c}{c} \\ \hline 0 & 0 & 1 \\ \hline \end{array}, E \right\} = \{\varnothing, \{c\}, E\}$$

$\alpha = 1$

$$T^*_{1}(E) = \left\{ \varnothing, \begin{array}{|c|c|c|} \multicolumn{1}{c}{a} & \multicolumn{1}{c}{b} & \multicolumn{1}{c}{c} \\ \hline 0 & 0 & 1 \\ \hline \end{array}, E \right\} = \{\varnothing, \{c\}, E\}$$

In this case the topologies have also been formed, with the same step of 0.1.

When the α-cuts are defined in this way, some authors prefer to speak of **semi-topologies**, justifying this by the fact that the $T^*_\alpha(E)$ are not found to be formed by any kind of meeting.

Finally, we will see a dual way of considering the α -cuts in relation with both criteria, beginning with the first of them.

$$(\mu_{\underset{\sim}{A}_j}(x) \leq \alpha) \Rightarrow (\mu_{A_j^{(\alpha)}}(x) = 1)$$
$$(\mu_{\underset{\sim}{A}_j}(x) > \alpha) \Rightarrow (\mu_{A_j^{(\alpha)}}(x) = 0)$$

$\alpha = 0$

$$T^*_0(E) = \left\{ \varnothing \, , \begin{array}{|c|c|c|} \multicolumn{1}{c}{a} & \multicolumn{1}{c}{b} & \multicolumn{1}{c}{c} \\ \hline 0 & 1 & 0 \\ \hline \end{array} , E \right\} = \{\varnothing, \{b\}, E\}$$

$\alpha = 0{,}1$

$$T^*_{0,1}(E) = \left\{ \varnothing \, , \begin{array}{|c|c|c|} \multicolumn{1}{c}{a} & \multicolumn{1}{c}{b} & \multicolumn{1}{c}{c} \\ \hline 1 & 1 & 0 \\ \hline \end{array} , E \right\} = \{\varnothing, \{a, b\}, E\}$$

$\alpha = 0{,}2$

$$T^*_{0,2}(E) = \left\{ \varnothing \, , \begin{array}{|c|c|c|} \multicolumn{1}{c}{a} & \multicolumn{1}{c}{b} & \multicolumn{1}{c}{c} \\ \hline 1 & 1 & 0 \\ \hline \end{array} , \begin{array}{|c|c|c|} \multicolumn{1}{c}{a} & \multicolumn{1}{c}{b} & \multicolumn{1}{c}{c} \\ \hline 0 & 1 & 0 \\ \hline \end{array} , \begin{array}{|c|c|c|} \multicolumn{1}{c}{a} & \multicolumn{1}{c}{b} & \multicolumn{1}{c}{c} \\ \hline 1 & 0 & 0 \\ \hline \end{array} , E \right\}$$

$$= \{\varnothing, \{a\}, \{b\}, \{a, b\}, E\}$$

$\alpha = 0{,}3$

$$T^*_{0,3}(E) = \left\{ \varnothing \, , \begin{array}{|c|c|c|} \multicolumn{1}{c}{a} & \multicolumn{1}{c}{b} & \multicolumn{1}{c}{c} \\ \hline 1 & 1 & 0 \\ \hline \end{array} , \begin{array}{|c|c|c|} \multicolumn{1}{c}{a} & \multicolumn{1}{c}{b} & \multicolumn{1}{c}{c} \\ \hline 0 & 1 & 0 \\ \hline \end{array} , \begin{array}{|c|c|c|} \multicolumn{1}{c}{a} & \multicolumn{1}{c}{b} & \multicolumn{1}{c}{c} \\ \hline 1 & 0 & 0 \\ \hline \end{array} , E \right\}$$

$$= \{\varnothing, \{a\}, \{b\}, \{a, b\}, E\}$$

$\alpha = 0{,}4$

$$T^*_{0,4}(E) = \left\{ \varnothing \, , \begin{array}{|c|c|c|} \multicolumn{1}{c}{a} & \multicolumn{1}{c}{b} & \multicolumn{1}{c}{c} \\ \hline 1 & 1 & 0 \\ \hline \end{array} , \begin{array}{|c|c|c|} \multicolumn{1}{c}{a} & \multicolumn{1}{c}{b} & \multicolumn{1}{c}{c} \\ \hline 1 & 0 & 0 \\ \hline \end{array} , \begin{array}{|c|c|c|} \multicolumn{1}{c}{a} & \multicolumn{1}{c}{b} & \multicolumn{1}{c}{c} \\ \hline 0 & 1 & 0 \\ \hline \end{array} , E \right\}$$

$$= \{\varnothing, \{a\}, \{b\}, \{a, b\}, E\}$$

$\alpha = 0{,}5$

$$T^*_{0{,}5}(E) = \left\{ \varnothing,\ \begin{array}{ccc} a & b & c \\ \hline \multicolumn{1}{|c|}{1} & \multicolumn{1}{c|}{1} & \multicolumn{1}{c|}{0} \\ \hline \end{array},\ \begin{array}{ccc} a & b & c \\ \hline \multicolumn{1}{|c|}{1} & \multicolumn{1}{c|}{0} & \multicolumn{1}{c|}{0} \\ \hline \end{array},\ \begin{array}{ccc} a & b & c \\ \hline \multicolumn{1}{|c|}{0} & \multicolumn{1}{c|}{1} & \multicolumn{1}{c|}{0} \\ \hline \end{array},\ E \right\}$$

$$= \{\varnothing, \{a\}, \{b\}, \{a, b\}, E\}$$

$\alpha = 0{,}6$

$$T^*_{0{,}6}(E) = \left\{ \varnothing,\ \begin{array}{ccc} a & b & c \\ \hline \multicolumn{1}{|c|}{1} & \multicolumn{1}{c|}{1} & \multicolumn{1}{c|}{0} \\ \hline \end{array},\ \begin{array}{ccc} a & b & c \\ \hline \multicolumn{1}{|c|}{1} & \multicolumn{1}{c|}{0} & \multicolumn{1}{c|}{0} \\ \hline \end{array},\ E \right\} = \{\varnothing, \{a\}, \{a, b\}, E\}$$

$\alpha = 0{,}7$

$$T^*_{0{,}7}(E) = \left\{ \varnothing,\ \begin{array}{ccc} a & b & c \\ \hline \multicolumn{1}{|c|}{1} & \multicolumn{1}{c|}{0} & \multicolumn{1}{c|}{0} \\ \hline \end{array},\ E \right\} = \{\varnothing, \{a\}, E\}$$

$\alpha = 0{,}8$

$$T^*_{0{,}8}(E) = \left\{ \varnothing,\ \begin{array}{ccc} a & b & c \\ \hline \multicolumn{1}{|c|}{1} & \multicolumn{1}{c|}{1} & \multicolumn{1}{c|}{0} \\ \hline \end{array},\ E \right\} = \{\varnothing, \{a, b\}, E\}$$

$\alpha = 0{,}9$

$$T^*_{0{,}9}(E) = \left\{ \varnothing,\ \begin{array}{ccc} a & b & c \\ \hline \multicolumn{1}{|c|}{1} & \multicolumn{1}{c|}{1} & \multicolumn{1}{c|}{0} \\ \hline \end{array},\ E \right\} = \{\varnothing, \{a, b\}, E\}$$

$\alpha = 1$

$$T^*_1(E) = \{\varnothing, E\}$$

In this case topologies are also obtained. And if to define the topologies we take:

$$\left(\mu_{\underset{\sim}{A}_j}(x) < \alpha\right) \Rightarrow \left(\mu_{A_j^{(\alpha)}}(x) = 1\right)$$
$$\left(\mu_{\underset{\sim}{A}_j}(x) \geq \alpha\right) \Rightarrow \left(\mu_{A_j^{(\alpha)}}(x) = 0\right)$$

the same occurs.

That said, a brief look at any of these ways of moving from uncertain topology to booleans allows the conclusion that neither focus A nor focus B fit. In this way it has been possible to verify that whatever the focus and whatever the way of performing the α -cuts, **it is not possible to achieve a general fit**.

5 Process of Obtaining of Boolean Topologies from a Referential of Fuzzy Subsets

When a referential of fuzzy subsets is considered (second focus), from which the opens which form the uncertain topology will be considered, we have seen that a topological space is not always obtained by using the technique of α -cuts. In continuation we will see a process to find topologies at distinct α levels, initially using the cut. Let us establish, for example:

$$(\mu_{\underset{\sim}{A}_j}(x) < \alpha) \Rightarrow (\mu_{A_j^{(\alpha)}}(x) = 0)$$
$$(\mu_{\underset{\sim}{A}_j}(x) \geq \alpha) \Rightarrow (\mu_{A_j^{(\alpha)}}(x) = 1),\ x \in \{a, b, c\}$$

This should allow the generation of an infinite number of boolean topologies (not forgetting that $\alpha \in [0.1]$) from an uncertain topology.

With the aim of giving a complete answer to however many questions this initiative may make, we propose to perform a decomposition by α -cuts of the fuzzy subsets taken as example up to now, using, as is our habit, the undecimal system. We therefore start from:

$$\underset{\sim}{P}_1 = \begin{array}{ccc} a & b & c \\ \hline .7 & .2 & .9 \end{array} \qquad \underset{\sim}{P}_2 = \begin{array}{ccc} a & b & c \\ \hline .3 & .9 & .7 \end{array} \qquad \underset{\sim}{P}_3 = \begin{array}{ccc} a & b & c \\ \hline .8 & .2 & .4 \end{array} \qquad \underset{\sim}{P}_4 = \begin{array}{ccc} a & b & c \\ \hline .2 & .9 & .8 \end{array}$$

If we begin the process making $\alpha = 1$, we see that the four fuzzy subsets result as **empty**.

We continue with $\propto \geq 0.9$:

$$\underset{\sim}{P}_1^{(0.9)} = \begin{array}{ccc} a & b & c \\ \hline 0 & 0 & 1 \end{array} \quad \underset{\sim}{P}_2^{(0.9)} = \begin{array}{ccc} a & b & c \\ \hline 0 & 1 & 0 \end{array} \quad \underset{\sim}{P}_3^{(0.9)} = \begin{array}{ccc} a & b & c \\ \hline 0 & 0 & 0 \end{array} \quad \underset{\sim}{P}_4^{(0.9)} = \begin{array}{ccc} a & b & c \\ \hline 0 & 1 & 0 \end{array}$$

We can see in this conversion that $\underset{\sim}{P}_3^{(0.9)}$ is empty and $\underset{\sim}{P}_2^{(0.9)}$ is the same as $\underset{\sim}{P}_4^{(0.9)}$.

With the aim of presenting comparable topological spaces at distinct levels we will use a way through which a result is found which automatically fulfils the axioms of the topology.

We will begin by showing the subsets which have the referential **a, b** and **c**.

$$F_a^{(0.9)} = \emptyset \qquad F_b^{(0.9)} = \{P_2^{(0.9)}, P_4^{(0.9)}\} \qquad F_c^{(0.9)} = \{P_1^{(0.9)}\}$$

and those which do not have them:

$$\overline{F_a^{(0,9}} = E \qquad \overline{F_b^{(0,9)}} = \{P_1^{(0,9)}, P_3^{(0,9)}\} \qquad \overline{F_c^{(0,9)}} = \{P_2^{(0,9)}, P_3^{(0,9)}, P_4^{(0,9)}\}$$

Next, all of the possible intersections are obtained:

$$G_1^{(0,9)} = F_a^{(0,9)} \cap F_b^{(0,9)} \cap F_c^{(0,9)} = \varnothing \qquad G_2^{(0,9)} = F_a^{(0,9)} \cap F_b^{(0,9)} \cap \overline{F_c^{(0,9)}} = \varnothing$$

$$G_3^{(0,9)} = F_a^{(0,9)} \cap \overline{F_b^{(0,9)}} \cap F_c^{(0,9)} = \varnothing \qquad G_4^{(0,9)} = \overline{F_a^{(0,9)}} \cap F_b^{(0,9)} \cap F_c^{(0,9)} = \varnothing$$

$$G_5^{(0,9)} = F_a^{(0,9)} \cap \overline{F_b^{(0,9)}} \cap \overline{F_c^{(0,9)}} = \varnothing$$

$$G_6^{(0,9)} = \overline{F_a^{(0,9}} \cap F_b^{(0,9)} \cap \overline{F_c^{(0,9)}} = \{P_2^{(0,9)}, P_4^{(0,9)}\} = \{\ \overset{a\;\;b\;\;c}{\boxed{0\,|\,1\,|\,0}}\ ,\ \overset{a\;\;b\;\;c}{\boxed{0\,|\,1\,|\,0}}\ \}$$

$$= \{\ \overset{a\;\;b\;\;c}{\boxed{1\,|\,0\,|\,0}}\ \}$$

$$G_7^{(0,9)} = \overline{F_a^{(0,9}} \cap \overline{F_b^{(0,9)}} \cap F_c^{(0,9)} = \{P_1^{(0,9)}\} = \{\ \overset{a\;\;b\;\;c}{\boxed{0\,|\,0\,|\,1}}\ \}$$

$$G_8^{(0,9)} = \overline{F_a^{(0,9}} \cap \overline{F_b^{(0,9)}} \cap \overline{F_c^{(0,9)}} = \{P_3^{(0,9)}\} = \{\ \overset{a\;\;b\;\;c}{\boxed{0\,|\,0\,|\,0}}\ \} = \varnothing$$

We can see an important question which has arisen in this α -cut, which, on the other hand, frequently occurs. In $G_6^{(0,9)}$, upon expressing two distinct fuzzy subsets, in the boolean way they are blended into only one. This substantially changes the conception of the result and the interpretation of the corresponding lattice, given that in maintaining the fuzzy form in the presentation of the subsets, those which are joined in $G_6^{(0,9)}$ would have been maintained differentiable. Let us see:

$$G_6^{(0,9)} = \{\ \overset{a\;\;\;b\;\;\;c}{\boxed{.3\,|\,.9\,|\,.7}}\ ,\ \overset{a\;\;\;b\;\;\;c}{\boxed{.2\,|\,.9\,|\,.8}}\ \}$$

Given that neither presentation falls in the form of the lattice and, therefore, in the topology, we propose to continue expressing the subsets in boolean form.

With this remark made, we **consider** the non-empty $G_i^{(0,9)}$, $i = 1, 2...$, which are $G_6^{(0,9)}$, $G_7^{(0,9)}$(as $G_8^{(0,9)}$ has resulted as empty) and **its unique union** $G_6^{(0,9)} \cup G_7^{(0,9)}$. Therefore, the topology at this level $\propto\ \geq 0.9$ is[4]:

$$T^{(0,9)}(E) = \{\varnothing, \{P_2^{(0,9)}, P_4^{(0,9)}\}, \{P_1^{(0,9)}\}, \{P_1^{(0,9)}, P_2^{(0,9)}, P_4^{(0,9)}\} = E\}$$

$$= \{\varnothing,\ \{\ \overset{a\;\;b\;\;c}{\boxed{0\,|\,1\,|\,0}}\ \},\ \{\ \overset{a\;\;b\;\;c}{\boxed{0\,|\,0\,|\,1}}\ \},\ \{\ \overset{a\;\;b\;\;c}{\boxed{0\,|\,0\,|\,1}}\ ,\ \overset{a\;\;b\;\;c}{\boxed{0\,|\,1\,|\,0}}\ \} = E\}$$

---

4 Taking into account that the empty $\varnothing$ is considered to always exist in any subset.

It results as easy to confirm that we find ourselves before a topology which can be expressed by one of the following lattices:

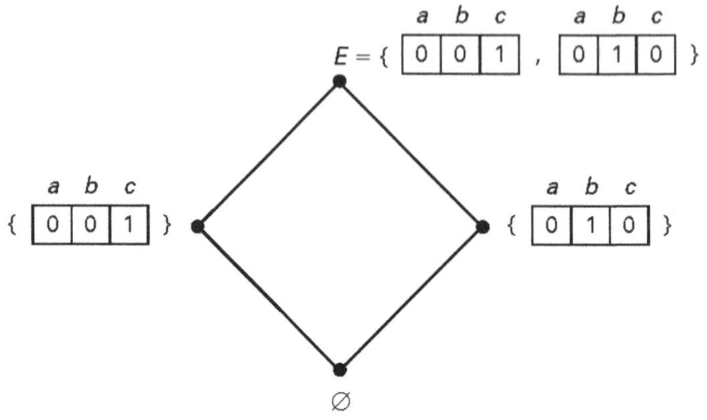

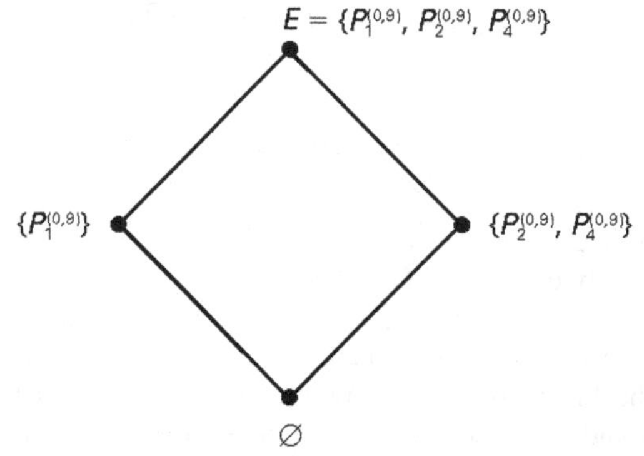

We move on to the level $\propto \geq 0.8$. The fuzzy subsets are converted into the following boolean subsets:

	a	*b*	*c*
$P_1^{(0,9)} =$	0	0	1
$P_2^{(0,9)} =$	0	1	0
$P_3^{(0,9)} =$	1	0	0
$P_4^{(0,9)} =$	0	1	1

We find:

$F_a^{(0,8)} = \{P_3^{(0,8)}\}$ $\quad F_b^{(0,8)} = \{P_2^{(0,8)}, P_4^{(0,8)}\}$ $\quad F_c^{(0,8)} = \{P_1^{(0,8)}, P_4^{(0,8)}\}$

$\overline{F_a^{(0,8)}} = \{P_1^{(0,8)}, P_2^{(0,8)}, P_4^{(0,8)}\}$ $\quad \overline{F_b^{(0,8)}} = \{P_1^{(0,8)}, P_3^{(0,8)}\}$ $\quad \overline{F_c^{(0,8)}} = \{P_2^{(0,8)}, P_3^{(0,8)}\}$

In continuation we achieve:

$$G_1^{(0,8)} = F_a^{(0,8)} \cap F_b^{(0,8)} \cap F_c^{(0,8)} = \varnothing \qquad G_2^{(0,8)} = F_a^{(0,8)} \cap F_b^{(0,8)} \cap \overline{F_c^{(0,8)}} = \varnothing$$

$$G_3^{(0,8)} = F_a^{(0,8)} \cap \overline{F_b^{(0,8)}} \cap F_c^{(0,8)} = \varnothing$$

$$G_4^{(0,8)} = \overline{F_a^{(0,8)}} \cap F_b^{(0,8)} \cap F_c^{(0,8)} = \{P_4^{(0,8)}\} = \left\{ \begin{array}{ccc} a & b & c \\ \hline \multicolumn{1}{|c|}{0} & \multicolumn{1}{c|}{1} & \multicolumn{1}{c|}{1} \\ \hline \end{array} \right\}$$

$$G_5^{(0,8)} = F_a^{(0,8)} \cap \overline{F_b^{(0,8)}} \cap \overline{F_c^{(0,8)}} = \{P_3^{(0,8)}\} = \left\{ \begin{array}{ccc} a & b & c \\ \hline \multicolumn{1}{|c|}{1} & \multicolumn{1}{c|}{0} & \multicolumn{1}{c|}{0} \\ \hline \end{array} \right\}$$

$$G_6^{(0,8)} = \overline{F_a^{(0,8)}} \cap F_b^{(0,8)} \cap \overline{F_c^{(0,8)}} = \{P_2^{(0,8)}\} = \left\{ \begin{array}{ccc} a & b & c \\ \hline \multicolumn{1}{|c|}{0} & \multicolumn{1}{c|}{1} & \multicolumn{1}{c|}{0} \\ \hline \end{array} \right\}$$

$$G_7^{(0,8)} = \overline{F_a^{(0,8)}} \cap \overline{F_b^{(0,8)}} \cap F_c^{(0,8)} = \{P_1^{(0,8)}\} = \left\{ \begin{array}{ccc} a & b & c \\ \hline \multicolumn{1}{|c|}{0} & \multicolumn{1}{c|}{0} & \multicolumn{1}{c|}{1} \\ \hline \end{array} \right\}$$

$$G_8^{(0,8)} = \overline{F_a^{(0,8)}} \cap \overline{F_b^{(0,8)}} \cap \overline{F_c^{(0,8)}} = \varnothing$$

The non-empty sets $G_i^{(0.8)}$, $i = 1, 2..., 8$ are $G_4^{(0.8)}$, $G_5^{(0.8)}$, $G_6^{(0.8)}$, $G_7^{(0.8)}$. These sets and all of their possible unions together with the empty form the subset $T^{(0.8)}(E)$, with its $2^4 = 16$ elements, which gives way to the following Boole lattice in whose vertices the subsets with the symbols $P_j^{(0.8)}$ $j = 1, 2, 3, 4$ have been placed.

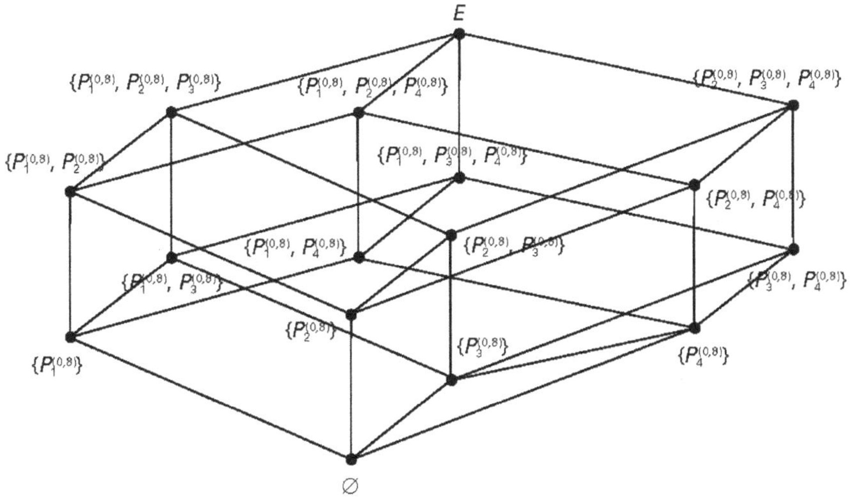

This level $\propto \geq 0.8$ allows us to make another important observation. The subsets which form $T^{(0.8)}(E)$ in each of the levels are formed by boolean subsets $P_1^{(\alpha)}, P_2^{(\alpha)}, P_3^{(\alpha)}, P_4^{(\alpha)}$ which do not always coincide when the cut is changed. In effect, comparing the levels $\propto \geq 0.9$ with $\propto \geq 0.8$ gives:

$$P_1^{(0,9)} = \begin{array}{|c|c|c|} \multicolumn{1}{c}{a} & \multicolumn{1}{c}{b} & \multicolumn{1}{c}{c} \\ \hline 0 & 0 & 1 \\ \hline \end{array} \qquad P_1^{(0,8)} = \begin{array}{|c|c|c|} \multicolumn{1}{c}{a} & \multicolumn{1}{c}{b} & \multicolumn{1}{c}{c} \\ \hline 0 & 0 & 1 \\ \hline \end{array} \qquad \text{coincide}$$

$$P_2^{(0,9)} = \begin{array}{|c|c|c|} \multicolumn{1}{c}{a} & \multicolumn{1}{c}{b} & \multicolumn{1}{c}{c} \\ \hline 0 & 1 & 0 \\ \hline \end{array} \qquad P_2^{(0,8)} = \begin{array}{|c|c|c|} \multicolumn{1}{c}{a} & \multicolumn{1}{c}{b} & \multicolumn{1}{c}{c} \\ \hline 0 & 1 & 0 \\ \hline \end{array} \qquad \text{coincide}$$

$$P_3^{(0,9)} = \begin{array}{|c|c|c|} \multicolumn{1}{c}{a} & \multicolumn{1}{c}{b} & \multicolumn{1}{c}{c} \\ \hline 0 & 0 & 0 \\ \hline \end{array} \qquad P_3^{(0,8)} = \begin{array}{|c|c|c|} \multicolumn{1}{c}{a} & \multicolumn{1}{c}{b} & \multicolumn{1}{c}{c} \\ \hline 1 & 0 & 0 \\ \hline \end{array} \qquad \text{do not coincide}$$

$$P_4^{(0,9)} = \begin{array}{|c|c|c|} \multicolumn{1}{c}{a} & \multicolumn{1}{c}{b} & \multicolumn{1}{c}{c} \\ \hline 0 & 1 & 0 \\ \hline \end{array} \qquad P_4^{(0,8)} = \begin{array}{|c|c|c|} \multicolumn{1}{c}{a} & \multicolumn{1}{c}{b} & \multicolumn{1}{c}{c} \\ \hline 0 & 1 & 1 \\ \hline \end{array} \qquad \text{do not coincide}$$

On this occasion, on moving from $\propto \geq 0.9$ to $\propto \geq 0.8$ the boolean subsets fit. But this will not always occur upon descending the levels $\propto \geq 0.7, 0.6, \ldots 0.1, 0$.

In the following we will see what occurs upon passing the cut $\propto \geq 0.7$, already previously dealt with. With a difference to before, the subsets resulting from the cut will be represented in a boolean way. Let:

$$P_1^{(0,7)} = \begin{array}{|c|c|c|} \multicolumn{1}{c}{a} & \multicolumn{1}{c}{b} & \multicolumn{1}{c}{c} \\ \hline 1 & 0 & 1 \\ \hline \end{array} \quad P_2^{(0,7)} = \begin{array}{|c|c|c|} \multicolumn{1}{c}{a} & \multicolumn{1}{c}{b} & \multicolumn{1}{c}{c} \\ \hline 0 & 1 & 1 \\ \hline \end{array} \quad P_3^{(0,7)} = \begin{array}{|c|c|c|} \multicolumn{1}{c}{a} & \multicolumn{1}{c}{b} & \multicolumn{1}{c}{c} \\ \hline 1 & 0 & 0 \\ \hline \end{array} \quad P_4^{(0,7)} = \begin{array}{|c|c|c|} \multicolumn{1}{c}{a} & \multicolumn{1}{c}{b} & \multicolumn{1}{c}{c} \\ \hline 0 & 1 & 1 \\ \hline \end{array}$$

Once again a coincidence arises, this time between $P_2^{(0.7)}$ and $P_4^{(0.7)}$.

In continuation we obtain:

$$F_a^{(0,7)} = \{P_1^{(0,7)}, P_3^{(0,7)}\} \qquad F_b^{(0,7)} = \{P_2^{(0,7)}, P_4^{(0,7)}\} \qquad F_c^{(0,7)} = \{P_1^{(0,7)}, P_2^{(0,7)}, P_4^{(0,7)}\}$$

and their compliments:

$$\overline{F_a^{(0,7)}} = \{P_2^{(0,7)}, P_4^{(0,7)}\} \qquad \overline{F_b^{(0,7)}} = \{P_1^{(0,7)}, P_3^{(0,7)}\} \qquad \overline{F_c^{(0,7)}} = \{P_3^{(0,7)}\}$$

In the following we list the subsets $G_i^{(0.7)}$, $i = 1, 2, \ldots 8$:

$$G_1^{(0,7)} = F_a^{(0,7)} \cap F_b^{(0,7)} \cap F_c^{(0,7)} = \varnothing \qquad G_2^{(0,7)} = F_a^{(0,7)} \cap F_b^{(0,7)} \cap \overline{F_c^{(0,7)}} = \varnothing$$

$$G_3^{(0,7)} = F_a^{(0,7)} \cap \overline{F_b^{(0,7)}} \cap F_c^{(0,7)} = \{P_1^{(0,7)}\} = \{ \begin{array}{|c|c|c|} \multicolumn{1}{c}{a} & \multicolumn{1}{c}{b} & \multicolumn{1}{c}{c} \\ \hline 1 & 0 & 1 \\ \hline \end{array} \}$$

$$G_4^{(0,7)} = \overline{F_a^{(0,7)}} \cap F_b^{(0,7)} \cap F_c^{(0,7)} = \{P_2^{(0,7)}, P_4^{(0,7)}\} = \{\overset{a\;b\;c}{\boxed{0\,|\,1\,|\,1}}\,,\ \overset{a\;b\;c}{\boxed{0\,|\,1\,|\,1}}\}$$

$$= \{\overset{a\;b\;c}{\boxed{0\,|\,1\,|\,1}}\}$$

$$G_5^{(0,7)} = F_a^{(0,7)} \cap \overline{F_b^{(0,7)}} \cap \overline{F_c^{(0,7)}} = \{P_3^{(0,7)}\} = \{\overset{a\;b\;c}{\boxed{1\,|\,0\,|\,0}}\}$$

$$G_6^{(0,7)} = \overline{F_a^{(0,7)}} \cap F_b^{(0,7)} \cap \overline{F_c^{(0,7)}} = \varnothing \qquad G_7^{(0,7)} = \overline{F_a^{(0,7)}} \cap \overline{F_b^{(0,7)}} \cap F_c^{(0,7)} = \varnothing$$

$$G_8^{(0,7)} = \overline{F_a^{(0,7)}} \cap \overline{F_b^{(0,7)}} \cap \overline{F_c^{(0,7)}} = \varnothing$$

We find that now, with a difference to that which occurred upon expressing the G_i subsets through fuzzy subsets, as we do this with booleans, the set $G_4^{(0.7)}$ consists of only one boolean subset, on not contemplating the nuance which allows the characteristic membership function in [0.1], belonging to fuzziness. Due to this, even when in both representations (fuzzy and boolean) the same non-empty subsets $G_3^{(0.7)}$, $G_4^{(0.7)}$, $G_5^{(0.7)}$ appear, their contents may vary, as occurs in the subset $G_4^{(0.7)}$. We express $T^{(0.7)}(E)$ in the two forms:

$$T^{(0,7)}(E) = \big\{\varnothing, \{P_1^{(0,7)}\}, \{P_3^{(0,7)}\}, \{P_2^{(0,7)}, P_4^{(0,7)}\}, \{P_1^{(0,7)}, P_3^{(0,7)}\},$$
$$\{P_1^{(0,7)}, P_2^{(0,7)}, P_4^{(0,7)}\}, \{P_2^{(0,7)}, P_3^{(0,7)}, P_4^{(0,7)}\}, E\big\}$$

$$= \big\{\varnothing, \{\overset{a\;b\;c}{\boxed{1\,|\,0\,|\,1}}\}, \{\overset{a\;b\;c}{\boxed{1\,|\,0\,|\,0}}\}, \{\overset{a\;b\;c}{\boxed{0\,|\,1\,|\,1}}\}, \{\overset{a\;b\;c}{\boxed{1\,|\,0\,|\,1}}\,,\ \overset{a\;b\;c}{\boxed{1\,|\,0\,|\,0}}\}$$
$$\{\overset{a\;b\;c}{\boxed{1\,|\,0\,|\,1}}\,,\ \overset{a\;b\;c}{\boxed{0\,|\,1\,|\,1}}\}, \{\overset{a\;b\;c}{\boxed{0\,|\,1\,|\,1}}\,,\ \overset{a\;b\;c}{\boxed{1\,|\,0\,|\,0}}\}, E\big\}$$

In continuation we include the corresponding lattice:

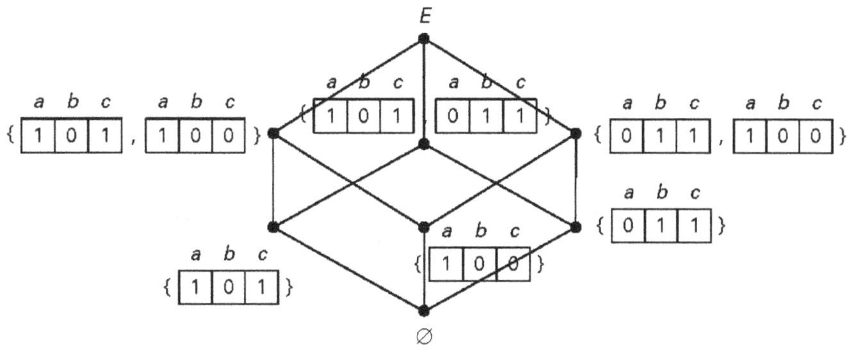

In this case we also find ourselves before a topological space.

Upon moving to the cuts $\propto \geq 0.6$, $\propto \geq 0.5$, the same boolean subsets are obtained as those found for $\propto \geq 0.7$, and so the process would be repeated in both levels and the topology would coincide with that corresponding to $\propto \geq 0.7$.

In the following we refer to the level $\propto \geq 0.4$. We have:

$$P_1^{(0,4)} = \begin{array}{|c|c|c|}\multicolumn{1}{c}{a}&\multicolumn{1}{c}{b}&\multicolumn{1}{c}{c}\\\hline 1&0&1\\\hline\end{array} \quad P_2^{(0,4)} = \begin{array}{|c|c|c|}\multicolumn{1}{c}{a}&\multicolumn{1}{c}{b}&\multicolumn{1}{c}{c}\\\hline 0&1&1\\\hline\end{array} \quad P_3^{(0,4)} = \begin{array}{|c|c|c|}\multicolumn{1}{c}{a}&\multicolumn{1}{c}{b}&\multicolumn{1}{c}{c}\\\hline 1&0&1\\\hline\end{array} \quad P_4^{(0,4)} = \begin{array}{|c|c|c|}\multicolumn{1}{c}{a}&\multicolumn{1}{c}{b}&\multicolumn{1}{c}{c}\\\hline 0&1&1\\\hline\end{array}$$

It results that:

$$F_a^{(0,4)} = \{P_1^{(0,4)}, P_3^{(0,4)}\} \qquad F_b^{(0,4)} = \{P_2^{(0,4)}, P_4^{(0,4)}\} \qquad F_c^{(0,4)} = \{P_1^{(0,4)}, P_2^{(0,4)}, P_3^{(0,4)}, P_4^{(0,4)}\}$$

The compliments are:

$$\overline{F_a^{(0,4)}} = \{P_2^{(0,4)}, P_4^{(0,4)}\} \qquad \overline{F_b^{(0,4)}} = \{P_1^{(0,4)}, P_3^{(0,4)}\} \qquad \overline{F_c^{(0,4)}} = \varnothing$$

We obtain:

$$G_1^{(0,4)} = F_a^{(0,4)} \cap F_b^{(0,4)} \cap F_c^{(0,4)} = \varnothing \qquad G_2^{(0,4)} = F_a^{(0,4)} \cap F_b^{(0,4)} \cap \overline{F_c^{(0,4)}} = \varnothing$$

$$G_3^{(0,4)} = F_a^{(0,4)} \cap \overline{F_b^{(0,4)}} \cap F_c^{(0,4)} = \{P_1^{(0,4)}, P_3^{(0,4)}\} = \{ \begin{array}{|c|c|c|}\multicolumn{1}{c}{a}&\multicolumn{1}{c}{b}&\multicolumn{1}{c}{c}\\\hline 1&0&1\\\hline\end{array}, \begin{array}{|c|c|c|}\multicolumn{1}{c}{a}&\multicolumn{1}{c}{b}&\multicolumn{1}{c}{c}\\\hline 1&0&1\\\hline\end{array} \}$$

$$= \{ \begin{array}{|c|c|c|}\multicolumn{1}{c}{a}&\multicolumn{1}{c}{b}&\multicolumn{1}{c}{c}\\\hline 1&0&1\\\hline\end{array} \}$$

$$G_4^{(0,4)} = \overline{F_a^{(0,4)}} \cap F_b^{(0,4)} \cap F_c^{(0,4)} = \{P_2^{(0,4)}, P_4^{(0,4)}\} = \{ \begin{array}{|c|c|c|}\multicolumn{1}{c}{a}&\multicolumn{1}{c}{b}&\multicolumn{1}{c}{c}\\\hline 0&1&1\\\hline\end{array}, \begin{array}{|c|c|c|}\multicolumn{1}{c}{a}&\multicolumn{1}{c}{b}&\multicolumn{1}{c}{c}\\\hline 0&1&1\\\hline\end{array} \}$$

$$= \{ \begin{array}{|c|c|c|}\multicolumn{1}{c}{a}&\multicolumn{1}{c}{b}&\multicolumn{1}{c}{c}\\\hline 0&1&1\\\hline\end{array} \}$$

As the non-empty subsets are $G_3^{(0.4)}$, $G_4^{(0.4)}$, the topology $T^{(0.4)}(E)$ will be:

$$T^{(0,4)}(E) = \{\varnothing, \{P_1^{(0,4)}, P_3^{(0,4)}\}, \{P_2^{(0,4)}, P_4^{(0,4)}\}, E\}$$

$$= \{\varnothing, \{ \begin{array}{|c|c|c|}\multicolumn{1}{c}{a}&\multicolumn{1}{c}{b}&\multicolumn{1}{c}{c}\\\hline 1&0&1\\\hline\end{array} \}, \{ \begin{array}{|c|c|c|}\multicolumn{1}{c}{a}&\multicolumn{1}{c}{b}&\multicolumn{1}{c}{c}\\\hline 0&1&1\\\hline\end{array} \}, \{ \begin{array}{|c|c|c|}\multicolumn{1}{c}{a}&\multicolumn{1}{c}{b}&\multicolumn{1}{c}{c}\\\hline 1&0&1\\\hline\end{array}, \begin{array}{|c|c|c|}\multicolumn{1}{c}{a}&\multicolumn{1}{c}{b}&\multicolumn{1}{c}{c}\\\hline 0&1&1\\\hline\end{array} \} = E\}$$

The corresponding lattice is:

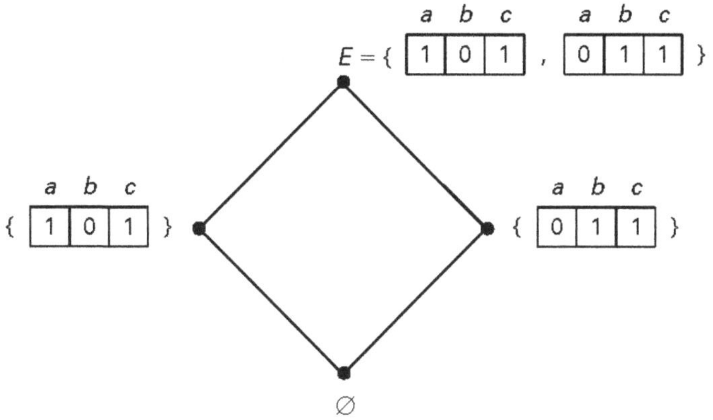

In continuation we refer to the level $\propto \geq 0.3$:

$$P_1^{(0,3)} = \begin{array}{ccc} a & b & c \\ \hline \boxed{1} & \boxed{0} & \boxed{1} \end{array} \quad P_2^{(0,3)} = \begin{array}{ccc} a & b & c \\ \hline \boxed{1} & \boxed{1} & \boxed{1} \end{array} \quad P_3^{(0,3)} = \begin{array}{ccc} a & b & c \\ \hline \boxed{1} & \boxed{0} & \boxed{1} \end{array} \quad P_4^{(0,3)} = \begin{array}{ccc} a & b & c \\ \hline \boxed{0} & \boxed{1} & \boxed{1} \end{array}$$

We have:

$$F_a^{(0,3)} = \{P_1^{(0,3)}, P_2^{(0,3)}, P_3^{(0,3)}\} \quad F_b^{(0,3)} = \{P_2^{(0,3)}, P_4^{(0,3)}\} \quad F_c^{(0,3)} = \{P_1^{(0,3)}, P_2^{(0,3)}, P_3^{(0,3)}, P_4^{(0,3)}\} = E$$

$$\overline{F_a^{(0,3)}} = \{P_4^{(0,3)}\} \quad \overline{F_b^{(0,3)}} = \{P_1^{(0,3)}, P_3^{(0,3)}\} \quad \overline{F_c^{(0,3)}} = \varnothing$$

Therefore:

$$G_1^{(0,3)} = F_a^{(0,3)} \cap F_b^{(0,3)} \cap F_c^{(0,3)} = \{P_2^{(0,3)}\} = \left\{ \begin{array}{ccc} a & b & c \\ \boxed{1} & \boxed{1} & \boxed{1} \end{array} \right\}$$

$$G_2^{(0,3)} = F_a^{(0,3)} \cap F_b^{(0,3)} \cap \overline{F_c^{(0,3)}} = \varnothing$$

$$G_3^{(0,3)} = F_a^{(0,3)} \cap \overline{F_b^{(0,3)}} \cap F_c^{(0,3)} = \{P_1^{(0,3)}, P_3^{(0,3)}\} = \left\{ \begin{array}{ccc} a & b & c \\ \boxed{1} & \boxed{0} & \boxed{1} \end{array}, \begin{array}{ccc} a & b & c \\ \boxed{1} & \boxed{0} & \boxed{1} \end{array} \right\}$$

$$= \left\{ \begin{array}{ccc} a & b & c \\ \boxed{1} & \boxed{0} & \boxed{1} \end{array} \right\}$$

$$G_4^{(0,3)} = \overline{F_a^{(0,3)}} \cap F_b^{(0,3)} \cap F_c^{(0,3)} = \{P_2^{(0,3)}\} = \left\{ \begin{array}{ccc} a & b & c \\ \boxed{0} & \boxed{1} & \boxed{1} \end{array} \right\}$$

$$G_5^{(0,3)} = F_a^{(0,3)} \cap \overline{F_b^{(0,3)}} \cap \overline{F_c^{(0,3)}} = \varnothing \qquad G_6^{(0,3)} = \overline{F_a^{(0,3)}} \cap F_b^{(0,3)} \cap \overline{F_c^{(0,3)}} = \varnothing$$

$$G_7^{(0,3)} = \overline{F_a^{(0,3)}} \cap \overline{F_b^{(0,3)}} \cap F_c^{(0,3)} = \varnothing \qquad G_8^{(0,3)} = \overline{F_a^{(0,3)}} \cap \overline{F_b^{(0,3)}} \cap \overline{F_c^{(0,3)}} = \varnothing$$

Having the non-empty subsets $G_1^{(0.3)}$, $G_3^{(0.3)}$, $G_4^{(0.3)}$, and so the topology will be:

$$T^{(0.3)}(E) = \{\varnothing, \{P_2^{(0.3)}\}, \{P_4^{(0.3)}\}, \{P_1^{(0.3)}, P_3^{(0.3)}\}, \{P_2^{(0.3)}, P_4^{(0.3)}\},$$
$$\{P_1^{(0.3)}, P_2^{(0.3)}, P_3^{(0.3)}\}, \{P_1^{(0.3)}, P_3^{(0.3)}, P_4^{(0.3)}\}, \{P_1^{(0.3)}, P_2^{(0.3)}, P_3^{(0.3)}, P_4^{(0.3)}\} = E\}$$

$$= \{\varnothing, \{\overset{a\;b\;c}{\boxed{1\,|\,1\,|\,1}}\}, \{\overset{a\;b\;c}{\boxed{0\,|\,1\,|\,1}}\}, \{\overset{a\;b\;c}{\boxed{1\,|\,0\,|\,1}}\}, \{\overset{a\;b\;c}{\boxed{1\,|\,1\,|\,1}}, \overset{a\;b\;c}{\boxed{0\,|\,1\,|\,1}}\}$$
$$\{\overset{a\;b\;c}{\boxed{1\,|\,0\,|\,1}}, \overset{a\;b\;c}{\boxed{1\,|\,1\,|\,1}}\}, \{\overset{a\;b\;c}{\boxed{1\,|\,0\,|\,1}}, \overset{a\;b\;c}{\boxed{0\,|\,1\,|\,1}}\},$$
$$\{\overset{a\;b\;c}{\boxed{1\,|\,0\,|\,1}}, \overset{a\;b\;c}{\boxed{1\,|\,1\,|\,1}}, \overset{a\;b\;c}{\boxed{0\,|\,1\,|\,1}}\} = E\}$$

We now present the lattice:

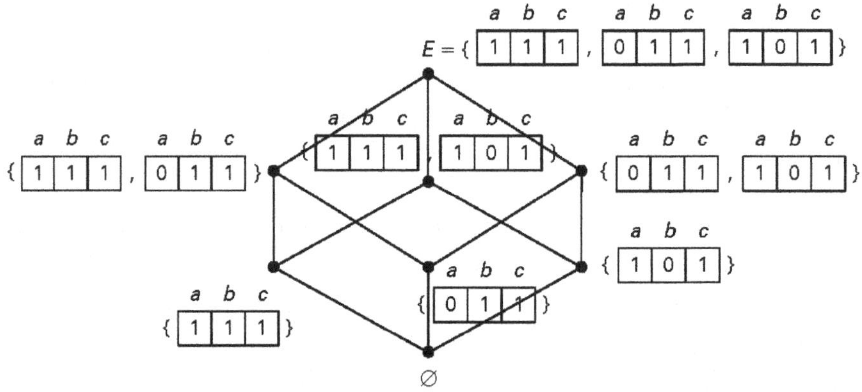

In this point we can say that this process of decomposition in $\propto$-cuts is finished, given that on making $\propto \geq 0.2$ we find that all of the boolean subsets $P_j^{(\alpha)}$, $j = 1, 2, 3, 4$, $\propto = [0.1]$ are converted into only one, in which all the values of the characteristic membership function are equal to the unit.

We have tried to conveniently visualise each of the boolean topologies by means of the corresponding lattice in order to, amongst other objectives, show that even though a fit takes place between some pairs of $\propto$ levels, a general fit is not produced for all the values of $\propto \in [0.1]$. Could this be important? Does it make sense to speak of fitting when the elements to fit change? Perhaps from a theoretical perspective. But we very much doubt that this will make economists and management specialists, who search for the best treatment for the problems which states and businesses pose, lose sleep. We believe that reaching this point has been an important step, but we are aware that it will not be the last. The door therefore remains open for those who wish to continue.

It has been possible to verify during this process of decomposition that the constituent elements of boolean topologies which arise from uncertain topology have sometimes been expressed considering the boolean subsets in a differing way even though they result as identical, as occurs with $P_2^{(0.9)}$ and $P_4^{(0.9)}$, for example. On other occasions we have only considered the elements of the "power set" for one of them when it coincides with another or others, such as the case of:

$$P_1^{(0,3)} = \begin{array}{ccc} a & b & c \\ \hline \multicolumn{1}{|c|}{1} & \multicolumn{1}{c|}{0} & \multicolumn{1}{c|}{1} \\ \hline \end{array} \qquad P_3^{(0,3)} = \begin{array}{ccc} a & b & c \\ \hline \multicolumn{1}{|c|}{1} & \multicolumn{1}{c|}{0} & \multicolumn{1}{c|}{1} \\ \hline \end{array}$$

to give an example. We have seen that the configuration of the lattices does not vary, but, what does change is its interpretation. So, which of these presentations is the most adequate? The answer is not universal. We will illustrate this statement with an example.

If each fuzzy subset $\underset{\sim}{P}_j$ describes, for example, a financial product through its qualities, characteristics or singularities (a, b, c in our case), the groupings between groups of products, elements of the "power set", **should explicitly show all of the products** which form them, independently of the fact that two or more of them have the same qualities at the level demanded. On the other hand, if one wishes to stress the characteristics, without giving importance to if they are possessed by one or more products, then, generally, in each element of the "power set" a single boolean subset must be taken into account, even though it describes more than one product. Therefore, the nature of the problem will dictate which of the two alternatives results as the most adequate in each situation or each problem posed.

We believe the interest for economics and management sciences to be able to have the theoretical and technical elements which arise from the adequate formulation of topological spaces unquestionable. Outlines derived from or connected with them have already been object of studies and efficient application in our field of knowledge. It is enough to remember among the first the so-called theory of clans and among the second the theory of affinities. But perhaps, up to now, these interesting findings have been found formulated without reference to the necessary links which unite them to a common core. We are convinced that this fact was not presided by ignorance, but by the lack of a basic structure with the sufficient solidity to be able to support these and other aspects of non-numerical mathematics of uncertainty, which has already allowed the elaboration of efficient models and algorithms.

We are conscious that our position in so much as to the dual possibility of fuzzification of topology is a bet as subject to criticism as many others may be. We assume with total humility the fact that it only deals with the first steps of a work and that, because of this, this is necessarily incomplete. In spite of this, we feel satisfied as our intuition shows us that the way undertaken can provide interesting developments in the future, which fill the important gaps that exist today in so much as the available instruments to face a complex world which is full of uncertainties.

6 Relationship between Uncertain Topological Spaces

In continuation we propose the development of an aspect of special interest when, like ourselves, the aim is the use of uncertain topology conceptualization in the field of economic-financial relationships. We are refering to the **functional application between topologies**. For this we will begin by defining the inverse fuzzy subset of a functional application **Γ**.

We insist on the necessity of making the **ideas of investment**[5] clear as here, in uncertainty, the two focuses followed now demand, we believe, explanatory comments of a distinct nature.

A) In the first instance, we will develop a focus from some referentials E_1 and E_2 whose elements are **referentials** of fuzzy subsets.

Therefore, we begin from the referentials E_1, E_2 and we assume a functional application **Γ** of E_1 coincides in E_2. If we have a fuzzy subset $\underset{\sim}{B} \subset E_2$ whose membership function is $\mu_{\underset{\sim}{B}}(y)$, $y \in E_2$, the inverse application $\mathbf{\Gamma}^{-1}$ of E_2 in E_1 gives an inverse fuzzy subset for the cited functional application **Γ** defined by:

$\forall x \in E_1$:

$$\mu_{\Gamma^{-1}\underset{\sim}{B}} = \mu_{\underset{\sim}{B}}(\Gamma(x))$$

We consider that this expression deserves an explanatory example, from the referentials:

$$E_1 = \{a, b, c, d, e, f, g\}$$

$$E_2 = \{\alpha, \beta, \gamma, \delta, \varepsilon\}$$

[5] We would like to point out as a reminder that in an **application Γ** of a set E_1 in a set E_2 all element $x \in E_1$ has only one image, but a $x \in E_2$ may have various sources. It is also said that **Γ** is a function which makes a correspondence with each x of the dominion E_1, one and the dominion E_2.

When **Γ** has the property that all elements of E_2 is an image of an element of E_1, it is said that **Γ** is an application over E_2 or that it is **epimorphic** (**surjective**).

When in an application **Γ** all elements $y \in E_2$ have as a maximum a source $x \in E_1$, it is said that a **monomorphic** or **univocal and invertible** application of E_1 in E_2 exists.

When **Γ** is at the same time epimorphic and monomorphic it is called **isomorphic** or **biunivocal application** of E_1 in E_2. In this supposition **Γ** establishes a **univocal and invertible** relationship. Only in this case can it be "accurately" said that a $\mathbf{\Gamma}^{-1}$ reciprocal or inverse to **Γ** exists. However, it must be taken into account that these terms are also used in a general sense, even though no isomorphism exists to define the set of all the sources of $y \in E_2$. Therefore, in this acceptation , $\mathbf{\Gamma}^{-1}$ is not "estrictu sensu" an application of E_1 in E_2, but a determined subset of E_1, which in certain cases may even be empty, corresponding to all subsets of E_2.

It is assumed that the functional application of E_1 in E_2 is that represented in the following graph:

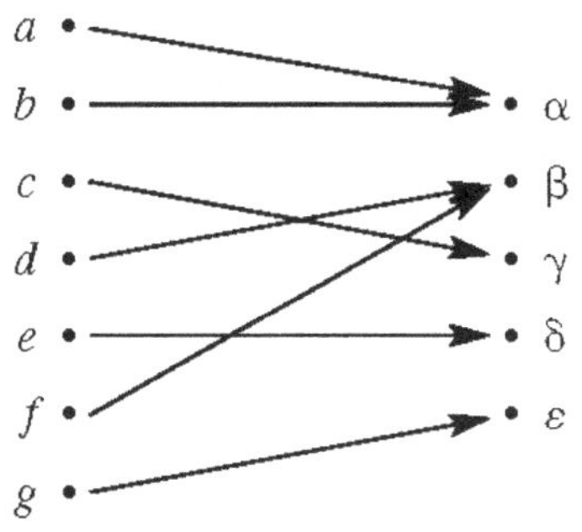

If the fuzzy subset $\underset{\sim}{B}$ is assumed as known:

$\underset{\sim}{B} =$

α	β	γ	δ	ε
.3	.7	.5	.9	.4

We have:

$$\mu_{\underset{\sim}{B}}(\Gamma(a)) = 0{,}3 \qquad \mu_{\underset{\sim}{B}}(\Gamma(b)) = 0{,}3 \qquad \mu_{\underset{\sim}{B}}(\Gamma(c)) = 0{,}5$$

$$\mu_{\underset{\sim}{B}}(\Gamma(d)) = 0{,}7 \qquad \mu_{\underset{\sim}{B}}(\Gamma(e)) = 0{,}9 \qquad \mu_{\underset{\sim}{B}}(\Gamma(f)) = 0{,}7 \qquad \mu_{\underset{\sim}{B}}(\Gamma(g)) = 0{,}4$$

We will therefore have as an inverse of $\underset{\sim}{B}$ for the application $\mathbf{\Gamma}$ the fuzzy subset:

$\Gamma^{-1}\underset{\sim}{B} =$

a	*b*	*c*	*d*	*e*	*f*	*g*
.3	.3	.5	.7	.9	.7	.4

In continuation we will set out the **induction** of a fuzzy subset by a functional application $\mathbf{\Gamma}$ from the fuzzy subset source of the application. As a starting point in this case we have a fuzzy subset $\underset{\sim}{A} \subset E_1$, with a membership function $\mu_{\underset{\sim}{A}}(x)$, $x \in E_1$. The image will be a fuzzy subset $\underset{\sim}{B}$ induced by a functional application $\mathbf{\Gamma}$ of E_1 in E_2 whose characteristic membership function is given in the following way:

$\forall\, y \in \underset{\sim}{B}$:

$$\mu_{\underset{\sim}{B}}(y) = \bigvee_{x \in \Gamma^{-1}(y)} \mu_{\underset{\sim}{A}}(x), \quad \text{si } \Gamma^{-1}(y) \neq \varnothing$$

$$= 0, \qquad \text{si } \Gamma^{-1}(y) = \varnothing$$

We believe that the development of an example beginning from the same premises as the previous can be clarifying. Now the process begins with $\underset{\sim}{A}$, maintaining the same functional application **Γ**. Therefore:

$$\underset{\sim}{A} = \begin{array}{|c|c|c|c|c|c|c|} \multicolumn{1}{c}{a} & \multicolumn{1}{c}{b} & \multicolumn{1}{c}{c} & \multicolumn{1}{c}{d} & \multicolumn{1}{c}{e} & \multicolumn{1}{c}{f} & \multicolumn{1}{c}{g} \\ \hline .3 & .3 & .5 & .7 & .9 & .7 & .4 \\ \hline \end{array}$$

The functional application gives:

$$\mu_{\underset{\sim}{B}}(\alpha) = 0{,}3 \vee 0{,}3 = 0{,}3 \qquad \mu_{\underset{\sim}{B}}(\beta) = 0{,}7 \vee 0{,}7 = 0{,}7 \qquad \mu_{\underset{\sim}{B}}(\gamma) = 0{,}5$$
$$\mu_{\underset{\sim}{B}}(\delta) = 0{,}9 \qquad \mu_{\underset{\sim}{B}}(\varepsilon) = 0{,}4$$

which is the fuzzy subset from which we would begin from in the inverse functional application, which is to say:

$$\Gamma\,\varnothing = \varnothing \quad \text{and} \quad \delta\, E = E$$

It can be seen that due to the fact of having taken the subset generated by the inverse of a functional application as an example, the obtaining of maximums always occurs comparing the same figures. This is a special case. Except in special cases the same does not occur when the fuzzy subset $\underset{\sim}{A}$ began from is original and all of the values of its characteristic membership function are known.

We will illustrate that which we have just stated with a new example of the same referentials and identical functional application but with a slightly different fuzzy subset $\underset{\sim}{A}^*$, the following:

$$\underset{\sim}{A}^* = \begin{array}{|c|c|c|c|c|c|c|} \multicolumn{1}{c}{a} & \multicolumn{1}{c}{b} & \multicolumn{1}{c}{c} & \multicolumn{1}{c}{d} & \multicolumn{1}{c}{e} & \multicolumn{1}{c}{f} & \multicolumn{1}{c}{g} \\ \hline .2 & .3 & .5 & .4 & .9 & .7 & .4 \\ \hline \end{array}$$

It will be seen that we have intentionally placed smaller figures respectively as values of the characteristic membership function of *a* and *d* than those of *b* and *f*. For the rest the fuzzy subset $\underset{\sim}{A}^*$ coincides with $\underset{\sim}{A}$. Upon doing it in this way the same image $\underset{\sim}{B}$ which was previously obtained should result, given that as a source *a* has *a* and *b* as those of β are *d* and *f*. In effect:

$$\underset{\sim}{B} = \begin{array}{|c|c|c|c|c|} \multicolumn{1}{c}{\alpha} & \multicolumn{1}{c}{\beta} & \multicolumn{1}{c}{\gamma} & \multicolumn{1}{c}{\delta} & \multicolumn{1}{c}{\varepsilon} \\ \hline .3 & .7 & .5 & .9 & .4 \\ \hline \end{array}$$

We believe that these examples can implicitly give an explanation of the criteria followed in order to obtain the inverse fuzzy subset for a functional application. In effect, when an element of E_2 is the image of two or more sources, elements of E_1, the value of the characteristic membership function of the image is the highest figure of all of the potential figures of the two or more sources. On the other hand, when the values of the characteristic membership function of the image are known, if this has various sources, upon not knowing the other value or values to assign to each one of the sources, for prudence the option is taken of giving all of them the same value, which is the maximum.

B) It seems justifiable to ask, in agreement with the logic of this work, how to explain the relationship between uncertain topological spaces which arises from a referential of fuzzy subsets. As we have previously done, we will present our proposal for the obtaining of the inverse of a functional application **Γ**.

We begin from two common subsets of **fuzzy subsets**, E_1 and E_2, and the functional application **Γ** of E_1 in E_2, is assumed as known. Given an element $\underset{\sim}{Q}_k$ of E_2, so that $\underset{\sim}{Q}_k \in E_2$, the inverse application $\mathbf{\Gamma}^{-1}$ of E_1 in E_2 gives an element $\mathbf{\Gamma}^{-1}\,\underset{\sim}{Q}_k \in E_1$ called inverse of the functional application **Γ** defined by:

$\forall \underset{\sim}{Q}_K \in E_2$:

$$\Gamma^{-1}\underset{\sim}{Q}_K = \underset{\sim}{Q}_K$$

We move on to develop a simple example. We have the following referentials:

$$E_1 = \{\underset{\sim}{P}_1, \underset{\sim}{P}_2, \underset{\sim}{P}_3, \underset{\sim}{P}_4\}$$
$$E_2 = \{\underset{\sim}{Q}_1, \underset{\sim}{Q}_2, \underset{\sim}{Q}_3\}$$

and the application **Γ** presented in the graph:

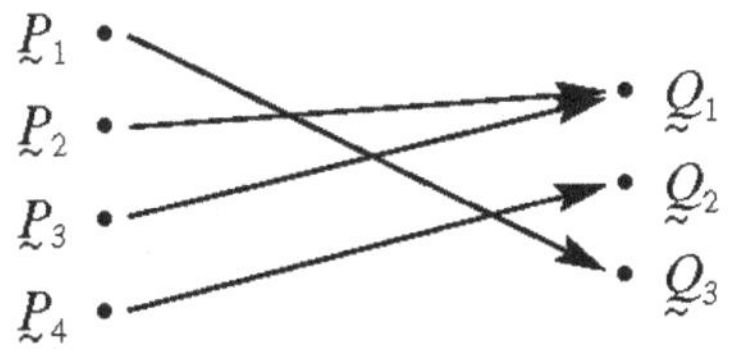

We assume that the fuzzy subsets $\underset{\sim}{Q}_k$, k = 1, 2, 3 are known

	a	b	c
$\underset{\sim}{Q}_1 =$	.6	.4	.9

	a	b	c
$\underset{\sim}{Q}_2 =$	.7	.8	.2

	a	b	c
$\underset{\sim}{Q}_3 =$	.3	.5	.8

With the definition accepted the previous application allows the obtention of:

$$\underset{\sim}{P}_1 = \Gamma^{-1}\underset{\sim}{Q}_3 = \underset{\sim}{Q}_3 = \begin{array}{ccc} a & b & c \\ \hline .3 & .5 & .8 \end{array}$$

$$\underset{\sim}{P}_2 = \Gamma^{-1}\underset{\sim}{Q}_1 = \underset{\sim}{Q}_1 = \begin{array}{ccc} a & b & c \\ \hline .6 & .4 & .9 \end{array}$$

$$\underset{\sim}{P}_3 = \Gamma^{-1}\underset{\sim}{Q}_1 = \underset{\sim}{Q}_1 = \begin{array}{ccc} a & b & c \\ \hline .6 & .4 & .9 \end{array}$$

$$\underset{\sim}{P}_4 = \Gamma^{-1}\underset{\sim}{Q}_2 = \underset{\sim}{Q}_2 = \begin{array}{ccc} a & b & c \\ \hline .7 & .8 & .2 \end{array}$$

We therefore have:

$$\underset{\sim}{P}_1 = \begin{array}{ccc} a & b & c \\ \hline .3 & .5 & .8 \end{array} \qquad \underset{\sim}{P}_2 = \begin{array}{ccc} a & b & c \\ \hline .6 & .4 & .9 \end{array} \qquad \underset{\sim}{P}_3 = \begin{array}{ccc} a & b & c \\ \hline .6 & .4 & .9 \end{array} \qquad \underset{\sim}{P}_4 = \begin{array}{ccc} a & b & c \\ \hline .7 & .8 & .2 \end{array}$$

All that we have just expounded easily allows us to see how the elements of the referential E_2 are generated by those of E_1 induced by a functional application $\mathbf{\Gamma}$. For this we will make:

$\forall \underset{\sim}{Q}_k \in E_2$:

$$\underset{\sim}{Q}_k = \bigvee_{\underset{\sim}{P}_j \in \Gamma^{-1}\underset{\sim}{Q}_k} \underset{\sim}{P}_j$$

We will see an example following the same referentials, identical application $\mathbf{\Gamma}$ and the same fuzzy subsets P_j, $j = 1, 2, 3, 4$ which were earlier object of findings and which we now consider as data in this new standing. So, a simple calculation is enough to find the same fuzzy subsets $\underset{\sim}{Q}_k$, k = 1, 2, 3, which were the starting point for the supposition of the inverse application $\mathbf{\Gamma}^{-1}$.

Now, we assume as known:

$$\underset{\sim}{P}_1 = \begin{array}{ccc} a & b & c \\ \hline .3 & .5 & .8 \end{array} \qquad \underset{\sim}{P}_2 = \begin{array}{ccc} a & b & c \\ \hline .6 & .4 & .9 \end{array} \qquad \underset{\sim}{P}_3 = \begin{array}{ccc} a & b & c \\ \hline .6 & .4 & .9 \end{array} \qquad \underset{\sim}{P}_4 = \begin{array}{ccc} a & b & c \\ \hline .7 & .8 & .2 \end{array}$$

It results:

$$\underset{\sim}{Q}_1 = \begin{array}{ccc} a & b & c \\ \hline \multicolumn{1}{|c|}{.6} & \multicolumn{1}{c|}{.4} & \multicolumn{1}{c|}{.9} \\ \hline \end{array} \ (\vee) \ \begin{array}{ccc} a & b & c \\ \hline \multicolumn{1}{|c|}{.6} & \multicolumn{1}{c|}{.4} & \multicolumn{1}{c|}{.9} \\ \hline \end{array} = \begin{array}{ccc} a & b & c \\ \hline \multicolumn{1}{|c|}{.6} & \multicolumn{1}{c|}{.4} & \multicolumn{1}{c|}{.9} \\ \hline \end{array}$$

$$\underset{\sim}{Q}_2 = \begin{array}{ccc} a & b & c \\ \hline \multicolumn{1}{|c|}{.7} & \multicolumn{1}{c|}{.8} & \multicolumn{1}{c|}{.2} \\ \hline \end{array}$$

$$\underset{\sim}{Q}_3 = \begin{array}{ccc} a & b & c \\ \hline \multicolumn{1}{|c|}{.3} & \multicolumn{1}{c|}{.5} & \multicolumn{1}{c|}{.8} \\ \hline \end{array}$$

But, leaving this example aside, if we begin from some source fuzzy subsets, distinct from each other, such as:

$$\underset{\sim}{P}_1 = \begin{array}{ccc} a & b & c \\ \hline \multicolumn{1}{|c|}{.4} & \multicolumn{1}{c|}{.8} & \multicolumn{1}{c|}{.2} \\ \hline \end{array} \quad \underset{\sim}{P}_2 = \begin{array}{ccc} a & b & c \\ \hline \multicolumn{1}{|c|}{.7} & \multicolumn{1}{c|}{.5} & \multicolumn{1}{c|}{.6} \\ \hline \end{array} \quad \underset{\sim}{P}_3 = \begin{array}{ccc} a & b & c \\ \hline \multicolumn{1}{|c|}{.1} & \multicolumn{1}{c|}{.9} & \multicolumn{1}{c|}{.4} \\ \hline \end{array} \quad \underset{\sim}{P}_4 = \begin{array}{ccc} a & b & c \\ \hline \multicolumn{1}{|c|}{.8} & \multicolumn{1}{c|}{.3} & \multicolumn{1}{c|}{.6} \\ \hline \end{array}$$

and we consider the same application $\mathbf{\Gamma}$, we would have[6]:

$$\underset{\sim}{Q}_1 = \underset{\sim}{P}_2 \ (\vee) \ \underset{\sim}{P}_3 = \begin{array}{ccc} a & b & c \\ \hline \multicolumn{1}{|c|}{.7} & \multicolumn{1}{c|}{.5} & \multicolumn{1}{c|}{.6} \\ \hline \end{array} \ (\vee) \ \begin{array}{ccc} a & b & c \\ \hline \multicolumn{1}{|c|}{.1} & \multicolumn{1}{c|}{.9} & \multicolumn{1}{c|}{.4} \\ \hline \end{array} = \begin{array}{ccc} a & b & c \\ \hline \multicolumn{1}{|c|}{.7} & \multicolumn{1}{c|}{.9} & \multicolumn{1}{c|}{.6} \\ \hline \end{array}$$

$$\underset{\sim}{Q}_2 = \underset{\sim}{P}_4 = \begin{array}{ccc} a & b & c \\ \hline \multicolumn{1}{|c|}{.8} & \multicolumn{1}{c|}{.3} & \multicolumn{1}{c|}{.6} \\ \hline \end{array}$$

$$\underset{\sim}{Q}_3 = \underset{\sim}{P}_1 = \begin{array}{ccc} a & b & c \\ \hline \multicolumn{1}{|c|}{.4} & \multicolumn{1}{c|}{.8} & \multicolumn{1}{c|}{.2} \\ \hline \end{array}$$

Let us see what happens when from the same suppositions the "power set" $P(E_1)$ is wished to be found from the inverse application $\mathbf{\Gamma}^{-1}\ Q(E_2)$.

The previous application $\mathbf{\Gamma}$ allows us to establish the following inverse functional application $\mathbf{\Gamma}^{-1}\ Q(E_2)$ in $P(E_1)$ in sagittate form, which we show in the following:

---

[6] Observe that, on this occasion, we have considered the "maximum" or "supremum" of the sources P2 and P3 to find the image of Q1.

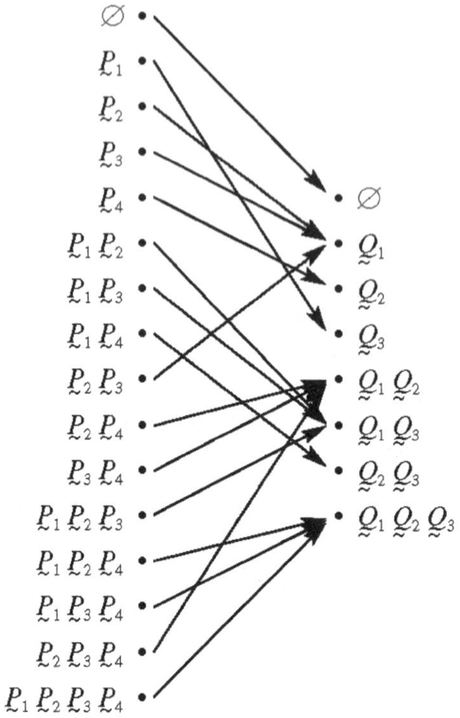

We have:

$$\Gamma^{-1}\varnothing = \varnothing \qquad \Gamma^{-1}\{\underset{\sim}{Q}_1\} = \big\{\{\underset{\sim}{P}_2\}, \{\underset{\sim}{P}_3\}, \{\underset{\sim}{P}_2, \underset{\sim}{P}_3\}\big\} \qquad \Gamma^{-1}\{\underset{\sim}{Q}_2\} = \{\underset{\sim}{P}_4\}$$

$$\Gamma^{-1}\{\underset{\sim}{Q}_3\} = \{\underset{\sim}{P}_1\} \qquad \Gamma^{-1}\{\underset{\sim}{Q}_1, \underset{\sim}{Q}_2\} = \big\{\{\underset{\sim}{P}_2, \underset{\sim}{P}_4\}, \{\underset{\sim}{P}_3, \underset{\sim}{P}_4\}, \{\underset{\sim}{P}_2, \underset{\sim}{P}_3, \underset{\sim}{P}_4\}\big\}$$

$$\Gamma^{-1}\{\underset{\sim}{Q}_1, \underset{\sim}{Q}_3\} = \big\{\{\underset{\sim}{P}_1, \underset{\sim}{P}_2\}, \{\underset{\sim}{P}_1, \underset{\sim}{P}_3\}, \{\underset{\sim}{P}_1, \underset{\sim}{P}_2, \underset{\sim}{P}_3\}\big\} \qquad \Gamma^{-1}\{\underset{\sim}{Q}_2, \underset{\sim}{Q}_3\} = \{\underset{\sim}{P}_1, \underset{\sim}{P}_4\}$$

$$\Gamma^{-1}\{\underset{\sim}{Q}_1, \underset{\sim}{Q}_2, \underset{\sim}{Q}_3\} = \big\{\{\underset{\sim}{P}_1, \underset{\sim}{P}_2, \underset{\sim}{P}_4\}, \{\underset{\sim}{P}_1, \underset{\sim}{P}_3, \underset{\sim}{P}_4\}, \{\underset{\sim}{P}_1, \underset{\sim}{P}_2, \underset{\sim}{P}_3, \underset{\sim}{P}_4\}\big\}$$

We do not stress on this point and if, on the other hand, we see how the established definition for the application $\mathbf{\Gamma}$ of E_1 in E_2 is also valid, with the adequate adaptations for a functional application between the respective "power sets". In effect:

Let:

$$\underset{\sim}{A}_j \in P(E_1)$$

The image of $\underset{\sim}{A}_j$ due to $\mathbf{\Gamma}$ is a functional application of $P(E_1)$ in $Q(E_2)$ so that:

$\forall\, \underset{\sim}{B}_k \in Q(E_2)$:

$$\underset{\sim}{B}_k = \bigvee_{\underset{\sim}{A}_j \in \Gamma^{-1}\underset{\sim}{B}_k} \underset{\sim}{A}_j, \quad \text{when} \quad \Gamma^{-1}\underset{\sim}{B}_k \neq \varnothing$$
$$= \varnothing, \qquad \text{when} \quad \Gamma^{-1}\underset{\sim}{B}_k = \varnothing$$

in which:

$$\Gamma^{-1}\underset{\sim}{B}_k = \{\underset{\sim}{A}_j / \Gamma\,\underset{\sim}{A}_j = \underset{\sim}{B}_k\}$$

We now reproduce the previous application $P(E_1)$ in $Q(E_2)$:

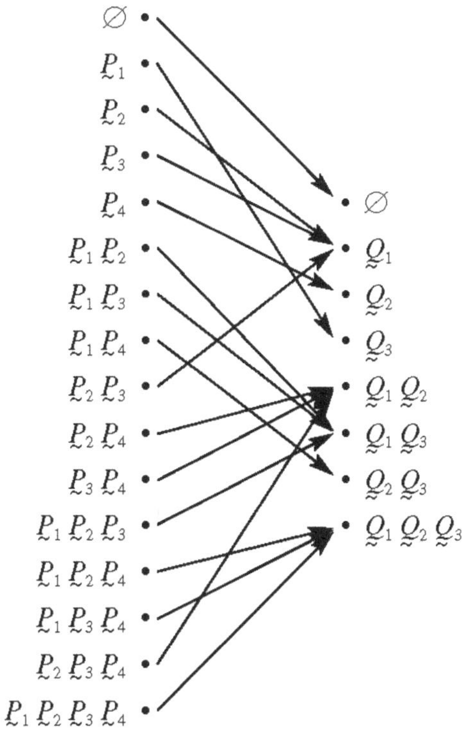

Being;

$\underset{\sim}{P}_1$ =

a	b	c
.3	.5	.8

$\underset{\sim}{P}_2$ =

a	b	c
.6	.4	.9

$\underset{\sim}{P}_3$ =

a	b	c
.6	.4	.9

$\underset{\sim}{P}_4$ =

a	b	c
.7	.8	.2

We will have:

$$\varnothing = \Gamma\,\varnothing$$

$\{\underset{\sim}{Q}_1\} = \vee\left\{\{\underset{\sim}{P}_2\}, \{\underset{\sim}{P}_3\}, \{\underset{\sim}{P}_2, \underset{\sim}{P}_3\}\right\} = \{\underset{\sim}{P}_2, \underset{\sim}{P}_3\}$ =

a	b	c
.6	.4	.9

$\{\underset{\sim}{Q}_2\} = \{\underset{\sim}{P}_4\}$ =

a	b	c
.7	.8	.2

$$\{\underset{\sim}{Q}_3\} = \{\underset{\sim}{P}_1\} = \begin{array}{|c|c|c|} \multicolumn{1}{c}{a} & \multicolumn{1}{c}{b} & \multicolumn{1}{c}{c} \\ \hline .3 & .5 & .8 \\ \hline \end{array}$$

$$\{\underset{\sim}{Q}_1, \underset{\sim}{Q}_2\} = \bigvee \{\{\underset{\sim}{P}_2, \underset{\sim}{P}_4\}, \{\underset{\sim}{P}_3, \underset{\sim}{P}_4\}, \{\underset{\sim}{P}_2, \underset{\sim}{P}_3, \underset{\sim}{P}_4\}\} = \{\underset{\sim}{P}_2, \underset{\sim}{P}_3, \underset{\sim}{P}_4\}$$

$$= \left\{ \begin{array}{|c|c|c|} \multicolumn{1}{c}{a} & \multicolumn{1}{c}{b} & \multicolumn{1}{c}{c} \\ \hline .6 & .4 & .9 \\ \hline \end{array} , \begin{array}{|c|c|c|} \multicolumn{1}{c}{a} & \multicolumn{1}{c}{b} & \multicolumn{1}{c}{c} \\ \hline .7 & .8 & .2 \\ \hline \end{array} \right\}$$

$$\{\underset{\sim}{Q}_1, \underset{\sim}{Q}_3\} = \bigvee \{\{\underset{\sim}{P}_1, \underset{\sim}{P}_2\}, \{\underset{\sim}{P}_1, \underset{\sim}{P}_3\}, \{\underset{\sim}{P}_1, \underset{\sim}{P}_2, \underset{\sim}{P}_3\}\} = \{\underset{\sim}{P}_1, \underset{\sim}{P}_2, \underset{\sim}{P}_3\}$$

$$= \left\{ \begin{array}{|c|c|c|} \multicolumn{1}{c}{a} & \multicolumn{1}{c}{b} & \multicolumn{1}{c}{c} \\ \hline .3 & .5 & .8 \\ \hline \end{array} , \begin{array}{|c|c|c|} \multicolumn{1}{c}{a} & \multicolumn{1}{c}{b} & \multicolumn{1}{c}{c} \\ \hline .6 & .4 & .9 \\ \hline \end{array} \right\}$$

$$\{\underset{\sim}{Q}_2, \underset{\sim}{Q}_3\} = \{\underset{\sim}{P}_1, \underset{\sim}{P}_4\} = \left\{ \begin{array}{|c|c|c|} \multicolumn{1}{c}{a} & \multicolumn{1}{c}{b} & \multicolumn{1}{c}{c} \\ \hline .3 & .5 & .8 \\ \hline \end{array} , \begin{array}{|c|c|c|} \multicolumn{1}{c}{a} & \multicolumn{1}{c}{b} & \multicolumn{1}{c}{c} \\ \hline .7 & .8 & .2 \\ \hline \end{array} \right\}$$

$$\{\underset{\sim}{Q}_1, \underset{\sim}{Q}_2, \underset{\sim}{Q}_3\} = \bigvee \{\{\underset{\sim}{P}_1, \underset{\sim}{P}_2, \underset{\sim}{P}_4\}, \{\underset{\sim}{P}_1, \underset{\sim}{P}_3, \underset{\sim}{P}_4\}, \{\underset{\sim}{P}_1, \underset{\sim}{P}_2, \underset{\sim}{P}_3, \underset{\sim}{P}_4\}\} = \{\underset{\sim}{P}_1, \underset{\sim}{P}_2, \underset{\sim}{P}_3, \underset{\sim}{P}_4\}$$

$$= \left\{ \begin{array}{|c|c|c|} \multicolumn{1}{c}{a} & \multicolumn{1}{c}{b} & \multicolumn{1}{c}{c} \\ \hline .3 & .5 & .8 \\ \hline \end{array} , \begin{array}{|c|c|c|} \multicolumn{1}{c}{a} & \multicolumn{1}{c}{b} & \multicolumn{1}{c}{c} \\ \hline .6 & .4 & .9 \\ \hline \end{array} , \begin{array}{|c|c|c|} \multicolumn{1}{c}{a} & \multicolumn{1}{c}{b} & \multicolumn{1}{c}{c} \\ \hline .7 & .8 & .2 \\ \hline \end{array} \right\}$$

which does nothing more than contrast the validity of the established concept.

In a general supposition in which all fuzzy subsets $\underset{\sim}{P}_j$ are distinct among themselves, the definition established for the induction of the elements of the "power set" of E_2, $Q(E_2)$ remains valid. Let us see this in an example in which we will consider the same referentials and the same functional application used up to now, only changing the fuzzy subset $\underset{\sim}{P}_3$ to make it distinct to $\underset{\sim}{P}_2$. In this way we begin from the following fuzzy subsets:

$$\underset{\sim}{P}_1 = \begin{array}{|c|c|c|} \multicolumn{1}{c}{a} & \multicolumn{1}{c}{b} & \multicolumn{1}{c}{c} \\ \hline .3 & .5 & .8 \\ \hline \end{array} \quad \underset{\sim}{P}_2 = \begin{array}{|c|c|c|} \multicolumn{1}{c}{a} & \multicolumn{1}{c}{b} & \multicolumn{1}{c}{c} \\ \hline .6 & .4 & .9 \\ \hline \end{array} \quad \underset{\sim}{P}_3 = \begin{array}{|c|c|c|} \multicolumn{1}{c}{a} & \multicolumn{1}{c}{b} & \multicolumn{1}{c}{c} \\ \hline .6 & .1 & .4 \\ \hline \end{array} \quad \underset{\sim}{P}_4 = \begin{array}{|c|c|c|} \multicolumn{1}{c}{a} & \multicolumn{1}{c}{b} & \multicolumn{1}{c}{c} \\ \hline .7 & .8 & .2 \\ \hline \end{array}$$

We have:

$$\varnothing = \Gamma \varnothing$$

$$\{\underset{\sim}{Q}_1\} = \bigvee \{\{\underset{\sim}{P}_2\}, \{\underset{\sim}{P}_3\}, \{\underset{\sim}{P}_2, \underset{\sim}{P}_3\}\} = \{\underset{\sim}{P}_2, \underset{\sim}{P}_3\} = \left\{ \begin{array}{|c|c|c|} \multicolumn{1}{c}{a} & \multicolumn{1}{c}{b} & \multicolumn{1}{c}{c} \\ \hline .6 & .4 & .9 \\ \hline \end{array} , \begin{array}{|c|c|c|} \multicolumn{1}{c}{a} & \multicolumn{1}{c}{b} & \multicolumn{1}{c}{c} \\ \hline .8 & .1 & .4 \\ \hline \end{array} \right\}$$

$$\{\underset{\sim}{Q}_2\} = \{\underset{\sim}{P}_4\} = \begin{array}{|c|c|c|}\multicolumn{1}{c}{a}&\multicolumn{1}{c}{b}&\multicolumn{1}{c}{c}\\\hline .7&.8&.2\\\hline\end{array}$$

$$\{\underset{\sim}{Q}_3\} = \{\underset{\sim}{P}_1\} = \begin{array}{|c|c|c|}\multicolumn{1}{c}{a}&\multicolumn{1}{c}{b}&\multicolumn{1}{c}{c}\\\hline .3&.5&.8\\\hline\end{array}$$

$$\{\underset{\sim}{Q}_1, \underset{\sim}{Q}_2\} = \vee \left\{ \{\underset{\sim}{P}_2, \underset{\sim}{P}_4\}, \{\underset{\sim}{P}_3, \underset{\sim}{P}_4\}, \{\underset{\sim}{P}_2, \underset{\sim}{P}_3, \underset{\sim}{P}_4\} \right\} = \{\underset{\sim}{P}_2, \underset{\sim}{P}_3, \underset{\sim}{P}_4\}$$

$$= \left\{ \begin{array}{|c|c|c|}\multicolumn{1}{c}{a}&\multicolumn{1}{c}{b}&\multicolumn{1}{c}{c}\\\hline .6&.4&.9\\\hline\end{array}, \begin{array}{|c|c|c|}\multicolumn{1}{c}{a}&\multicolumn{1}{c}{b}&\multicolumn{1}{c}{c}\\\hline .8&.1&.4\\\hline\end{array}, \begin{array}{|c|c|c|}\multicolumn{1}{c}{a}&\multicolumn{1}{c}{b}&\multicolumn{1}{c}{c}\\\hline .7&.8&.2\\\hline\end{array} \right\}$$

$$\{\underset{\sim}{Q}_1, \underset{\sim}{Q}_3\} = \vee \left\{ \{\underset{\sim}{P}_1, \underset{\sim}{P}_2\}, \{\underset{\sim}{P}_1, \underset{\sim}{P}_3\}, \{\underset{\sim}{P}_1, \underset{\sim}{P}_2, \underset{\sim}{P}_3\} \right\} = \{\underset{\sim}{P}_1, \underset{\sim}{P}_2, \underset{\sim}{P}_3\}$$

$$= \left\{ \begin{array}{|c|c|c|}\multicolumn{1}{c}{a}&\multicolumn{1}{c}{b}&\multicolumn{1}{c}{c}\\\hline .3&.5&.8\\\hline\end{array}, \begin{array}{|c|c|c|}\multicolumn{1}{c}{a}&\multicolumn{1}{c}{b}&\multicolumn{1}{c}{c}\\\hline .6&.4&.9\\\hline\end{array}, \begin{array}{|c|c|c|}\multicolumn{1}{c}{a}&\multicolumn{1}{c}{b}&\multicolumn{1}{c}{c}\\\hline .8&.1&.4\\\hline\end{array} \right\}$$

$$\{\underset{\sim}{Q}_2, \underset{\sim}{Q}_3\} = \{\underset{\sim}{P}_1, \underset{\sim}{P}_4\}\} = \left\{ \begin{array}{|c|c|c|}\multicolumn{1}{c}{a}&\multicolumn{1}{c}{b}&\multicolumn{1}{c}{c}\\\hline .3&.5&.8\\\hline\end{array}, \begin{array}{|c|c|c|}\multicolumn{1}{c}{a}&\multicolumn{1}{c}{b}&\multicolumn{1}{c}{c}\\\hline .7&.8&.2\\\hline\end{array} \right\}$$

$$\{\underset{\sim}{Q}_1, \underset{\sim}{Q}_2, \underset{\sim}{Q}_3\} = \vee \left\{ \{\underset{\sim}{P}_1, \underset{\sim}{P}_2, \underset{\sim}{P}_4\}, \{\underset{\sim}{P}_1, \underset{\sim}{P}_3, \underset{\sim}{P}_4\}, \{\underset{\sim}{P}_1, \underset{\sim}{P}_2, \underset{\sim}{P}_3, \underset{\sim}{P}_4\} \right\} = \{\underset{\sim}{P}_1, \underset{\sim}{P}_2, \underset{\sim}{P}_3, \underset{\sim}{P}_4\}$$

$$= \left\{ \begin{array}{|c|c|c|}\multicolumn{1}{c}{a}&\multicolumn{1}{c}{b}&\multicolumn{1}{c}{c}\\\hline .3&.5&.8\\\hline\end{array}, \begin{array}{|c|c|c|}\multicolumn{1}{c}{a}&\multicolumn{1}{c}{b}&\multicolumn{1}{c}{c}\\\hline .6&.4&.9\\\hline\end{array}, \begin{array}{|c|c|c|}\multicolumn{1}{c}{a}&\multicolumn{1}{c}{b}&\multicolumn{1}{c}{c}\\\hline .8&.1&.4\\\hline\end{array}, \begin{array}{|c|c|c|}\multicolumn{1}{c}{a}&\multicolumn{1}{c}{b}&\multicolumn{1}{c}{c}\\\hline .7&.8&.2\\\hline\end{array} \right\}$$

We consider that the generation of elements of the referential or of the "power set" by the direct and inverse functional applications has been sufficiently shown with these simple examples.

We will now move on to state some simple properties which we centre on in this second focus, for when we consider their adaptation to the supposition of referentials which are those of the fuzzy subsets as immediate.

1. Given a functional application of $P(E_1)$ in $Q(E_2)$, $\mathbf{\Gamma}$, the following is fulfilled:

$\forall \underset{\sim}{B}_k \in Q(E_2)$:

$$\Gamma^{-1} \overline{\underset{\sim}{B}_k} \subset \overline{\Gamma^{-1} \underset{\sim}{B}_k}$$

We will confirm this in the previous application:

$$\Gamma^{-1}\,\overline{\varnothing} = \Gamma^{-1}\{\underset{\sim}{Q}_1, \underset{\sim}{Q}_2, \underset{\sim}{Q}_3\} = \{\underset{\sim}{P}_1, \underset{\sim}{P}_2, \underset{\sim}{P}_3, \underset{\sim}{P}_4\} \qquad \overline{\Gamma^{-1}\varnothing} = \{\underset{\sim}{P}_1, \underset{\sim}{P}_2, \underset{\sim}{P}_3, \underset{\sim}{P}_4\}$$

$$\Gamma^{-1}\,\overline{\{\underset{\sim}{Q}_1\}} = \Gamma^{-1}\{\underset{\sim}{Q}_2, \underset{\sim}{Q}_3\} = \{\underset{\sim}{P}_1, \underset{\sim}{P}_4\} \qquad \overline{\Gamma^{-1}\{\underset{\sim}{Q}_1\}} = \overline{\{\underset{\sim}{P}_2, \underset{\sim}{P}_3\}} = \{\underset{\sim}{P}_1, \underset{\sim}{P}_4\}$$

$$\Gamma^{-1}\,\overline{\{\underset{\sim}{Q}_2\}} = \Gamma^{-1}\{\underset{\sim}{Q}_1, \underset{\sim}{Q}_3\} = \{\underset{\sim}{P}_1, \underset{\sim}{P}_2, \underset{\sim}{P}_3\} \qquad \overline{\Gamma^{-1}\{\underset{\sim}{Q}_2\}} = \overline{\{\underset{\sim}{P}_4\}} = \{\underset{\sim}{P}_1, \underset{\sim}{P}_2, \underset{\sim}{P}_3\}$$

$$\Gamma^{-1}\,\overline{\underset{\sim}{Q}_3} = \Gamma^{-1}\{\underset{\sim}{Q}_1, \underset{\sim}{Q}_2\} = \{\underset{\sim}{P}_2, \underset{\sim}{P}_3, \underset{\sim}{P}_4\} \qquad \overline{\Gamma^{-1}\underset{\sim}{Q}_3} = \overline{\{\underset{\sim}{P}_1\}} = \{\underset{\sim}{P}_2, \underset{\sim}{P}_3, \underset{\sim}{P}_4\}$$

$$\Gamma^{-1}\,\overline{\{\underset{\sim}{Q}_1, \underset{\sim}{Q}_2\}} = \Gamma^{-1}\{\underset{\sim}{Q}_3\} = \{\underset{\sim}{P}_1\} \qquad \overline{\Gamma^{-1}\{\underset{\sim}{Q}_1, \underset{\sim}{Q}_2\}} = \overline{\{\underset{\sim}{P}_2, \underset{\sim}{P}_3, \underset{\sim}{P}_4\}} = \{\underset{\sim}{P}_1\}$$

$$\Gamma^{-1}\,\overline{\{\underset{\sim}{Q}_1, \underset{\sim}{Q}_3\}} = \Gamma^{-1}\{\underset{\sim}{Q}_2\} = \{\underset{\sim}{P}_4\} \qquad \overline{\Gamma^{-1}\{\underset{\sim}{Q}_1, \underset{\sim}{Q}_3\}} = \overline{\{\underset{\sim}{P}_1, \underset{\sim}{P}_2, \underset{\sim}{P}_3\}} = \{\underset{\sim}{P}_4\}$$

$$\Gamma^{-1}\,\overline{\{\underset{\sim}{Q}_2, \underset{\sim}{Q}_3\}} = \Gamma^{-1}\{\underset{\sim}{Q}_1\} = \{\underset{\sim}{P}_2, \underset{\sim}{P}_3\} \qquad \overline{\Gamma^{-1}\{\underset{\sim}{Q}_2, \underset{\sim}{Q}_3\}} = \overline{\{\underset{\sim}{P}_1, \underset{\sim}{P}_4\}} = \{\underset{\sim}{P}_2, \underset{\sim}{P}_3\}$$

$$\Gamma^{-1}\,\overline{\{\underset{\sim}{Q}_1, \underset{\sim}{Q}_2, \underset{\sim}{Q}_3\}} = \Gamma^{-1}\varnothing = \varnothing \qquad \overline{\Gamma^{-1}\{\underset{\sim}{Q}_1, \underset{\sim}{Q}_2, \underset{\sim}{Q}_3\}} = \overline{\{\underset{\sim}{P}_1, \underset{\sim}{P}_2, \underset{\sim}{P}_3, \underset{\sim}{P}_4\}} = \varnothing$$

The fulfillment of the proposed property is confirmed.

2. If we have a functional application $\boldsymbol{\Gamma}$ of $P(E_1)$ in $Q(E_2)$, the following is fulfilled:

$\forall \underset{\sim}{A}_j \in P(E_1)$:

$$\Gamma\,\overline{\underset{\sim}{A}_j} \supset \overline{\Gamma\,\underset{\sim}{A}_j}$$

We move on to the corresponding confirmation in our repeated example.

$$\Gamma\,\overline{\varnothing} = \Gamma\,E = \{\underset{\sim}{Q}_1, \underset{\sim}{Q}_2, \underset{\sim}{Q}_3\} \qquad \overline{\Gamma\varnothing} = \overline{\varnothing} = \{\underset{\sim}{Q}_1, \underset{\sim}{Q}_2, \underset{\sim}{Q}_3\}$$

$$\Gamma\,\overline{\{\underset{\sim}{P}_1\}} = \Gamma\,\{\underset{\sim}{P}_2, \underset{\sim}{P}_3, \underset{\sim}{P}_4\} = \{\underset{\sim}{Q}_1, \underset{\sim}{Q}_2\} \qquad \overline{\Gamma\,\{\underset{\sim}{P}_1\}} = \overline{\{\underset{\sim}{Q}_3\}} = \{\underset{\sim}{Q}_1, \underset{\sim}{Q}_2\}$$

$$\Gamma\,\overline{\{\underset{\sim}{P}_2\}} = \Gamma\,\{\underset{\sim}{P}_1, \underset{\sim}{P}_3, \underset{\sim}{P}_4\} = \{\underset{\sim}{Q}_1, \underset{\sim}{Q}_2, \underset{\sim}{Q}_3\} \qquad \overline{\Gamma\,\{\underset{\sim}{P}_2\}} = \overline{\{\underset{\sim}{Q}_1\}} = \{\underset{\sim}{Q}_2, \underset{\sim}{Q}_3\}$$

$$\Gamma\,\overline{\{\underset{\sim}{P}_3\}} = \Gamma\,\{\underset{\sim}{P}_1, \underset{\sim}{P}_2, \underset{\sim}{P}_4\} = \{\underset{\sim}{Q}_1, \underset{\sim}{Q}_2, \underset{\sim}{Q}_3\} \qquad \overline{\Gamma\,\{\underset{\sim}{P}_3\}} = \overline{\{\underset{\sim}{Q}_1\}} = \{\underset{\sim}{Q}_2, \underset{\sim}{Q}_3\}$$

$$\Gamma\,\overline{\{\underset{\sim}{P}_4\}} = \Gamma\,\{\underset{\sim}{P}_1, \underset{\sim}{P}_2, \underset{\sim}{P}_3\} = \{\underset{\sim}{Q}_1, \underset{\sim}{Q}_3\} \qquad \overline{\Gamma\,\{\underset{\sim}{P}_4\}} = \overline{\{\underset{\sim}{Q}_2\}} = \{\underset{\sim}{Q}_1, \underset{\sim}{Q}_3\}$$

$$\Gamma\,\overline{\{\underset{\sim}{P}_1, \underset{\sim}{P}_2\}} = \Gamma\,\{\underset{\sim}{P}_3, \underset{\sim}{P}_4\} = \{\underset{\sim}{Q}_1, \underset{\sim}{Q}_2\} \qquad \overline{\Gamma\,\{\underset{\sim}{P}_1, \underset{\sim}{P}_2\}} = \overline{\{\underset{\sim}{Q}_1, \underset{\sim}{Q}_3\}} = \{\underset{\sim}{Q}_2\}$$

$$\Gamma\,\overline{\{\underset{\sim}{P}_1, \underset{\sim}{P}_3\}} = \Gamma\,\{\underset{\sim}{P}_2, \underset{\sim}{P}_4\} = \{\underset{\sim}{Q}_1, \underset{\sim}{Q}_2\} \qquad \overline{\Gamma\,\{\underset{\sim}{P}_1, \underset{\sim}{P}_3\}} = \overline{\{\underset{\sim}{Q}_1, \underset{\sim}{Q}_3\}} = \{\underset{\sim}{Q}_2\}$$

$$\Gamma\,\overline{\{\underset{\sim}{P}_1, \underset{\sim}{P}_4\}} = \Gamma\,\{\underset{\sim}{P}_2, \underset{\sim}{P}_3\} = \{\underset{\sim}{Q}_1\} \qquad \overline{\Gamma\,\{\underset{\sim}{P}_1, \underset{\sim}{P}_4\}} = \overline{\{\underset{\sim}{Q}_2, \underset{\sim}{Q}_3\}} = \{\underset{\sim}{Q}_1\}$$

$$\Gamma\,\overline{\{\underset{\sim}{P}_2, \underset{\sim}{P}_3\}} = \Gamma\,\{\underset{\sim}{P}_1, \underset{\sim}{P}_4\} = \{\underset{\sim}{Q}_2, \underset{\sim}{Q}_3\} \qquad \overline{\Gamma\,\{\underset{\sim}{P}_2, \underset{\sim}{P}_3\}} = \overline{\{\underset{\sim}{Q}_1\}} = \{\underset{\sim}{Q}_2, \underset{\sim}{Q}_3\}$$

$$\Gamma\,\overline{\{\underset{\sim}{P}_2, \underset{\sim}{P}_4\}} = \Gamma\,\{\underset{\sim}{P}_1, \underset{\sim}{P}_3\} = \{\underset{\sim}{Q}_1, \underset{\sim}{Q}_3\} \qquad \overline{\Gamma\,\{\underset{\sim}{P}_2, \underset{\sim}{P}_4\}} = \overline{\{\underset{\sim}{Q}_1, \underset{\sim}{Q}_2\}} = \{\underset{\sim}{Q}_3\}$$

$$\Gamma\,\overline{\{\underset{\sim}{P}_3, \underset{\sim}{P}_4\}} = \Gamma\,\{\underset{\sim}{P}_1, \underset{\sim}{P}_2\} = \{\underset{\sim}{Q}_1, \underset{\sim}{Q}_3\} \qquad \overline{\Gamma\,\{\underset{\sim}{P}_3, \underset{\sim}{P}_4\}} = \overline{\{\underset{\sim}{Q}_1, \underset{\sim}{Q}_2\}} = \{\underset{\sim}{Q}_3\}$$

$$\Gamma\,\overline{\{\underset{\sim}{P}_1, \underset{\sim}{P}_2, \underset{\sim}{P}_3\}} = \Gamma\,\{\underset{\sim}{P}_4\} = \{\underset{\sim}{Q}_2\} \qquad \overline{\Gamma\,\{\underset{\sim}{P}_1, \underset{\sim}{P}_2, \underset{\sim}{P}_3\}} = \overline{\{\underset{\sim}{Q}_1, \underset{\sim}{Q}_3\}} = \{\underset{\sim}{Q}_2\}$$

$\Gamma\,\{\overline{\underset{\sim}{P}_1, \underset{\sim}{P}_2, \underset{\sim}{P}_4}\} = \Gamma\,\{\underset{\sim}{P}_3\} = \{\underset{\sim}{Q}_1\}$

$\Gamma\,\{\overline{\underset{\sim}{P}_1, \underset{\sim}{P}_3, \underset{\sim}{P}_4}\} = \Gamma\,\{\underset{\sim}{P}_2\} = \{\underset{\sim}{Q}_1\}$

$\Gamma\,\{\overline{\underset{\sim}{P}_2, \underset{\sim}{P}_3, \underset{\sim}{P}_4}\} = \Gamma\,\{\underset{\sim}{P}_1\} = \{\underset{\sim}{Q}_3\}$

$\Gamma\,\{\overline{\underset{\sim}{P}_1, \underset{\sim}{P}_2, \underset{\sim}{P}_3, \underset{\sim}{P}_4}\} = \Gamma\,\varnothing = \varnothing$

$\overline{\Gamma\,\{\underset{\sim}{P}_1, \underset{\sim}{P}_2, \underset{\sim}{P}_4\}} = \overline{\{\underset{\sim}{Q}_1, \underset{\sim}{Q}_3, \underset{\sim}{Q}_3\}} = \varnothing$

$\overline{\Gamma\,\{\underset{\sim}{P}_1, \underset{\sim}{P}_3, \underset{\sim}{P}_4\}} = \overline{\{\underset{\sim}{Q}_1, \underset{\sim}{Q}_3, \underset{\sim}{Q}_3\}} = \varnothing$

$\overline{\Gamma\,\{\underset{\sim}{P}_2, \underset{\sim}{P}_3, \underset{\sim}{P}_4\}} = \overline{\{\underset{\sim}{Q}_1, \underset{\sim}{Q}_2\}} = \{\underset{\sim}{Q}_3\}$

$\overline{\Gamma\,\{\underset{\sim}{P}_1, \underset{\sim}{P}_2, \underset{\sim}{P}_3, \underset{\sim}{P}_4\}} = \overline{\{\underset{\sim}{Q}_1, \underset{\sim}{Q}_2, \underset{\sim}{Q}_3\}} = \varnothing$

3. Once again we consider the two referentials E_1 and E_2 with their corresponding "power sets". The following is fulfilled:

$\forall\, \underset{\sim}{B}_k, \underset{\sim}{B}_l \in Q(E_2)$:

$$(\underset{\sim}{B}_k \subset \underset{\sim}{B}_l) \Rightarrow (\Gamma^{-1}\underset{\sim}{B}_k \subset \Gamma^{-1}\underset{\sim}{B}_l)$$

We move on to the corresponding confirmation.

$\{\underset{\sim}{Q}_1\} \subset \{\underset{\sim}{Q}_1, \underset{\sim}{Q}_2\}$

$\{\underset{\sim}{Q}_1\} \subset \{\underset{\sim}{Q}_1, \underset{\sim}{Q}_3\}$

$\{\underset{\sim}{Q}_1\} \subset \{\underset{\sim}{Q}_1, \underset{\sim}{Q}_2, \underset{\sim}{Q}_3\}$

$\{\underset{\sim}{Q}_2\} \subset \{\underset{\sim}{Q}_1, \underset{\sim}{Q}_2\}$

$\{\underset{\sim}{Q}_2\} \subset \{\underset{\sim}{Q}_2, \underset{\sim}{Q}_3\}$

$\{\underset{\sim}{Q}_2\} \subset \{\underset{\sim}{Q}_1, \underset{\sim}{Q}_2, \underset{\sim}{Q}_3\}$

$\{\underset{\sim}{Q}_3\} \subset \{\underset{\sim}{Q}_1, \underset{\sim}{Q}_3\}$

$\{\underset{\sim}{Q}_3\} \subset \{\underset{\sim}{Q}_2, \underset{\sim}{Q}_3\}$

$\{\underset{\sim}{Q}_3\} \subset \{\underset{\sim}{Q}_1, \underset{\sim}{Q}_2, \underset{\sim}{Q}_3\}$

$\{\underset{\sim}{Q}_1, \underset{\sim}{Q}_2\} \subset \{\underset{\sim}{Q}_1, \underset{\sim}{Q}_2, \underset{\sim}{Q}_3\}$

$\{\underset{\sim}{Q}_1, \underset{\sim}{Q}_3\} \subset \{\underset{\sim}{Q}_1, \underset{\sim}{Q}_2, \underset{\sim}{Q}_3\}$

$\{\underset{\sim}{Q}_2, \underset{\sim}{Q}_3\} \subset \{\underset{\sim}{Q}_1, \underset{\sim}{Q}_2, \underset{\sim}{Q}_3\}$

$\{\underset{\sim}{P}_2, \underset{\sim}{P}_3\} \subset \{\underset{\sim}{P}_2, \underset{\sim}{P}_3, \underset{\sim}{P}_4\}$

$\{\underset{\sim}{P}_2, \underset{\sim}{P}_3\} \subset \{\underset{\sim}{P}_1, \underset{\sim}{P}_2, \underset{\sim}{P}_3\}$

$\{\underset{\sim}{P}_2, \underset{\sim}{P}_3\} \subset \{\underset{\sim}{P}_1, \underset{\sim}{P}_2, \underset{\sim}{P}_3, \underset{\sim}{P}_4\}$

$\{\underset{\sim}{P}_4\} \subset \{\underset{\sim}{P}_2, \underset{\sim}{P}_3, \underset{\sim}{P}_4\}$

$\{\underset{\sim}{P}_4\} \subset \{\underset{\sim}{P}_1, \underset{\sim}{P}_4\}$

$\{\underset{\sim}{P}_4\} \subset \{\underset{\sim}{P}_1, \underset{\sim}{P}_2, \underset{\sim}{P}_3, \underset{\sim}{P}_4\}$

$\{\underset{\sim}{P}_1\} \subset \{\underset{\sim}{P}_1, \underset{\sim}{P}_2, \underset{\sim}{P}_3\}$

$\{\underset{\sim}{P}_1\} \subset \{\underset{\sim}{P}_1, \underset{\sim}{P}_4\}$

$\{\underset{\sim}{P}_1\} \subset \{\underset{\sim}{P}_1, \underset{\sim}{P}_2, \underset{\sim}{P}_3, \underset{\sim}{P}_4\}$

$\{\underset{\sim}{P}_2, \underset{\sim}{P}_3, \underset{\sim}{P}_4\} \subset \{\underset{\sim}{P}_1, \underset{\sim}{P}_2, \underset{\sim}{P}_3, \underset{\sim}{P}_4\}$

$\{\underset{\sim}{P}_1, \underset{\sim}{P}_2, \underset{\sim}{P}_3\} \subset \{\underset{\sim}{P}_1, \underset{\sim}{P}_2, \underset{\sim}{P}_3, \underset{\sim}{P}_4\}$

$\{\underset{\sim}{P}_1, \underset{\sim}{P}_4\} \subset \{\underset{\sim}{P}_1, \underset{\sim}{P}_2, \underset{\sim}{P}_3, \underset{\sim}{P}_4\}$

Once again the verification confirms the stated property.

4. Given a functional application $\mathbf{\Gamma}$ of $P(E_1)$ in $Q(E_2)$:

$\forall\, \underset{\sim}{A}_j \in P(E_1)$:

$$\underset{\sim}{A}_j \subset \Gamma^{-1}\,(\Gamma\,\underset{\sim}{A}_j)$$

We use the same functional application to contrast this property. We have:

$$\{\underset{\sim}{P}_1\} \subset \Gamma^{-1}\{\underset{\sim}{Q}_3\} = \{\underset{\sim}{P}_1\}$$
$$\{\underset{\sim}{P}_2\} \subset \Gamma^{-1}\{\underset{\sim}{Q}_1\} = \{\underset{\sim}{P}_2, \underset{\sim}{P}_3\}$$
$$\{\underset{\sim}{P}_3\} \subset \Gamma^{-1}\{\underset{\sim}{Q}_1\} = \{\underset{\sim}{P}_2, \underset{\sim}{P}_3\}$$
$$\{\underset{\sim}{P}_4\} \subset \Gamma^{-1}\{\underset{\sim}{Q}_2\} = \{\underset{\sim}{P}_4\}$$

$$\{\underset{\sim}{P}_1, \underset{\sim}{P}_2\} \subset \Gamma^{-1}\{\underset{\sim}{Q}_1, \underset{\sim}{Q}_3\} = \{\underset{\sim}{P}_1, \underset{\sim}{P}_2, \underset{\sim}{P}_3\}$$
$$\{\underset{\sim}{P}_1, \underset{\sim}{P}_3\} \subset \Gamma^{-1}\{\underset{\sim}{Q}_1, \underset{\sim}{Q}_3\} = \{\underset{\sim}{P}_1, \underset{\sim}{P}_2, \underset{\sim}{P}_3\}$$
$$\{\underset{\sim}{P}_1, \underset{\sim}{P}_4\} \subset \Gamma^{-1}\{\underset{\sim}{Q}_2, \underset{\sim}{Q}_3\} = \{\underset{\sim}{P}_1, \underset{\sim}{P}_4\}$$
$$\{\underset{\sim}{P}_2, \underset{\sim}{P}_3\} \subset \Gamma^{-1}\{\underset{\sim}{Q}_1\} = \{\underset{\sim}{P}_2, \underset{\sim}{P}_3\}$$
$$\{\underset{\sim}{P}_2, \underset{\sim}{P}_4\} \subset \Gamma^{-1}\{\underset{\sim}{Q}_1, \underset{\sim}{Q}_2\} = \{\underset{\sim}{P}_2, \underset{\sim}{P}_3, \underset{\sim}{P}_4\}$$
$$\{\underset{\sim}{P}_3, \underset{\sim}{P}_4\} \subset \Gamma^{-1}\{\underset{\sim}{Q}_1, \underset{\sim}{Q}_2\} = \{\underset{\sim}{P}_2, \underset{\sim}{P}_3, \underset{\sim}{P}_4\}$$

$$\{\underset{\sim}{P}_1, \underset{\sim}{P}_2, \underset{\sim}{P}_3\} \subset \Gamma^{-1}\{\underset{\sim}{Q}_1, \underset{\sim}{Q}_3\} = \{\underset{\sim}{P}_1, \underset{\sim}{P}_2, \underset{\sim}{P}_3\}$$
$$\{\underset{\sim}{P}_1, \underset{\sim}{P}_2, \underset{\sim}{P}_4\} \subset \Gamma^{-1}\{\underset{\sim}{Q}_1, \underset{\sim}{Q}_2, \underset{\sim}{Q}_3\} = \{\underset{\sim}{P}_1, \underset{\sim}{P}_2, \underset{\sim}{P}_3, \underset{\sim}{P}_4\}$$
$$\{\underset{\sim}{P}_1, \underset{\sim}{P}_3, \underset{\sim}{P}_4\} \subset \Gamma^{-1}\{\underset{\sim}{Q}_1, \underset{\sim}{Q}_2, \underset{\sim}{Q}_3\} = \{\underset{\sim}{P}_1, \underset{\sim}{P}_2, \underset{\sim}{P}_3, \underset{\sim}{P}_4\}$$
$$\{\underset{\sim}{P}_2, \underset{\sim}{P}_3, \underset{\sim}{P}_4\} \subset \Gamma^{-1}\{\underset{\sim}{Q}_1, \underset{\sim}{Q}_2\} = \{\underset{\sim}{P}_2, \underset{\sim}{P}_3, \underset{\sim}{P}_4\}$$

$$\{\underset{\sim}{P}_1, \underset{\sim}{P}_2, \underset{\sim}{P}_3, \underset{\sim}{P}_4\} \subset \Gamma^{-1}\{\underset{\sim}{Q}_1, \underset{\sim}{Q}_2, \underset{\sim}{Q}_3\} = \{\underset{\sim}{P}_1, \underset{\sim}{P}_2, \underset{\sim}{P}_3, \underset{\sim}{P}_4\}$$

The property is fulfilled for all the elements of the "power set" $P(E_1)$.

We have put forward only a few of the many properties which allow the construction of the "power set" $Q(E_2)$ as induction of the functional application $\boldsymbol{\Gamma}$ upon choosing the greater of the elements of the "power set" of $P(E_1)$ as an image of its elements.

7 Continuous Functional Application in Uncertainty

In continuation we will see an interesting concept which may open interesting **perspectives**. We are refering to the notion of "fuzzy-continuous functional application", also called "*F*-continuous" (fuzzy-continuous).

It is said that an application $\boldsymbol{\Gamma}$ of E_1 in E_2 is "F-continuous" if the inverse of all open-$\boldsymbol{\Gamma}_2$ in E_2 is an open-$\boldsymbol{\Gamma}_1$ in E_1.

The amplitude of this definition allows efficient developments as much from the focus from one type of referentials as from the other. In continuation we will see the form which the continuous functional application acquires in both cases. For this we will be using simple but sufficiently revealing examples.

A) In the first focus two referentials E_1 and E_2 are considered, such as the following:

$$E_1 = \{a, b, c, d\}$$
$$E_2 = \{\alpha, \beta, \gamma\}$$

and the existence of a functional application $\mathbf{\Gamma}$ of E_1 in E_2 is assumed:

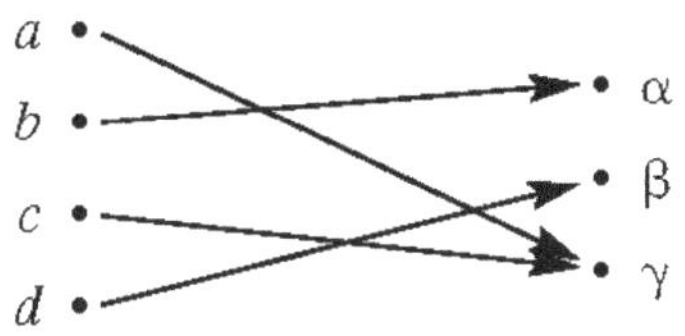

In the same way an uncertain topology E_1, $T_1(E_1)$ and an uncertain topology E_2, $T_2(E_2)$ are taken, which we represent through the following lattices:

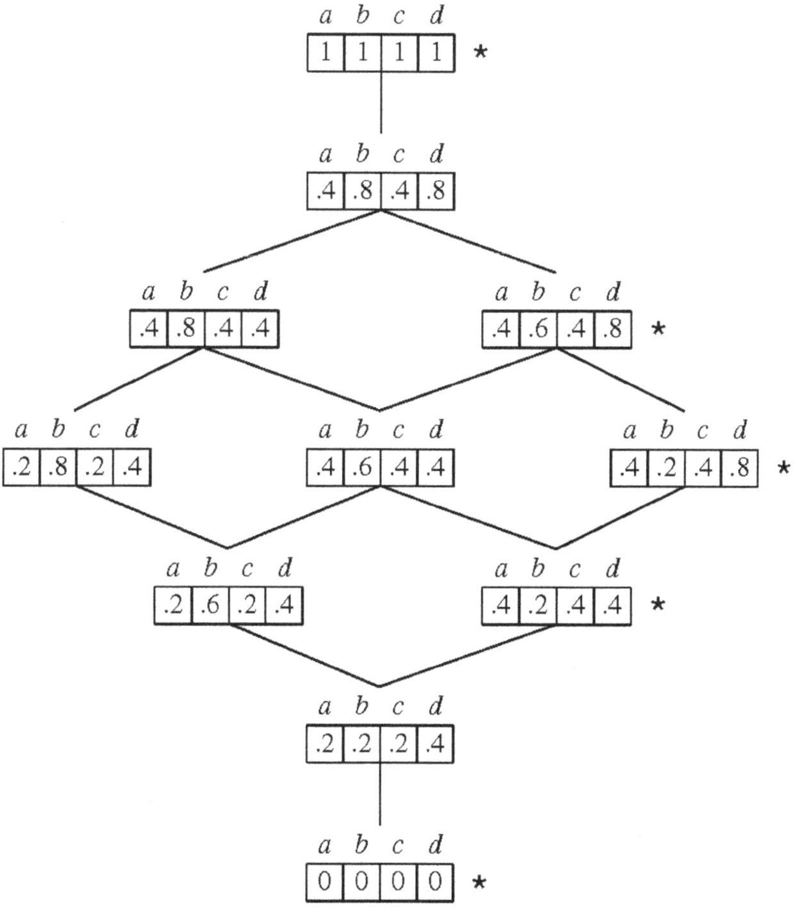

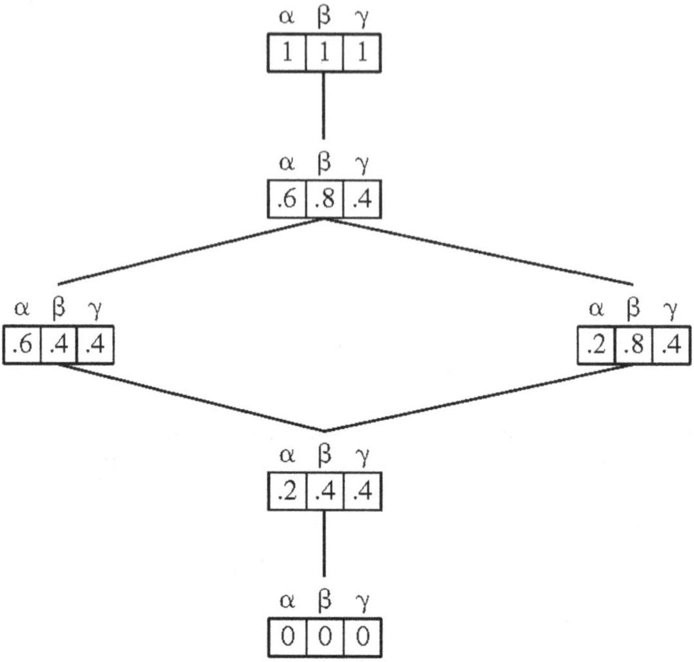

To confirm if the previous functional application is F-continuous it is enough to verify the fulfillment of the following condition:

$$\forall\, x \in E_1,\ \underset{\sim}{B}_k \in T_2(E_2):$$

$$\mu_{\Gamma^{-1}\underset{\sim}{B}_k} = \mu_{\underset{\sim}{B}_k}(\Gamma(x))$$

which we do in the following way of sagittate representation:

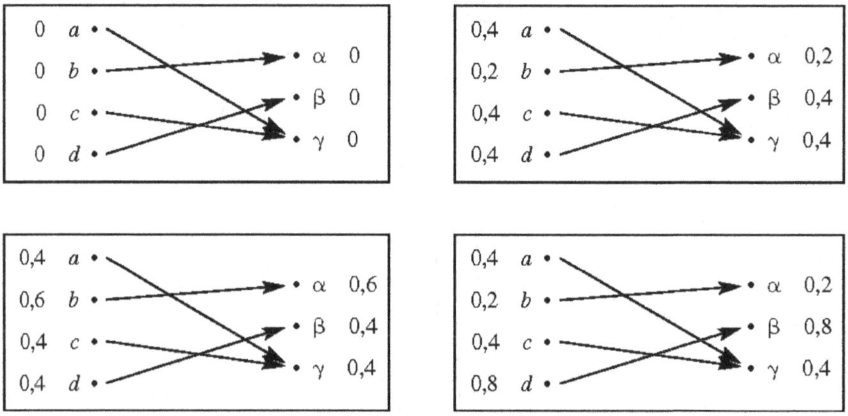

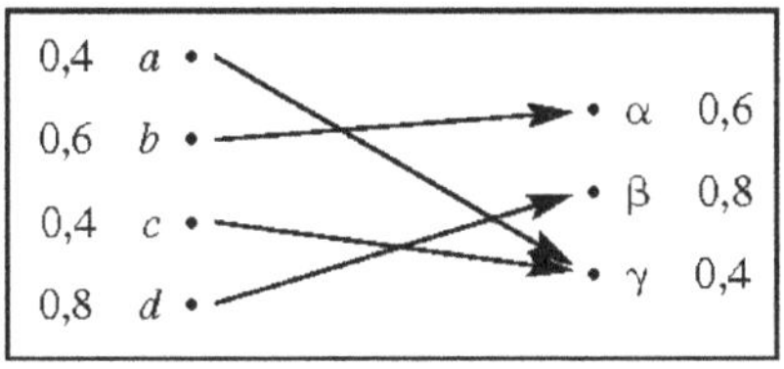

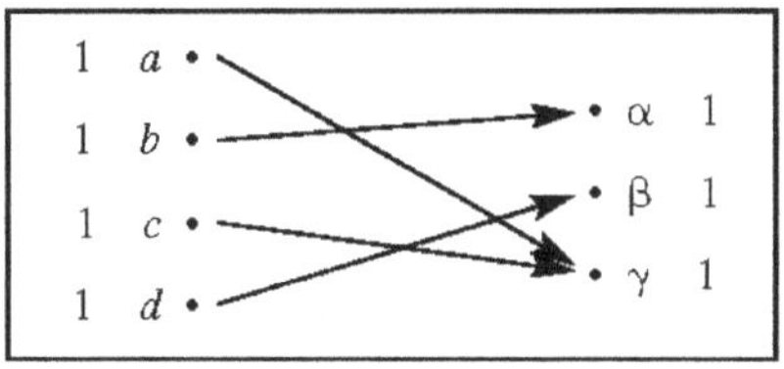

In effect, the functional application is F-continuous, having taken into account that each fuzzy subset $\mathbf{\Gamma}^{-1}$ $\underset{\sim}{B}_k$ obtained by means of the condition demanded corresponding to an open-$\mathbf{\Gamma}_1$ of the topological space E, $T_1(E_1)$. We hope that this example has been enough to show the F-continuity from the focus undertaken and that we are now able to, stating some properties, establish the existing relationship between them[7].

If two uncertain topologies are known, $T_1(E_1)$ over (E_1) and $T_2(E_2)$ over (E_2), we can state the four following properties:

1. The functional application $\mathbf{\Gamma}$ is F-continuous.
2. The inverse of every closed fuzzy subset $\underset{\sim}{B}_k \in T_2(E_2)$ is a closed fuzzy subset $\underset{\sim}{A}_j \in T_1(E_1)$.
3. For every fuzzy subset $\underset{\sim}{A}_j \in P(E_1)$, the inverse of each neighbourhood of $\mathbf{\Gamma}\underset{\sim}{A}_j$ is a neighbourhood of $\underset{\sim}{A}_j$.
4. For every fuzzy subset $\underset{\sim}{A}_j \in P(E_1)$ and every neighbourhood $\underset{\sim}{V}$ of $\mathbf{\Gamma}\,\underset{\sim}{A}_j$, there exists a neighbourhood $\underset{\sim}{W}$ of $\underset{\sim}{A}_j$, so that $\mathbf{\Gamma}\,\underset{\sim}{W} \subset \underset{\sim}{V}$.

Therefore, between them, we may establish the following relationships:

a) (Property 1) $\Leftrightarrow$ (Property 2)

In effect, Due to the definition of an inverse fuzzy subset for a functional application $\mathbf{\Gamma}$, one has:

$\forall x \in E_1$:

$$\mu_{\Gamma^{-1}\underset{\sim}{B}}(x) = \mu_{\underset{\sim}{B}}(\Gamma(x))$$

and also:

$\forall x \in E_1$:

$$\begin{aligned}\mu_{\Gamma^{-1}\bar{\underset{\sim}{B}}_k}(x) &= \mu_{\bar{\underset{\sim}{B}}_k}(\Gamma(x))\\ &= 1 - \mu_{\underset{\sim}{B}_k}(\Gamma(x))\\ &= 1 - \mu_{\Gamma^{-1}\underset{\sim}{B}_k}(x)\\ &= \mu_{\overline{\Gamma^{-1}\underset{\sim}{B}_k}}(x)\end{aligned}$$

[7] For the development of this part of the epigraph we have based our work on Kaufmann, A.: *Introduction à la théorie des sous-ensambles flous à l'usage des ingénieurs. 4. Compléments et nouvelles applications*. Ed. Masson. París, 1977, pages 102-106.

Therefore:

$\forall \underset{\sim}{B}_k \subset E_2$:

$$\Gamma^{-1} \overline{\underset{\sim}{B}_k} = \overline{\Gamma^{-1} \underset{\sim}{B}_k}$$

b) (Property 3) $\Leftrightarrow$ (Property 4).

We will first check that (Property 3) $\Rightarrow$ (Property 4) and then that (Property 4) $\Rightarrow$ (Property 3).

For the first of these it is enough to say that with $\mathbf{\Gamma}^{-1}\ \underset{\sim}{V}$ being a neighbourhood of $\underset{\sim}{A}_j$ one has:

$$\Gamma \underset{\sim}{W} = \Gamma (\Gamma^{-1} \underset{\sim}{V}) \subset \underset{\sim}{W}$$

in which $\underset{\sim}{W} = \mathbf{\Gamma}^{-1}\ \underset{\sim}{V}$.

For the second, we will begin from the affirmation that $\underset{\sim}{V}$ is a neighbourhood of $\mathbf{\Gamma}\ \underset{\sim}{A}_j$,. A neighbourhood $\underset{\sim}{W}$ of $\underset{\sim}{A}_j$ therefore exists so that $\mathbf{\Gamma}\ \underset{\sim}{W} \subset \underset{\sim}{V}$. In this way:

$$\Gamma^{-1} (\Gamma \underset{\sim}{W}) \subset \Gamma^{-1} \underset{\sim}{V}$$

But with:

$$\underset{\sim}{W} \subset \Gamma^{-1} (\Gamma \underset{\sim}{W})$$

it results that $\mathbf{\Gamma}^{-1}\ \underset{\sim}{V}$. is a neighbourhood of $\underset{\sim}{A}_j$.

(Property 1) $\Rightarrow$ (Property 3).

At first glance we can see that the inverse is not fulfilled, which implies that the neighbourhoods do not provide enough elements for the fuzzy-continuous concept. Having said that, if $\mathbf{\Gamma}$ is F-continuous one has $\underset{\sim}{A}_j \in P(E)$ and $\underset{\sim}{V}$ is a neighbourhood of $\mathbf{\Gamma}\ \underset{\sim}{A}_j$, therefore $\underset{\sim}{V}$ contains an open neighbourhood $\underset{\sim}{W}$ of $\mathbf{\Gamma}\ \underset{\sim}{A}_j$.

But as:

$$\Gamma \underset{\sim}{A}_j \subset \underset{\sim}{W} \subset \underset{\sim}{V}$$

then:

$$\Gamma^{-1} (\Gamma \underset{\sim}{A}_j) \subset \Gamma^{-1} \underset{\sim}{W} \subset \Gamma^{-1} \underset{\sim}{V}$$

But:

$$\underset{\sim}{A}_j \subset \Gamma^{-1} (\Gamma \underset{\sim}{A}_j)$$

and $\mathbf{\Gamma}^{-1}\ \underset{\sim}{W}$ is an open. Therefore $\mathbf{\Gamma}^{-1}\ \underset{\sim}{V}$ is a neighbourhood of $\underset{\sim}{A}_j$.

B) Even when we have not finished with the possibilities which F-continuous applications offer, we have taken the decision of leaving here the line followed from the first focus to move on to the possibilities of the second. For this we will use the referentials:

$$E_1 = \{\underset{\sim}{P}_1, \underset{\sim}{P}_2, \underset{\sim}{P}_3, \underset{\sim}{P}_4\}$$
$$E_2 = \{\underset{\sim}{Q}_1, \underset{\sim}{Q}_2, \underset{\sim}{Q}_3\}$$

and the following fuzzy topologies of E_1 and E_2:

$$T_1(E_1) = \left\{\varnothing, \{\underset{\sim}{P}_1\}, \{\underset{\sim}{P}_3\}, \{\underset{\sim}{P}_1, \underset{\sim}{P}_3\}, \{\underset{\sim}{P}_2, \underset{\sim}{P}_4\}, \{\underset{\sim}{P}_1, \underset{\sim}{P}_2, \underset{\sim}{P}_4\}, \{\underset{\sim}{P}_2, \underset{\sim}{P}_3, \underset{\sim}{P}_4\}, E_1\right\}$$
$$T_2(E_2) = \left\{\varnothing, \{\underset{\sim}{Q}_3\}, \{\underset{\sim}{Q}_1, \underset{\sim}{Q}_2\}, E_2\right\}$$

which we present in the corresponding Boole lattices:

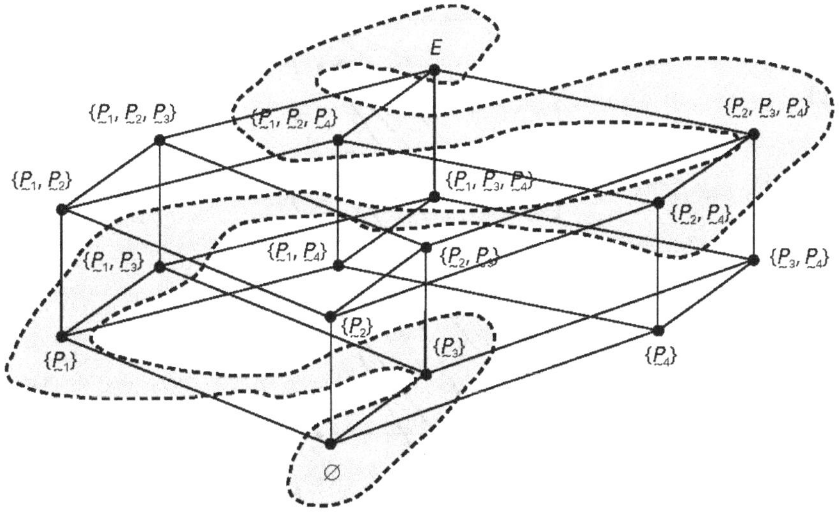

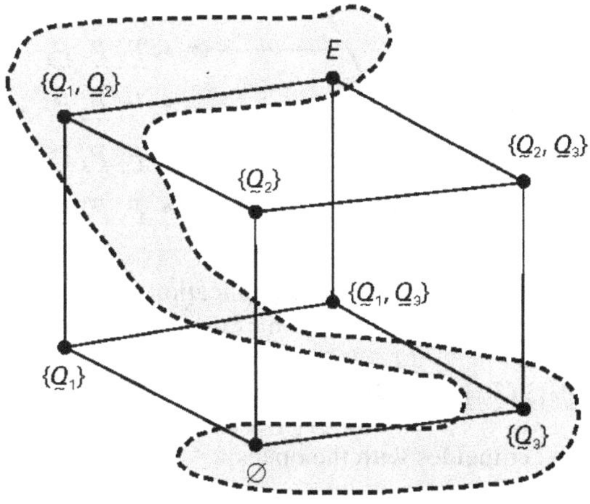

When it is known all topology may be expressed through its interior application δ, or also by its adherent application Γ. We remind ourselves to this respect that the interior $\underset{\sim}{A}_j \in T(E)$ is the greatest open-T contained in $\underset{\sim}{A}_j$. In our topologies this would be:

$$\delta\varnothing = \varnothing \quad \delta\{\underset{\sim}{P}_1\} = \{\underset{\sim}{P}_1\} \quad \delta\{\underset{\sim}{P}_2\} = \varnothing \quad \delta\{\underset{\sim}{P}_3\} = \{\underset{\sim}{P}_3\} \quad \delta\{\underset{\sim}{P}_4\} = \varnothing$$

$$\delta\{\underset{\sim}{P}_1, \underset{\sim}{P}_2\} = \{\underset{\sim}{P}_1\} \quad \delta\{\underset{\sim}{P}_1, \underset{\sim}{P}_3\} = \{\underset{\sim}{P}_1, \underset{\sim}{P}_3\} \quad \delta\{\underset{\sim}{P}_1, \underset{\sim}{P}_4\} = \{\underset{\sim}{P}_1\}$$

$$\delta\{\underset{\sim}{P}_2, \underset{\sim}{P}_3\} = \{\underset{\sim}{P}_3\} \quad \delta\{\underset{\sim}{P}_2, \underset{\sim}{P}_4\} = \{\underset{\sim}{P}_2, \underset{\sim}{P}_4\} \quad \delta\{\underset{\sim}{P}_3, \underset{\sim}{P}_4\} = \{\underset{\sim}{P}_3\}$$

$$\delta\{\underset{\sim}{P}_1, \underset{\sim}{P}_2, \underset{\sim}{P}_3\} = \{\underset{\sim}{P}_1, \underset{\sim}{P}_3\} \quad \delta\{\underset{\sim}{P}_1, \underset{\sim}{P}_2, \underset{\sim}{P}_4\} = \{\underset{\sim}{P}_1, \underset{\sim}{P}_2, \underset{\sim}{P}_4\}$$

$$\delta\{\underset{\sim}{P}_1, \underset{\sim}{P}_3, \underset{\sim}{P}_4\} = \{\underset{\sim}{P}_1, \underset{\sim}{P}_3\} \quad \delta\{\underset{\sim}{P}_2, \underset{\sim}{P}_3, \underset{\sim}{P}_4\} = \{\underset{\sim}{P}_2, \underset{\sim}{P}_3, \underset{\sim}{P}_4\} \quad \delta E_1 = E_1$$

with its representation through the following graph:

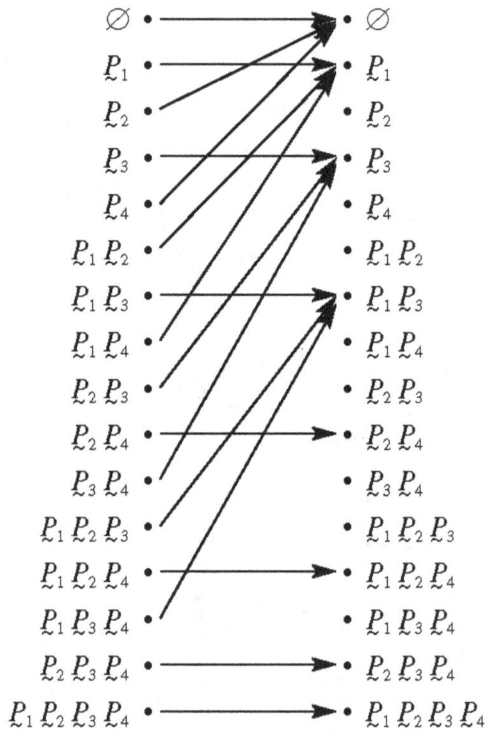

The presentation through the adherent application demands that the closed-Ts are found by means of the complementation of the open-Ts. They are:

$$C_e = \{\varnothing, \{\underset{\sim}{P}_1\}, \{\underset{\sim}{P}_3\}, \{\underset{\sim}{P}_1, \underset{\sim}{P}_3\}, \{\underset{\sim}{P}_2, \underset{\sim}{P}_4\}, \{\underset{\sim}{P}_1, \underset{\sim}{P}_2, \underset{\sim}{P}_4\}, \{\underset{\sim}{P}_2, \underset{\sim}{P}_3, \underset{\sim}{P}_4\}, E_1\}$$

which, exceptionally, coincides with the opens.

As the adherence of $\underset{\sim}{A}_j \in T(E)$ is the smallest closed-T which $\underset{\sim}{A}_j$ contains, we will have:

$$\Gamma \varnothing = \varnothing \quad \Gamma\{\underset{\sim}{P}_1\} = \{\underset{\sim}{P}_1\} \quad \Gamma\{\underset{\sim}{P}_2\} = \{\underset{\sim}{P}_2, \underset{\sim}{P}_4\} \quad \Gamma\{\underset{\sim}{P}_3\} = \{\underset{\sim}{P}_3\}$$

$$\Gamma\{\underset{\sim}{P}_4\} = \{\underset{\sim}{P}_2, \underset{\sim}{P}_4\} \quad \Gamma\{\underset{\sim}{P}_1, \underset{\sim}{P}_2\} = \{\underset{\sim}{P}_1, \underset{\sim}{P}_2, \underset{\sim}{P}_4\} \quad \Gamma\{\underset{\sim}{P}_1, \underset{\sim}{P}_3\} = \{\underset{\sim}{P}_1, \underset{\sim}{P}_3\}$$

$$\Gamma\{\underset{\sim}{P}_1, \underset{\sim}{P}_4\} = \{\underset{\sim}{P}_1, \underset{\sim}{P}_2, \underset{\sim}{P}_4\} \quad \Gamma\{\underset{\sim}{P}_2, \underset{\sim}{P}_3\} = \{\underset{\sim}{P}_2, \underset{\sim}{P}_3, \underset{\sim}{P}_4\} \quad \Gamma\{\underset{\sim}{P}_2, \underset{\sim}{P}_4\} = \{\underset{\sim}{P}_2, \underset{\sim}{P}_4\}$$

$$\Gamma\{\underset{\sim}{P}_3, \underset{\sim}{P}_4\} = \{\underset{\sim}{P}_2, \underset{\sim}{P}_3, \underset{\sim}{P}_4\} \quad \Gamma\{\underset{\sim}{P}_1, \underset{\sim}{P}_2, \underset{\sim}{P}_3\} = E_1 \quad \Gamma\{\underset{\sim}{P}_1, \underset{\sim}{P}_2, \underset{\sim}{P}_4\} = \{\underset{\sim}{P}_1, \underset{\sim}{P}_2, \underset{\sim}{P}_4\}$$

$$\Gamma\{\underset{\sim}{P}_1, \underset{\sim}{P}_3, \underset{\sim}{P}_4\} = E_1 \quad \Gamma\{\underset{\sim}{P}_2, \underset{\sim}{P}_3, \underset{\sim}{P}_4\} = \{\underset{\sim}{P}_2, \underset{\sim}{P}_3, \underset{\sim}{P}_4\} \quad \Gamma E_1 = E_1$$

its representation in graph form is the following:

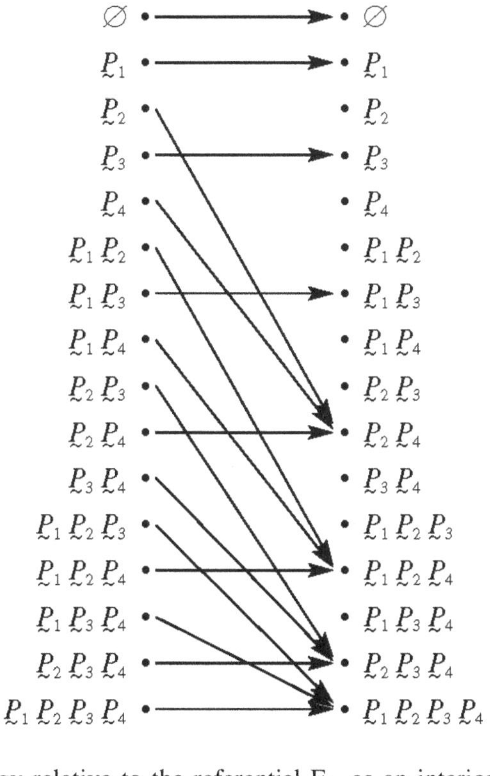

For the topology relative to the referential E_2, as an interior application δ we will have:

$$\delta \varnothing = \varnothing \quad \delta\{\underset{\sim}{Q}_1\} = \varnothing \quad \delta\{\underset{\sim}{Q}_2\} = \varnothing \quad \delta\{\underset{\sim}{Q}_3\} = \{\underset{\sim}{Q}_3\}$$

$$\delta\{\underset{\sim}{Q}_1, \underset{\sim}{Q}_2\} = \{\underset{\sim}{Q}_1, \underset{\sim}{Q}_2\} \quad \delta\{\underset{\sim}{Q}_1, \underset{\sim}{Q}_3\} = \{\underset{\sim}{Q}_3\} \quad \delta\{\underset{\sim}{Q}_2, \underset{\sim}{Q}_3\} = \{\underset{\sim}{Q}_3\} \quad \delta E_2 = E_2$$

with the following graph:

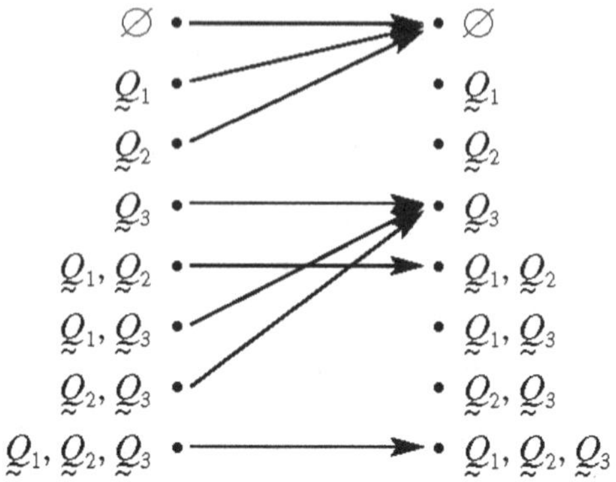

As the closed-Ts are:

$$C_e = \{\varnothing, \{\underset{\sim}{Q}_3\}, \{\underset{\sim}{Q}_1, \underset{\sim}{Q}_2\}, E_2\}$$

the adherent application $\mathbf{\Gamma}$ will be:

$$\Gamma\varnothing = \varnothing \quad \Gamma\{\underset{\sim}{Q}_1\} = \{\underset{\sim}{Q}_1, \underset{\sim}{Q}_2\} \quad \Gamma\{\underset{\sim}{Q}_2\} = \{\underset{\sim}{Q}_1, \underset{\sim}{Q}_2\} \quad \Gamma\{\underset{\sim}{Q}_3\} = \{\underset{\sim}{Q}_3\}$$
$$\Gamma\{\underset{\sim}{Q}_1, \underset{\sim}{Q}_2\} = \{\underset{\sim}{Q}_1, \underset{\sim}{Q}_2\} \quad \Gamma\{\underset{\sim}{Q}_1, \underset{\sim}{Q}_3\} = E_2 \quad \Gamma\{\underset{\sim}{Q}_2, \underset{\sim}{Q}_3\} = E_2 \quad \Gamma E_2 = E_2$$

whose representation in graph form is:

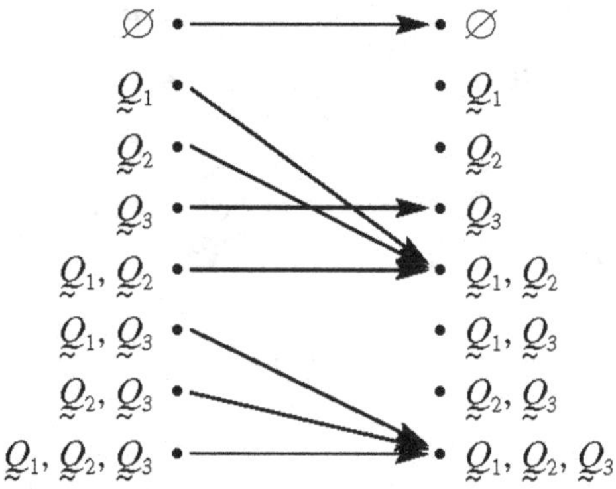

Replacing the Latin letters for P_i and the Greek for $\underset{\sim}{Q}_{k_\varnothing}$.

After this brief reminder, we will continue in our process establishing the same functional application of E_1 in E_2 of the first focus, substituting the Latin letters for P_i and the Greek for $Q_{k_\varnothing}$. Therefore:

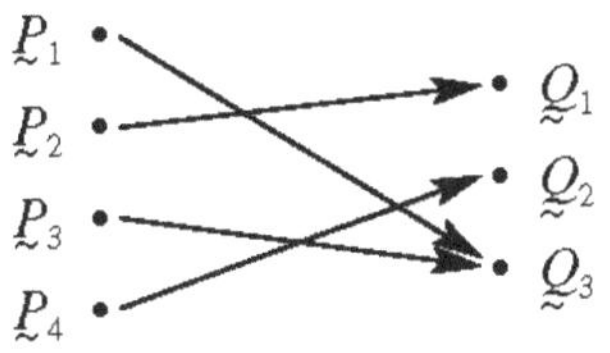

The following application of $P(E_1)$ in $Q(E_2)$ is obtained:

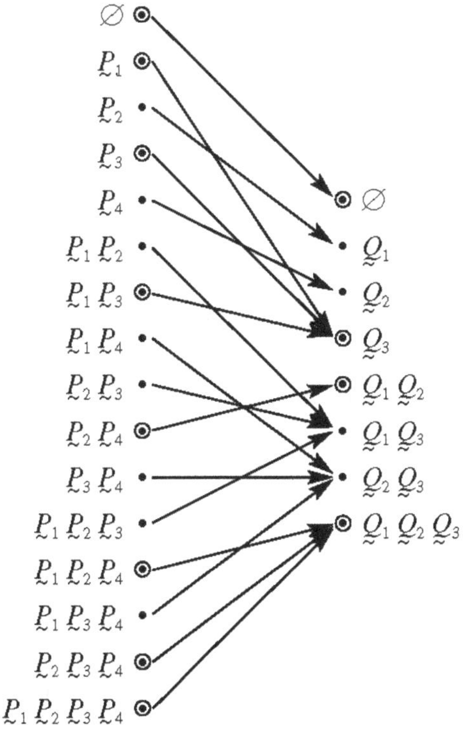

We will now move on to confirm if this application **Γ** is F-continuous. For this we will represent each of the fuzzy subsets of the referentials E_1 and E_2 by means of the values of the characteristic membership function. We assume that they are the following:

$$\underset{\sim}{P}_1 = \begin{array}{|c|c|c|} \multicolumn{1}{c}{a} & \multicolumn{1}{c}{b} & \multicolumn{1}{c}{c} \\ \hline .4 & .8 & .6 \\ \hline \end{array} \quad \underset{\sim}{P}_2 = \begin{array}{|c|c|c|} \multicolumn{1}{c}{a} & \multicolumn{1}{c}{b} & \multicolumn{1}{c}{c} \\ \hline .7 & .2 & .9 \\ \hline \end{array} \quad \underset{\sim}{P}_3 = \begin{array}{|c|c|c|} \multicolumn{1}{c}{a} & \multicolumn{1}{c}{b} & \multicolumn{1}{c}{c} \\ \hline .4 & .8 & .6 \\ \hline \end{array} \quad \underset{\sim}{P}_4 = \begin{array}{|c|c|c|} \multicolumn{1}{c}{a} & \multicolumn{1}{c}{b} & \multicolumn{1}{c}{c} \\ \hline .6 & .8 & .9 \\ \hline \end{array}$$

$$\underset{\sim}{Q}_1 = \begin{array}{|c|c|c|} \multicolumn{1}{c}{a} & \multicolumn{1}{c}{b} & \multicolumn{1}{c}{c} \\ \hline .7 & .2 & .9 \\ \hline \end{array} \quad \underset{\sim}{Q}_2 = \begin{array}{|c|c|c|} \multicolumn{1}{c}{a} & \multicolumn{1}{c}{b} & \multicolumn{1}{c}{c} \\ \hline .6 & .8 & .9 \\ \hline \end{array} \quad \underset{\sim}{Q}_3 = \begin{array}{|c|c|c|} \multicolumn{1}{c}{a} & \multicolumn{1}{c}{b} & \multicolumn{1}{c}{c} \\ \hline .4 & .8 & .6 \\ \hline \end{array}$$

Let us see if the inverse of all open-T_2s are open-T_1s.

$$\Gamma^{-1}\varnothing = \Gamma^{-1}\{ \begin{array}{|c|c|c|} \multicolumn{1}{c}{a} & \multicolumn{1}{c}{b} & \multicolumn{1}{c}{c} \\ \hline 0 & 0 & 0 \\ \hline \end{array} \} = \varnothing = \{ \begin{array}{|c|c|c|} \multicolumn{1}{c}{a} & \multicolumn{1}{c}{b} & \multicolumn{1}{c}{c} \\ \hline 0 & 0 & 0 \\ \hline \end{array} \}$$

$$\Gamma^{-1}\{\underset{\sim}{Q}_3\} = \Gamma^{-1}\{ \begin{array}{|c|c|c|} \multicolumn{1}{c}{a} & \multicolumn{1}{c}{b} & \multicolumn{1}{c}{c} \\ \hline .4 & .8 & .6 \\ \hline \end{array} \} = \{\{\underset{\sim}{P}_1\}, \{\underset{\sim}{P}_3\}, \{\underset{\sim}{P}_1, \underset{\sim}{P}_3\}\} = \{ \begin{array}{|c|c|c|} \multicolumn{1}{c}{a} & \multicolumn{1}{c}{b} & \multicolumn{1}{c}{c} \\ \hline .4 & .8 & .6 \\ \hline \end{array} \}$$

$$\Gamma^{-1}\{\underset{\sim}{Q}_1, \underset{\sim}{Q}_2\} = \Gamma^{-1}\{ \begin{array}{|c|c|c|} \multicolumn{1}{c}{a} & \multicolumn{1}{c}{b} & \multicolumn{1}{c}{c} \\ \hline .7 & .2 & .9 \\ \hline \end{array}, \begin{array}{|c|c|c|} \multicolumn{1}{c}{a} & \multicolumn{1}{c}{b} & \multicolumn{1}{c}{c} \\ \hline .6 & .8 & .9 \\ \hline \end{array} \} = \{\underset{\sim}{P}_2, \underset{\sim}{P}_4\} = \{ \begin{array}{|c|c|c|} \multicolumn{1}{c}{a} & \multicolumn{1}{c}{b} & \multicolumn{1}{c}{c} \\ \hline .7 & .2 & .9 \\ \hline \end{array}, \begin{array}{|c|c|c|} \multicolumn{1}{c}{a} & \multicolumn{1}{c}{b} & \multicolumn{1}{c}{c} \\ \hline .6 & .8 & .9 \\ \hline \end{array} \}$$

$$\Gamma^{-1}\{\underset{\sim}{Q}_1, \underset{\sim}{Q}_2, \underset{\sim}{Q}_3\} = \Gamma^{-1}\{ \begin{array}{|c|c|c|} \multicolumn{1}{c}{a} & \multicolumn{1}{c}{b} & \multicolumn{1}{c}{c} \\ \hline .7 & .2 & .9 \\ \hline \end{array}, \begin{array}{|c|c|c|} \multicolumn{1}{c}{a} & \multicolumn{1}{c}{b} & \multicolumn{1}{c}{c} \\ \hline .6 & .8 & .9 \\ \hline \end{array}, \begin{array}{|c|c|c|} \multicolumn{1}{c}{a} & \multicolumn{1}{c}{b} & \multicolumn{1}{c}{c} \\ \hline .4 & .8 & .6 \\ \hline \end{array} \} = \{\{\underset{\sim}{P}_1, \underset{\sim}{P}_2, \underset{\sim}{P}_4\},$$

$$\{\underset{\sim}{P}_2, \underset{\sim}{P}_3, \underset{\sim}{P}_4\}, \{\underset{\sim}{P}_1, \underset{\sim}{P}_2, \underset{\sim}{P}_3, \underset{\sim}{P}_4\}\} = \{ \begin{array}{|c|c|c|} \multicolumn{1}{c}{a} & \multicolumn{1}{c}{b} & \multicolumn{1}{c}{c} \\ \hline .4 & .8 & .6 \\ \hline \end{array}, \begin{array}{|c|c|c|} \multicolumn{1}{c}{a} & \multicolumn{1}{c}{b} & \multicolumn{1}{c}{c} \\ \hline .7 & .2 & .9 \\ \hline \end{array}, \begin{array}{|c|c|c|} \multicolumn{1}{c}{a} & \multicolumn{1}{c}{b} & \multicolumn{1}{c}{c} \\ \hline .6 & .8 & .9 \\ \hline \end{array} \}$$

Having seen the results obtained the conclusion is reached that the previous application if F-continuous.

In this focus it may also result as interesting to show the relationships between F-continuous functional applications within previously shown fuzzy topologies. For this we once again assume the availability of two fuzzy topologies $T(E_1)$ and $T(E_2)$ in which the four following properties are given:

1. An F-continuous application exists between them
2. The inverse of all elements $\underset{\sim}{B}_k \in T(E_2)$ is an element $\underset{\sim}{A}_j \in T(E_1)$.
3. For all elements $\underset{\sim}{A}_j \in P(E_1)$ it is fulfilled that the inverse of each neighbour of $\mathbf{\Gamma}\, \underset{\sim}{A}_j$ is a neighbour of $\underset{\sim}{A}_j$.
4. For all elements $\underset{\sim}{A}_j \in P(E_1)$ and each neighbourhood $\underset{\sim}{V}$ of $\mathbf{\Gamma}\, \underset{\sim}{A}_j$, there exists a neighbourhood $\underset{\sim}{W}$ of $\underset{\sim}{A}_j$ so that $\mathbf{\Gamma}\, \underset{\sim}{W} \subset \underset{\sim}{V}$.

Once again the following relationships between them may be established, seen from the first focus:

a) (Property 1) $\Leftrightarrow$ (Property 2)

"The existence of an F-continuous functional application gives that the inverse of all elements of the topology $T(E_2)$ is an element of the topology $T(E_1)$ and vice versa". This is an immediate consequence which the already confirmed property fulfils of:

$\forall \underset{\sim}{B}_k \in Q(E_2)$:

$$\Gamma^{-1}\,\overline{\underset{\sim}{B}_k} = \overline{\Gamma^{-1}\,\underset{\sim}{B}_k}$$

b) (Property 3) $\Leftrightarrow$ (Property 4)

"When for all elements $\underset{\sim}{A}_j \in P(E_1)$ the inverse of each neighbour $\boldsymbol{\Gamma}\,\underset{\sim}{A}_j$ is a neighbour of $\underset{\sim}{A}_j$ there exists a neighbourhood $\underset{\sim}{W}$ of $\underset{\sim}{A}_j$ such as $\boldsymbol{\Gamma}\,\underset{\sim}{W} \subset \underset{\sim}{V}$, in which $\underset{\sim}{V}$ is a neighbourhood of $\boldsymbol{\Gamma}\,\underset{\sim}{A}_j$ and vice versa"

We will now move on to develop an example starting from the topologies $T(E_1)$ and $T(E_2)$ which have served us up to now.

$$T(E_1) = \{\emptyset, \{\underset{\sim}{P}_1\}, \{\underset{\sim}{P}_3\}, \{\underset{\sim}{P}_1, \underset{\sim}{P}_3\}, \{\underset{\sim}{P}_2, \underset{\sim}{P}_4\}, \{\underset{\sim}{P}_1, \underset{\sim}{P}_2, \underset{\sim}{P}_4\}, \{\underset{\sim}{P}_2, \underset{\sim}{P}_3, \underset{\sim}{P}_4\}, E_1\}$$
$$T(E_2) = \{\emptyset, \{\underset{\sim}{Q}_2\}, \{\underset{\sim}{Q}_1, \underset{\sim}{Q}_2\}, E_2\}$$

Now an element of $P(E_1)$ such as $\{\underset{\sim}{P}_1, \underset{\sim}{P}_3\}$ will be considered and we will indicate its neighbourhood in the corresponding Boole lattice.

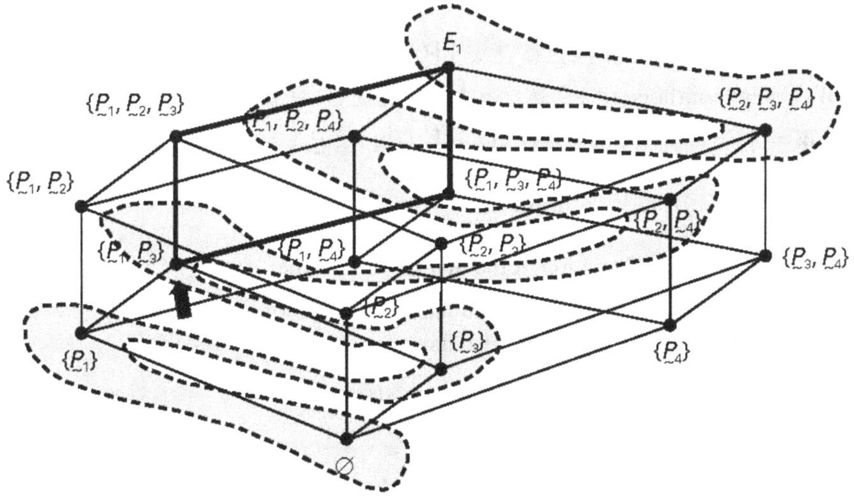

As $\mathbf{\Gamma}\{\underset{\sim}{P}_1, \underset{\sim}{P}_3\}$ is $\{\underset{\sim}{Q}_3\}$, we should search for the neighbours of this element, which, in our case, are $\{\underset{\sim}{Q}_1, \underset{\sim}{Q}_3\}$, $\{\underset{\sim}{Q}_2, \underset{\sim}{Q}_3\}$, E_2, shown in the following Boole lattice:

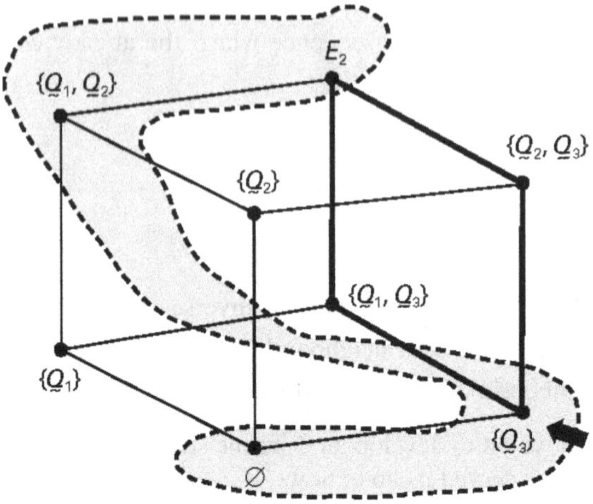

Let us see which are its inverses:

$$\Gamma^{-1}\{\underset{\sim}{Q}_1, \underset{\sim}{Q}_3\} = \{\underset{\sim}{P}_1, \underset{\sim}{P}_2, \underset{\sim}{P}_3\}, \qquad \text{is a neighbour of } \{\underset{\sim}{P}_1, \underset{\sim}{P}_3\}$$
$$\Gamma^{-1}\{\underset{\sim}{Q}_2, \underset{\sim}{Q}_3\} = \{\underset{\sim}{P}_1, \underset{\sim}{P}_3, \underset{\sim}{P}_4\}, \qquad \text{is a neighbour of } \{\underset{\sim}{P}_1, \underset{\sim}{P}_3\}$$
$$\Gamma^{-1} E_2 = \{\underset{\sim}{P}_1, \underset{\sim}{P}_2, \underset{\sim}{P}_3, \underset{\sim}{P}_4\}, \qquad \text{is a neighbour of } \{\underset{\sim}{P}_1, \underset{\sim}{P}_3\}$$

And also:

$$\underset{\sim}{V} = \text{neighbourhood of } \mathbf{\Gamma}\{\underset{\sim}{P}_1, \underset{\sim}{P}_3\} = \left\{\{\underset{\sim}{Q}_1, \underset{\sim}{Q}_3\}, \{\underset{\sim}{Q}_2, \underset{\sim}{Q}_3\}, E_2\right\}$$
$$\underset{\sim}{W} = \text{neighbourhood of } \{\underset{\sim}{P}_1, \underset{\sim}{P}_3\} = \left\{\{\underset{\sim}{P}_1, \underset{\sim}{P}_2, \underset{\sim}{P}_3\}, \{\underset{\sim}{P}_1, \underset{\sim}{P}_3, \underset{\sim}{P}_4\}, E_1\right\}$$
$$\Gamma \underset{\sim}{W} = \left\{\Gamma\{\underset{\sim}{P}_1, \underset{\sim}{P}_2, \underset{\sim}{P}_3\} = \{\underset{\sim}{Q}_1, \underset{\sim}{Q}_3\}, \Gamma\{\underset{\sim}{P}_1, \underset{\sim}{P}_3, \underset{\sim}{P}_4\} = \{\underset{\sim}{Q}_2, \underset{\sim}{Q}_3\}, \Gamma E_1 = E_2\right\}$$

In this way $\mathbf{\Gamma}\, \underset{\sim}{W} \subset \underset{\sim}{V}$ is fulfilled.

We believe the confirmation of the inverse to be unnecessary.

It can be seen that our confirmation has been based on the fact that $\mathbf{\Gamma}^{-1}\, \underset{\sim}{V}$ is a neighbourhood of $\{\underset{\sim}{P}_1, \underset{\sim}{P}_3\}$ and we also have:

$$\Gamma \underset{\sim}{W} = \Gamma(\Gamma^{-1} \underset{\sim}{V}) \subset \underset{\sim}{W}$$

in which $\underset{\sim}{W} = \mathbf{\Gamma}^{-1}\, \underset{\sim}{V}$.

In so much as the inverse proposal, with $\underset{\sim}{V}$ being a neighbourhood of $\mathbf{\Gamma}\{\underset{\sim}{P}_1, \underset{\sim}{P}_3\}$, there exists a neighbourhood $\underset{\sim}{W}$ of $\{\underset{\sim}{P}_1, \underset{\sim}{P}_3\}$ such as:

$$\Gamma \underset{\sim}{W} \subset \underset{\sim}{V}$$

Therefore:

$$\Gamma^{-1}(\Gamma \underset{\sim}{W}) \subset \Gamma^{-1} \underset{\sim}{V}$$

But as:

$$\underset{\sim}{W} \subset \Gamma^{-1}(\Gamma \underset{\sim}{W})$$

it results that:

$$\Gamma^{-1} \underset{\sim}{V} \text{ is a neighbourhood of } \{\underset{\sim}{P}_1, \underset{\sim}{P}_3\}$$

(Property 1) $\Rightarrow$ (Property 3).

"If an F-continuous functional application exists, the inverse of each neighbour of $\Gamma \underset{\sim}{A}_j$ is a neighbour of $\underset{\sim}{A}_j \in P(E_1)$."

In this case it is necessary to remember the existence of a relationship in one sense but not in the inverse. It is this way as a consequence of the neighbourhoods not having the capacity to determine an F-continuous functional application.

We will continue with the same F-continuous functional application of our example, to choose as an element of $P(E_1)$, $\{\underset{\sim}{P}_1, \underset{\sim}{P}_3\}$, as we have done previously.

We have the following:

$$\Gamma \{\underset{\sim}{P}_1, \underset{\sim}{P}_3\} = \{\underset{\sim}{Q}_3\}$$

and its neighbourhood as we have seen is:

$$\text{Neighbourhood } \{\underset{\sim}{Q}_3\} = \{\{\underset{\sim}{Q}_1, \underset{\sim}{Q}_3\}, \{\underset{\sim}{Q}_2, \underset{\sim}{Q}_3\}. E_2\}$$

As we already know the inverses are:

$$\Gamma^{-1} \{\underset{\sim}{Q}_1, \underset{\sim}{Q}_3\} = \{\underset{\sim}{P}_1, \underset{\sim}{P}_2, \underset{\sim}{P}_3\}$$
$$\Gamma^{-1} \{\underset{\sim}{Q}_2, \underset{\sim}{Q}_3\} = \{\underset{\sim}{P}_1, \underset{\sim}{P}_3, \underset{\sim}{P}_4\}$$
$$\Gamma^{-1} E_2 = E_1$$

All are neighbours of $\{\underset{\sim}{P}_1, \underset{\sim}{P}_3\}$

Even when the panorama presented up to now is obligatorily incomplete, we believe that it will be enough to state the conglomeration of knowledge of uncertain topology for, based on this, the construction of operative techniques capable of dealing with complex structures and relationships between distinct systems with rapid mutations. To complete our offer, a "flash" is necessary on **a class of application** in which we will later make a base to achieve good formulation of the affinities. We are refering to **homeomorphism** in uncertain topologies.

8 Homeomorphic Applications in Uncertainty

We have studied, in the **determinist field**, the notion of homeomorphism as a bijective application in a set E_1 over a set E_2 which also induces a bijection in the respective "power sets" $P(E_1)$ over $Q(E_2)$ and which defines a biunivocal correspondence between the **opens** of E_1 and the **opens** of E_2. We have also said that, therefore, the **topological spaces** $(E_1, T_1(E_1))$ and $(E_2, T_2(E_2))$ are homeomorphic.

In general this conception of homeomorphism remains when we place ourselves in the **field of uncertainty,** whatever the focus taken into consideration, although, maybe, for greater commodity we adopt as a notion of ***F*- homeomorphism** that which conceives fuzzy homeomorphism ϕ as that bijective application of E_1 over E_2 in which as much ϕ as ϕ^{-1} are F-continuous. It therefore occurs that the **opens** of $\boldsymbol{E_1}$ (respectively the closes) are transformed into **opens** of $\boldsymbol{E_2}$ (respectively into closes) and vice versa. Upon dealing with a bijective application, there should be:

$$\text{cardinal } E_1 = \text{ cardinal } E_2$$

In the most superficial way, although without affecting the rigour which is necessary, we will expound, as we have done up to now, the most significant elements of each of the focuses, using an example which allows for easy comparison.

A) We begin from the existence of the two referentials E_1 and E_2 which have the same cardinal:

$$E_1 = \{a, b, c, d\}$$
$$E_2 = \{\alpha, \beta, \gamma, \delta\}$$

with a bijective application $\boldsymbol{\Gamma}$, which is a copy of that used in our deterministic description:

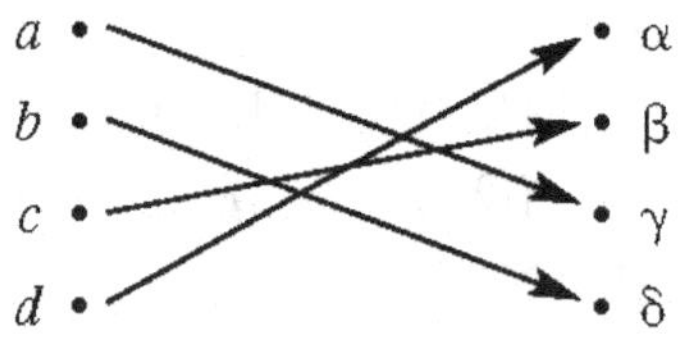

Let us now consider the topology $T_1(E_1)$:

$$T_1(E_1) = \left\{ \varnothing, \begin{array}{cccc} a & b & c & d \\ \hline \multicolumn{1}{|c|}{.2} & \multicolumn{1}{c|}{.2} & \multicolumn{1}{c|}{.2} & \multicolumn{1}{c|}{.4} \\ \hline \end{array}, \begin{array}{cccc} a & b & c & d \\ \hline \multicolumn{1}{|c|}{.6} & \multicolumn{1}{c|}{.3} & \multicolumn{1}{c|}{.2} & \multicolumn{1}{c|}{.4} \\ \hline \end{array}, \begin{array}{cccc} a & b & c & d \\ \hline \multicolumn{1}{|c|}{.2} & \multicolumn{1}{c|}{.2} & \multicolumn{1}{c|}{.4} & \multicolumn{1}{c|}{.4} \\ \hline \end{array}, \begin{array}{cccc} a & b & c & d \\ \hline \multicolumn{1}{|c|}{.7} & \multicolumn{1}{c|}{.5} & \multicolumn{1}{c|}{.2} & \multicolumn{1}{c|}{.5} \\ \hline \end{array}, \right.$$

$$= \begin{array}{cccc} a & b & c & d \\ \hline \multicolumn{1}{|c|}{.6} & \multicolumn{1}{c|}{.3} & \multicolumn{1}{c|}{.4} & \multicolumn{1}{c|}{.4} \\ \hline \end{array}, \begin{array}{cccc} a & b & c & d \\ \hline \multicolumn{1}{|c|}{.2} & \multicolumn{1}{c|}{.2} & \multicolumn{1}{c|}{.8} & \multicolumn{1}{c|}{.4} \\ \hline \end{array}, \begin{array}{cccc} a & b & c & d \\ \hline \multicolumn{1}{|c|}{.7} & \multicolumn{1}{c|}{.5} & \multicolumn{1}{c|}{.4} & \multicolumn{1}{c|}{.5} \\ \hline \end{array}, \begin{array}{cccc} a & b & c & d \\ \hline \multicolumn{1}{|c|}{.6} & \multicolumn{1}{c|}{.3} & \multicolumn{1}{c|}{.8} & \multicolumn{1}{c|}{.4} \\ \hline \end{array},$$

$$\left. = \begin{array}{cccc} a & b & c & d \\ \hline \multicolumn{1}{|c|}{.7} & \multicolumn{1}{c|}{.5} & \multicolumn{1}{c|}{.8} & \multicolumn{1}{c|}{.5} \\ \hline \end{array}, E \right\}$$

Represented by the following lattice:

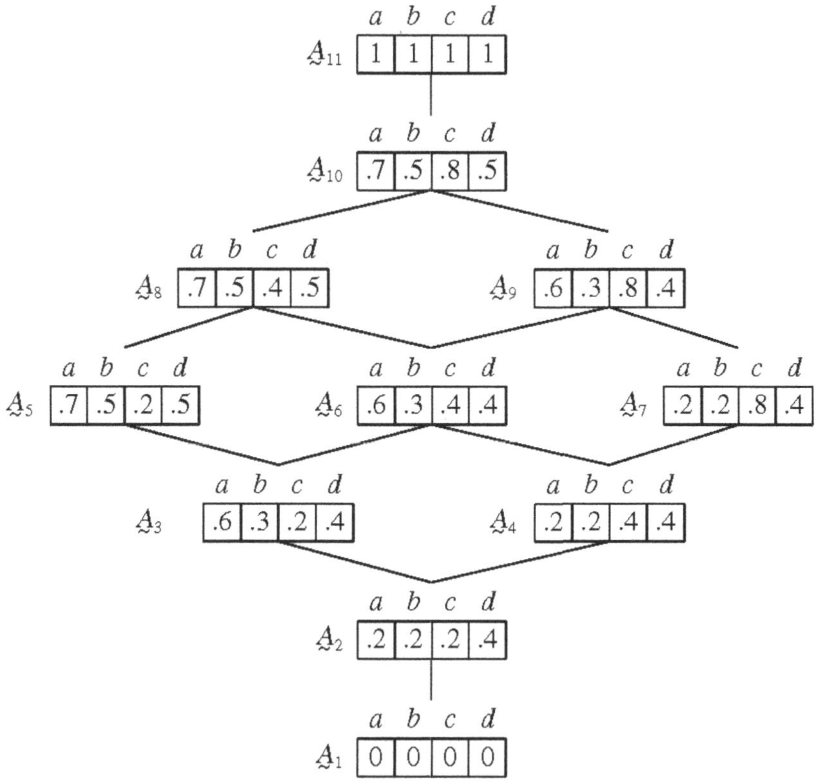

As a result the closes are:

$$C_e^{(1)}(E_1) = \Big\{ \varnothing, \begin{array}{cccc} a & b & c & d \\ \hline .3 & .5 & .2 & .5 \end{array}, \begin{array}{cccc} a & b & c & d \\ \hline .3 & .5 & .6 & .5 \end{array}, \begin{array}{cccc} a & b & c & d \\ \hline .4 & .7 & .2 & .6 \end{array}, \begin{array}{cccc} a & b & c & d \\ \hline .3 & .5 & .8 & .5 \end{array},$$

$$= \begin{array}{cccc} a & b & c & d \\ \hline .4 & .7 & .6 & .6 \end{array}, \begin{array}{cccc} a & b & c & d \\ \hline .8 & .8 & .2 & .6 \end{array}, \begin{array}{cccc} a & b & c & d \\ \hline .4 & .7 & .8 & .6 \end{array}, \begin{array}{cccc} a & b & c & d \\ \hline .8 & .8 & .6 & .6 \end{array},$$

$$, \begin{array}{cccc} a & b & c & d \\ \hline .8 & .8 & .8 & .6 \end{array}, E_1 \Big\}$$

To find a topology of E_2, $T_2(E_2)$ which allows homeomorphism with $T_1(E_1)$ it is enough to obtain the fuzzy subsets whose values of the characteristic membership function are established according to the bijective application **Γ**. It will be, in our case:

$$T_2(E_2) = \left\{ \emptyset, \overset{\alpha\ \ \beta\ \ \gamma\ \ \delta}{\boxed{.4\,|\,.2\,|\,.2\,|\,.2}}, \overset{\alpha\ \ \beta\ \ \gamma\ \ \delta}{\boxed{.4\,|\,.2\,|\,.6\,|\,.3}}, \overset{\alpha\ \ \beta\ \ \gamma\ \ \delta}{\boxed{.4\,|\,.4\,|\,.2\,|\,.2}}, \overset{\alpha\ \ \beta\ \ \gamma\ \ \delta}{\boxed{.5\,|\,.2\,|\,.7\,|\,.5}}, \right.$$

$$= \overset{\alpha\ \ \beta\ \ \gamma\ \ \delta}{\boxed{.4\,|\,.4\,|\,.6\,|\,.3}}, \overset{\alpha\ \ \beta\ \ \gamma\ \ \delta}{\boxed{.4\,|\,.8\,|\,.2\,|\,.2}}, \overset{\alpha\ \ \beta\ \ \gamma\ \ \delta}{\boxed{.5\,|\,.4\,|\,.7\,|\,.5}}, \overset{\alpha\ \ \beta\ \ \gamma\ \ \delta}{\boxed{.4\,|\,.8\,|\,.6\,|\,.3}},$$

$$\left. = \overset{\alpha\ \ \beta\ \ \gamma\ \ \delta}{\boxed{.5\,|\,.8\,|\,.7\,|\,.5}}, E_2 \right\}$$

With the following lattice representation:

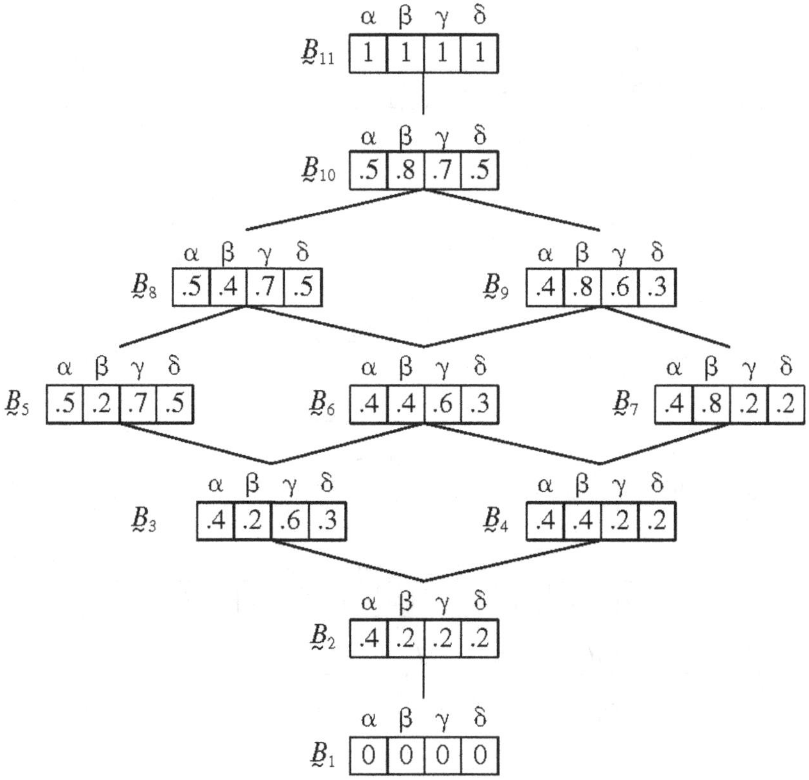

It becomes evident that the uncertain topological spaces (E_1, $T_1(E_1)$) and (E_2, $T_2(E_2)$) are homeomorphic.

On this occasion we have searched for an uncertain topological space starting from one known, which gave way to a homeomorphism through a bijective functional application. The case in which one would like to know if two uncertain topologies are homeomorphic is also frequent. Without disdaining the importance of this standing we will not insist on this subject, from the focus undertaken up to now, to move on to that which takes some referentials E_1 and E_2 of fuzzy subsets as a starting point.

B) We propose the development of a process from an illustrating example which allows us to show the form that a homeomorphism acquires when the fuzzyfication collects the content in the own elements of the following referential sets:

$$E_1 = \{\underset{\sim}{P}_1, \underset{\sim}{P}_2, \underset{\sim}{P}_3, \underset{\sim}{P}_4\}$$
$$E_2 = \{\underset{\sim}{Q}_1, \underset{\sim}{Q}_2, \underset{\sim}{Q}_3, \underset{\sim}{Q}_4\}$$

and a functional application $\boldsymbol{\Gamma}$ which, to aid comparison, we maintain as that used in the previous focus with the necessary change of the referentials. We represent this in the following:

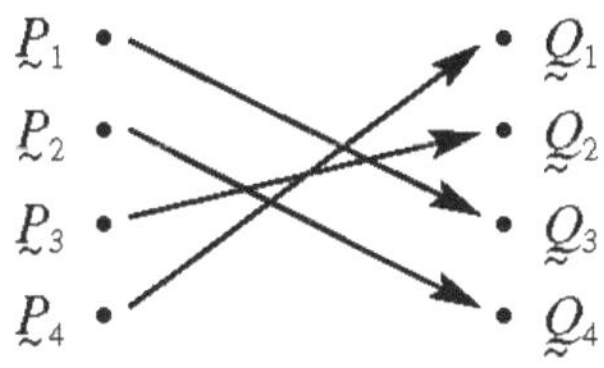

In the deterministic development we have seen that this bijective application also induces a bijective function between the "power sets" $P(E_1)$ and $Q(E_2)$.

We will now consider a topology $T_1(E_1)$ such as:

$$T_1(E_1) = \left\{\varnothing, \{\underset{\sim}{P}_1\}, \{\underset{\sim}{P}_3\}, \{\underset{\sim}{P}_2, \underset{\sim}{P}_4\}, \{\underset{\sim}{P}_1, \underset{\sim}{P}_3\}, \{\underset{\sim}{P}_1, \underset{\sim}{P}_2, \underset{\sim}{P}_4\}, \{\underset{\sim}{P}_2, \underset{\sim}{P}_3, \underset{\sim}{P}_4\}, E_1\right\}$$

This topology, the source of the application $P(E_1) \rightarrow Q(E_2)$, provides the following image:

$$T_2(E_2) = \left\{\varnothing, \{\underset{\sim}{Q}_3\}, \{\underset{\sim}{Q}_2\}, \{\underset{\sim}{Q}_1, \underset{\sim}{Q}_4\}, \{\underset{\sim}{Q}_2, \underset{\sim}{Q}_3\}, \{\underset{\sim}{Q}_1, \underset{\sim}{Q}_3, \underset{\sim}{Q}_4\}, \{\underset{\sim}{Q}_1, \underset{\sim}{Q}_2, \underset{\sim}{Q}_4\}, E_2\right\}$$

It is rapidly verified that a bijection between $T_1(E_1)$ and $T_2(E_2)$ is established in the induced application $P(E_1) \rightarrow Q(E_2)$

$$
\begin{aligned}
\varnothing &\longrightarrow \varnothing\\
\underset{\sim}{P}_1 &\longrightarrow \underset{\approx}{Q}_3\\
\underset{\sim}{P}_3 &\longrightarrow \underset{\approx}{Q}_2\\
\underset{\sim}{P}_2, \underset{\sim}{P}_4 &\longrightarrow \underset{\approx}{Q}_1, \underset{\approx}{Q}_4\\
\underset{\sim}{P}_1, \underset{\sim}{P}_3 &\longrightarrow \underset{\approx}{Q}_2, \underset{\approx}{Q}_3\\
\underset{\sim}{P}_1, \underset{\sim}{P}_2, \underset{\sim}{P}_4 &\longrightarrow \underset{\approx}{Q}_1, \underset{\approx}{Q}_3, \underset{\approx}{Q}_4\\
\underset{\sim}{P}_2, \underset{\sim}{P}_3, \underset{\sim}{P}_4 &\longrightarrow \underset{\approx}{Q}_1, \underset{\approx}{Q}_2, \underset{\approx}{Q}_4\\
E_1 &\longrightarrow E_2
\end{aligned}
$$

If these two topologies are represented by means of the corresponding Boole lattices it can be seen, as it could be in no other way, that they are homeomorphic. In effect:

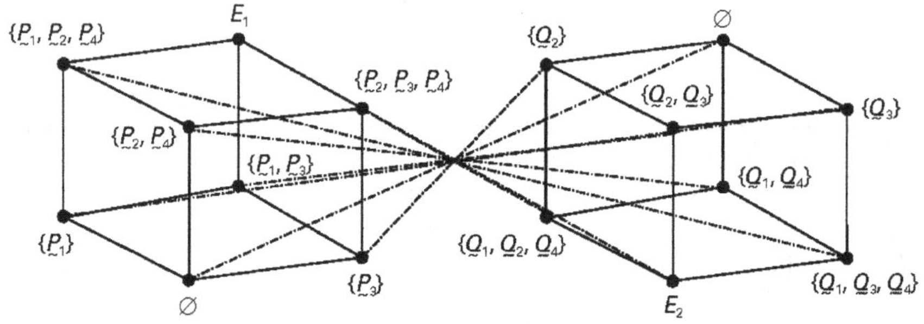

The superposition of both lattices gives a single lattice:

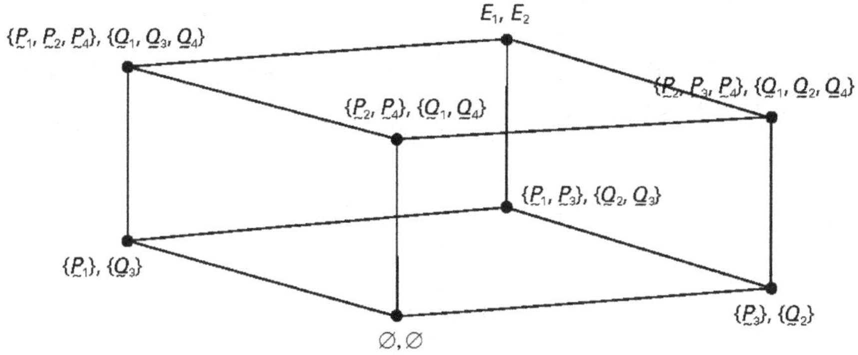

The formation of a single lattice from the two bijective structures, associating in this way elements of the "power set" of the two sets, provides ample possibilities in related processes, some of which have already been developed in the last decades[8] and we sense that also in many unexplored up to now, at least in that which we have been able to ascertain. On this path we believe that our authentic ambition lies: offering the most solid base possible in order to construct new operative techniques on this for the treatment of a wide range of economics and management phenomena in a context of growing uncertainty.

[8] A significant example of that which we have just stated we have in the representation of the affinities by means of Galois lattices, as can be confirmed in the development of the third and final part of this work.

Part III

Operative Techniques for Economy and Management

Chapter 5
Technical Elements with Topological Support

1 A Topology Named Clan

With the aim of advancing in the construction of technical elements capable of being used in the solution of economics and management problems, we will begin with the already presented axiom of the topology which we reproduce in continuation:

1. $\varnothing \in T(E)$
2. $E \in T(E)$
3. $(A_j \in T(E), A_k \in T(E)) \Rightarrow (A_j \cap A_k) \in T(E)$
4. $(A_j \in T(E), A_k \in T(E)) \Rightarrow (A_j \cup A_k) \in T(E)$

If we add a fifth axiom to these four:

5. $(A_j \in T(E)) \Rightarrow (\overline{A}_j \in T(E))$

we find ourselves before a particular topology in which all of the opens are closes and that, in addition, if the topology contains $\underset{\sim}{A}_j$ it also contains $\overline{\underset{\sim}{A}}_j$.

Now it does not result as difficult to state that it is not necessary to define this type of topology with these 5 axioms. It is enough to do this with the following three:

1. $E \in T(E)$
2. $(A_j \in T(E)) \Rightarrow (\overline{A}_j \in T(E))$
3. $(A_j \in T(E) \text{ y } A_k \in T(E)) \Rightarrow (A_j \cup A_k) \in T(E)$

as from these properties it is easy to deduce:

4. $\varnothing \in T(E)$
5. $(A_j \in T(E) \text{ y } A_k \in T(E)) \Rightarrow (A_j \cap A_k) \in T(E)$

In effect:

If $E \in T(E)$, with $\overline{E} = \varnothing$ for the second of these axioms the fourth is also fulfilled.

J. Gil Aluja & A.M. Gil Lafuente: Towards an Advanced Modelling, STUDFUZZ 276, pp. 187–215.
springerlink.com

Also, if:

$$(A_j, A_k \in T(E)) \Rightarrow (A_j \cup A_k) \in T(E)$$

as:

$$\overline{A_j}, \overline{A_k} \in T(E)$$

for the second axiom it may be written:

$$\overline{A_j \cup A_k} \in T(E)$$

and for the theorem of De Morgan:

$$\overline{A_j \cup A_k} \in \overline{A_j} \cap \overline{A_k}$$

As we have repeatedly mentioned, as all $A_j \in T(E)$ involves that all $A_j \in T(E)$, and also the intersections of the A_j and A_k upon equally belonging to $T(E)$, also belong to the intersection of its compliments $A_j \cap A_k$. Which is to say:

$$A_j \cap A_k \in T(E)$$

We will move on to an example, beginning from the referential:

$$E = \{a, b, c, d\}$$

The following subset of $P(E)$ is a topology of this nature:

$$\{\varnothing, \{a\}, \{b\}, \{a, b\}, \{c, d\}, \{a, c, d\}, \{b, c, d\}, E\}$$

In the figures which we present in continuation this topology in $P(E)$ is situated and in the following the Boole sub-lattice that is formed is shown in the most visible way. We have evidently confirmed the fulfillment of its axioms "a priori".

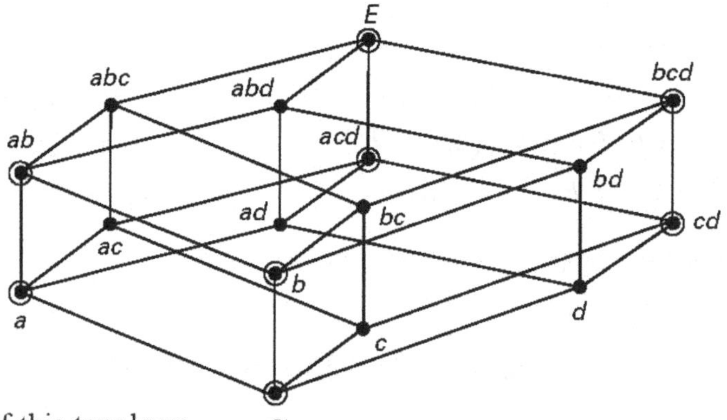

O Elements of this topology ∅

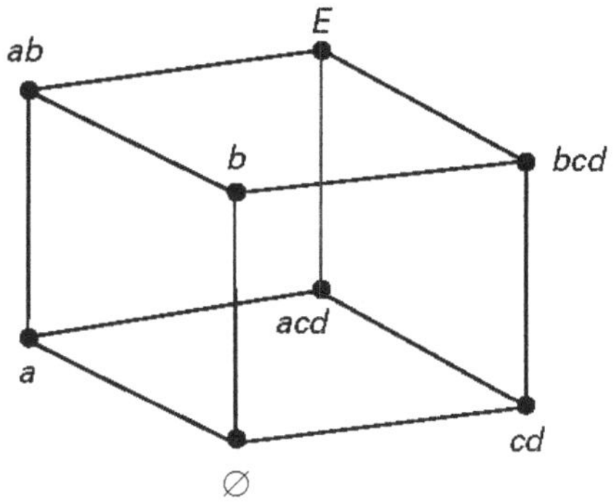

Sub-lattice of the Boole lattice which forms this topology.

It has been seen in the block dedicated to ordinary topology that the elements of $P(E)$ which are not opens are obtained in δ if $A_l \in P(E)$, the interior of A_l is the largest open contained in A_l. The same occurs for $\mathbf{\Gamma}$ with the smallest close of A_l.

The following figure shows the interior application δ of $T(E)$.

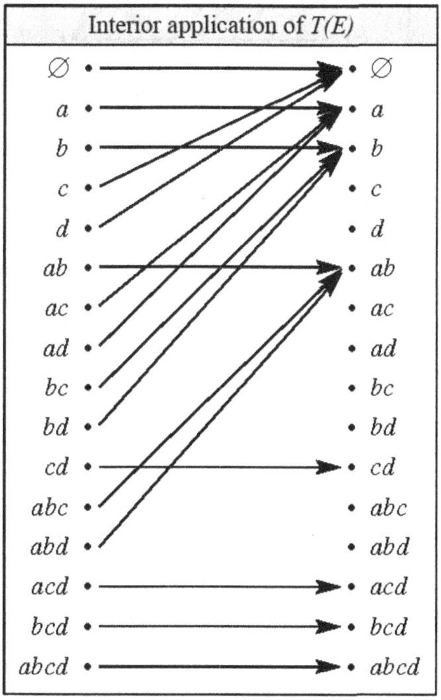

The adherent application of $T(E)$ is found by the formula so often repeated:

$$\Gamma A_j = \overline{\delta \overline{A_j}}$$

We reproduce this in continuation:

Adherence application of $T(E)$

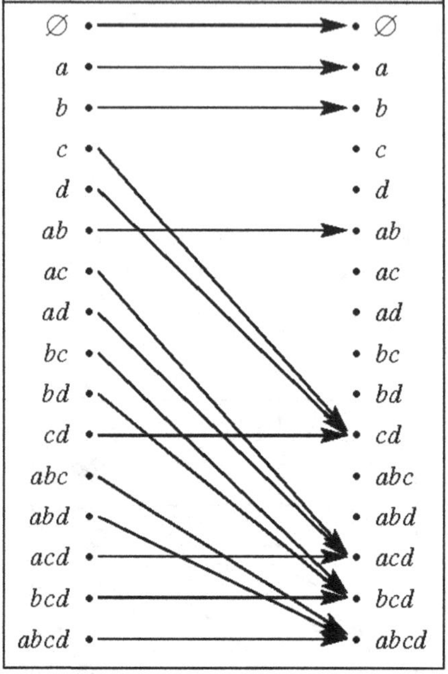

We will confirm that all of the opens are also closes.

$$\overline{\delta\, \overline{a}} = \overline{\delta\, bcd} = \overline{bcd} = a$$
$$\overline{\delta\, \overline{b}} = \overline{\delta\, acd} = \overline{acd} = b$$
$$\overline{\delta\, \overline{ab}} = \overline{\delta\, cd} = \overline{cd} = ab$$
$$\overline{\delta\, \overline{cd}} = \overline{\delta\, ab} = \overline{ab} = cd$$
$$\overline{\delta\, \overline{acd}} = \overline{\delta\, b} = \overline{b} = acd$$
$$\overline{\delta\, \overline{bcd}} = \overline{\delta\, a} = \overline{a} = bcd$$

We may now ask how a topology of this nature can be generated. There evidently exist various ways capable of taking us to this topological space. But the example relative to the grouping of financial products deliberately developed from pretopologies has provided us with an excellent idea which may well be useful for our purposes.

Let us assume as known a family such as:

$$F = \{A_1, A_2, \ldots, A_r\}$$
$$A_j \in P(E) \qquad j = 1, 2, \ldots, r \qquad A_j \neq \varnothing$$

in which each group of financial products whose parts all have common attributes among themselves is expressed by A_j, $i = 1, 2, \ldots, r$.

Beginning from F all of the possible intersections are performed taking the A_j^* of F:

$$A_1^* \cap A_2^* \cap \ldots \cap A_r^*$$

in which $A_j^* = A_j$ or $\overline{A}_j$.

The non empty groupings obtained and all of the possible unions between them are considered, $\varnothing$ is added and a $T(E)$ is obtained which fulfils the axioms established for a topology in which the opens coincide with the closes.

Let us look at a simple example:

Let us assume:

$$E = \{a, b, c, d, e, f, g\} \quad \text{y} \quad F = \big\{\{a, c, d, e, f\} = A_1, \quad \{b, c, d, g\} = A_2\big\}$$

We obtain:

$$A_1 \cap A_2 = \{c, d\} \quad A_1 \cap \overline{A}_2 = \{a, e, f\} \quad \overline{A}_1 \cap A_2 = \{b, g\} \quad \overline{A}_1 \cap \overline{A}_2 = \varnothing$$

which are the groupings obtained from F. We therefore have as a topology given rise to by F:

$$T(F) = \big\{\varnothing, \{c, d\}, \{a, e, f\}, \{b, g\}, \{a, c, d, e, f\}, \{b, c, d, g\}, \{a, b, e, f, g,\}, E\big\}$$

In the following figure this topology given rise to by F has been represented:

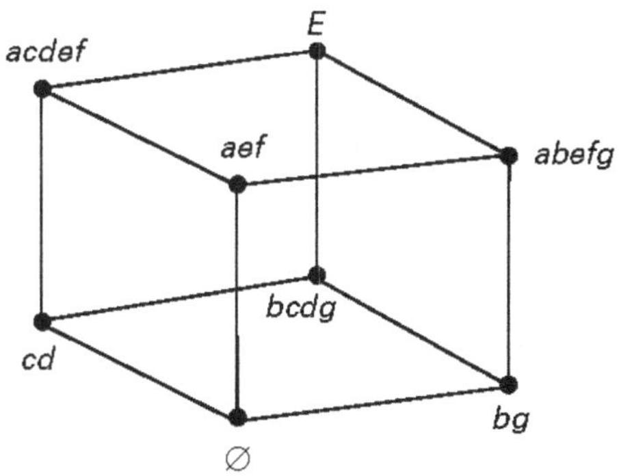

2 The Use of These Topologies in the Area of Management

The topology in which the opens coincide with the closes has given rise to the **concept of clan**.

The **notion of clan** has resulted as extremely useful for the **treatment of information** when the data is found collected in file form[1]. **A file** is a finite set E of specific registers, documents, formulae, programmes and information of a diverse nature.

In continuation we perform a brief outline of the main elements which configure a **clan**, adapted to the necessities of the treatment of the information.

A property P susceptible to being possessed by at least one register of file E is called **key**. A file may contain a set of keys:

$$C = \{P_1, P_2, \ldots, P_n\}$$

in such a way that all register should have at least one property P_i.

The pair formed by a file is designated as (E, C) and the set of keys C.

Also known as:

$$A_j = f(P_j) \qquad j = 1, 2, \ldots, n$$

the part $A_j \in P(E)$ which has the property P_j.

The family of the non empty parts of E which have the property P_1, the property P_2, etc. are called:

$$F = \{A_1, A_2, \ldots, A_n\} = \{f(P_1), f(P_2), \ldots, f(P_n)\}$$

A file (E, C) from a set of keys C may also be designated by the pair (E, F). In this way we arrive at the concept of topological space. A clan is therefore a topology in which the opens and the closes coincide.

From F a clan which is designated by $K(E, F)$ can arise.

Let us take a look at an example, beginning from the referential:

$$E = \{a, b, c, d, e, f, g, h, i, j, k, l\}$$

and the set of keys:

$$C = \{P_1, P_2, P_3\}$$

Let us establish:

$$A_1 = f(P_1) = \{a, c, e, f, g, h, k, l\}$$
$$A_2 = f(P_2) = \{b, c, d, e, f, l\}$$
$$A_3 = f(P_3) = \{d, h, i, j, l\}$$

---

[1] Kaufmann, A., and Gil Aluja, J.: Técnicas operativas de gestión para el tratamiento de la incertidumbre. Ed. Hispano-Europea. Barcelona, 1987, pages. 239-245. Courtillot, M.: Structure canonique des fichiers. AIER-AFCET, vol. 7, January 1973, pages. 2-15.

We obtain:

$$\overline{A}_1 = f(\overline{P}_1) = \{b, d, i, j\}$$
$$\overline{A}_2 = f(P_2) = \{a, g, h, i, j, k\}$$
$$\overline{A}_3 = f(P_3) = \{a, b, c, e, f, g, k\}$$

From which results:

$$f(P_1 \Delta P_2 \Delta P_3) = A_1 \cap A_2 \cap A_3 = \{l\}$$
$$f(P_1 \Delta P_2 \Delta \overline{P}_3) = A_1 \cap A_2 \cap \overline{A}_3 = \{c, e, f\}$$
$$f(P_1 \Delta \overline{P}_2 \Delta P_3) = A_1 \cap \overline{A}_2 \cap A_3 = \{h\}$$
$$f(P_1 \Delta \overline{P}_2 \Delta \overline{P}_3) = A_1 \cap \overline{A}_2 \cap \overline{A}_3 = \{a, g, k\}$$
$$f(\overline{P}_1 \Delta P_2 \Delta P_3) = \overline{A}_1 \cap A_2 \cap A_3 = \{d\}$$
$$f(\overline{P}_1 \Delta P_2 \Delta \overline{P}_3) = \overline{A}_1 \cap A_2 \cap \overline{A}_3 = \{b\}$$
$$f(\overline{P}_1 \Delta \overline{P}_2 \Delta P_3) = \overline{A}_1 \cap \overline{A}_2 \cap A_3 = \{i, j\}$$
$$f(\overline{P}_1 \Delta \overline{P}_2 \Delta \overline{P}_3) = \overline{A}_1 \cap \overline{A}_2 \cap \overline{A}_3 = \varnothing$$

In clan theory these groupings are called **min-terms** or **atoms**.

In this way seven **non-empty atoms** of the file are obtained, which form a set D of the following atoms:

$$D(F) = \big\{\{1\}, \{c, e, f\}, \{h\}, \{a, g, k\}, \{d\}, \{b\}, \{i, j\}\big\}$$

from which if the empty $\varnothing$ and all of the possible unions are added **the clan** is obtained.

Let us see another supposition with a lower number of referential elements. Let:

$$E = \{a, b, c, d, e\}$$

and the set of keys:

$$C = \{P_1, P_2, P_3)$$

We assume that the following contents of A_j^*, $j = 1, 2, 3$ are established in agreement with the nature of the problem:

$$A_1 = f(P_1) = \{a, b\}$$
$$A_2 = f(P_2) = \{a, b, c\}$$
$$A_3 = f(P_3) = \{d, e\}$$
$$\overline{A}_1 = f(\overline{P}_1) = \{c, d, e\}$$
$$\overline{A}_2 = f(\overline{P}_2) = \{d, e\}$$
$$\overline{A}_3 = f(\overline{P}_3) = \{a, b, c\}$$

The corresponding atoms are obtained:

$$f(P_1 \,\Delta\, P_2 \,\Delta\, P_3) = \varnothing$$
$$f(P_1 \,\Delta\, P_2 \,\Delta\, \overline{P}_3) = \{a, b\}$$
$$f(P_1 \,\Delta\, \overline{P}_2 \,\Delta\, P_3) = \varnothing$$
$$f(P_1 \,\Delta\, \overline{P}_2 \,\Delta\, \overline{P}_3) = \varnothing$$
$$f(\overline{P}_1 \,\Delta\, P_2 \,\Delta\, P_3) = \varnothing$$
$$f(\overline{P}_1 \,\Delta\, P_2 \,\Delta\, \overline{P}_3) = \{c\}$$
$$f(\overline{P}_1 \,\Delta\, \overline{P}_2 \,\Delta\, P_3) = \{d, e\}$$
$$f(\overline{P}_1 \,\Delta\, \overline{P}_2 \,\Delta\, \overline{P}_3) = \varnothing$$

Therefore:

$$D(F) = \big\{\{a, b\}, \{c\}, \{d, e\}\big\}$$

And from here the clan which we present in continuation results:

$$K(F) = \big\{\varnothing, \{a, b\}, \{c\}, \{d, e\}, \{a, b, c\}, \{c, d, e\}, \{a, b, d, e\}, E\big\}$$

which can be represented by means of the following sub-lattice of the Boole lattice corresponding to the "power set":

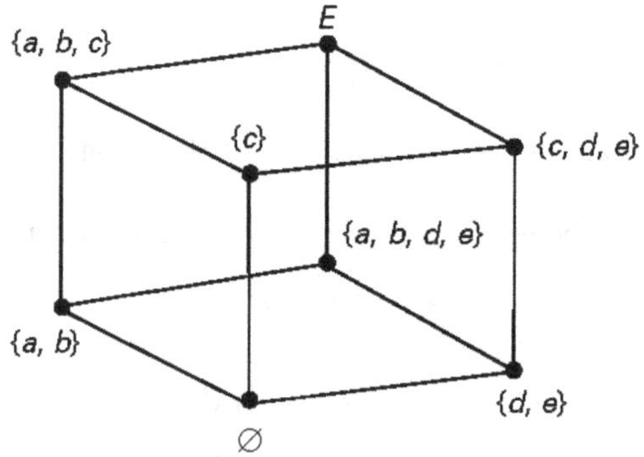

These theoretical and technical elements have been repeatedly used for the solution of some approaches which resulted as difficult to formalise in other ways. By way of example we mention the model followed for a selection of financial products[2], the most general application in the field of investment[3] or work, which gave place to the **Mapclan** model for the grouping of products[4].

---

[2] Gil Lafuente, A. M.: Nuevas estrategias para el análisis financiero en la empresa. Ed. Ariel. Barcelona, 2001, pages. 350-356.

[3] Gil Aluja, J.: Investment in uncertainty. Kluwer Academic Publishers, Boston, London and Dordrecht, 1999, pages. 115-123.

[4] Gil Aluja, J.: «MAPCLAN, model for assembling products by means of Clans.» Proceedings of the Third International Conference on Modelling and Simulation Ms' 97. Melbourne, 29-31 October 1997, pages. 496-504.

There are certain kinds of economics and management problems in which a distinct approach can be performed. The starting point is a key and the atom to which it corresponds is wanted to be obtained.

To illustrate that which we have just stated, we will begin with a key such as:

$$P = (P_1 \,\Delta\, \overline{P}_2 \,\Delta\, P_3) \,\nabla\, \left(\overline{P_1 \,\Delta\, \overline{P}_2}\right)$$

Its meaning is obvious: it deals with obtaining the properties P_1 and P_3 and the property **non-** $\boldsymbol{P_2}$ and/or that the property P_1 does not exist and neither that of property non-P_2.

It is normal to transform this key into the form of boolean variables in scientific and technical medium. In our example this would be:

$$x = x_1\, \overline{x_2}\, x_3 \,\dot{+}\, \overline{x_1\, \overline{x_2}} = x_1\, \overline{x_2}\, x_3 \,\dot{+}\, \overline{x_1}\, x_2$$

Even though it may be well-known by many, we have not resisted the temptation of reproducing an example which has become a classic in the literature on this subject[5]. It deals with the solving of a problem of the choice of a car. A set of 5 models is considered:

$$E = \{M_1, M_2, M_3, M_4, M_5\}$$

The decision maker takes into account the following properties:

P_1: 4 doors	$\overline{P}_1$: 2 doors
$P_2 \leq 5$ CV	$\overline{P}_2$: >5 CV
P_3: power steering	$\overline{P}_3$: no power steering
P_4: luxury interior	$\overline{P}_4$: standard interior

The examination of the catalogue of each of the models gives:

$$f(P_1) = \{M_1, M_2, M_4\}$$
$$f(P_2) = \{M_2, M_3, M_4, M_5\}$$
$$f(P_3) = \{M_3, M_5\}$$
$$f(P_4) = \{M_1, M_2\}$$

With the aim of "organising" this information it is normal to use the matrix which shows the correspondence between car models and properties:

	P_1	P_2	P_3	P_4	$\overline{P}_1$	$\overline{P}_2$	$\overline{P}_3$	$\overline{P}_4$
M_1	1			1		1	1	
M_2	1	1		1			1	
M_3		1	1		1			1
M_4	1	1					1	1
M_5		1	1		1			1

[5] Kaufmann, A., y Gil Aluja, J.: Técnicas especiales para la gestión de expertos. Ed. Milladoiro. Santiago de Compostela, 1993, pages 73-74.

The min-terms are:

$$
\begin{array}{llll}
f(1234) = \varnothing & f(123\bar{4}) = \varnothing & f(12\bar{3}4) = \{M_2\} & f(12\overline{34}) = \{M_4\} \\
f(1\bar{2}34) = \varnothing & f(1\bar{2}3\bar{4}) = \varnothing & f(1\overline{23}4) = \{M_1\} & f(1\overline{234}) = \varnothing \\
f(\bar{1}234) = \varnothing & f(\bar{1}23\bar{4}) = \{M_3, M_5\} & f(\bar{1}2\bar{3}4) = \varnothing & f(\bar{1}2\overline{34}) = \varnothing \\
f(\overline{12}34) = \varnothing & f(\overline{12}3\bar{4}) = \varnothing & f(\overline{123}4) = \varnothing & f(\overline{1234}) = \varnothing
\end{array}
$$

Therefore the **non-empty** min-terms or atoms are: $\{M_1\}$,$\{M_2\}$,$\{M_4\}$, $\{M_3, M_5\}$. As we have repeatedly signalled, these atoms, their possible unions and the empty form a clan, which is to say a topology.

This topology can be represented by means of a Boole lattice such as:

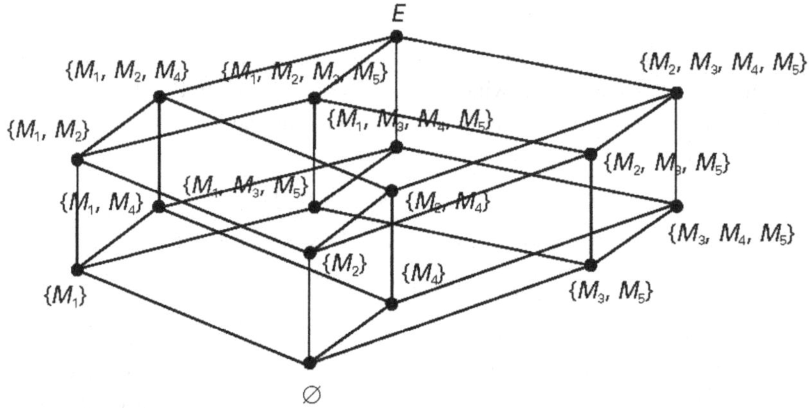

We do not believe it necessary to insist that this lattice is at the same time a Boole sub-lattice of the elements of the "power set" $P(E)$.

For this we assume that the model or models which have the following property are wished to be found:

$$P = \bar{P}_1 \,\Delta\, P_2 \,\Delta\, P_3 \,\Delta\, \bar{P}_4$$

which is to say a car with 2 doors (not 4 doors), with a power equal or less to 5 CV, that it has power steering and a standard interior (not luxury).

Observing the min-terms $\{M_3, M_5\}$ must satisfy the following:

$$
\begin{aligned}
\bar{P}_1 \,\Delta\, P_2 \,\Delta\, P_3 \,\Delta\, \bar{P}_4 &= \{M_3, M_5\} \cap \{M_2, M_3, M_4, M_5\} \\
&\quad \cap \{M_3, M_5\} \cap \{M_3, M_4, M_5\} \\
&= \{M_3, M_5\}
\end{aligned}
$$

It is evident that this outline is not restricted to the clans which coincide with the min-terms, but that any condition with **y** or with **y/o** may be established. In this way we assume that only the following is demanded:

$$(P_1 \Delta P_2) \nabla \overline{P}_4$$

We obtain:

$$\begin{aligned}(P_1 \Delta P_2) \Delta \overline{P}_4 &= (\{M_1, M_2, M_4\} \cap \{M_2, M_3, M_4, M_5\}) \\ &\quad \cup \{M_3, M_4, M_5\} \\ &= \{M_2, M_4\} \cup \{M_3, M_4, M_5\} \\ &= \{M_2, M_3, M_4, M_5\}\end{aligned}$$

Therefore, all the models satisfy the conditions demanded except M_1.

The example reproduced and others which we could present show that if one has 1) the file split into atoms, and 2) a procedure which allows the atoms to be clearly stated, any key which is desired may be obtained.

The use of clans for the solution of the problems of analysis and decision in the field of economics and management has been frequent in recent years. But, maybe, when it has merited greater attention from researchers and business men, it has been as a result of the extension of the method of clans with the use of fuzzy variables, which is to say [0, 1]. It therefore results as interesting to move on to a generalisation of our proposal with the use of fuzzy logic.

3 Clans in Uncertainty

The axiomatic shown for the definition of this type of topology and consequently of a clan are also maintained when an extension to the fuzzy subsets takes place. Here also, as it could not be in any other way, the subsets which we consider could be fuzzy subsets of a referential E or the same fuzzy subsets which make up the elements of the referential, as we have seen in the two focuses of uncertain topological spaces.

A) We will briefly see how the first of these ways is conceived. Starting from a non-fuzzy subset of fuzzy subsets $\underset{\sim}{A}_j^{(i)} \in L^E$, $L = [0, 1]$, $i = 1, 2, \ldots, r$. These fuzzy subsets will be defined in the following way:

A referential is assumed as given:

$$E = \{x_1, x_2, \ldots, x_n\}$$

and also another referential:

$$C = \{P_1, P_2, \ldots, P_m\}$$

Between these there exists a relationship so that if $x_i \in E$, when it has a value of the characteristic function P_j with the level μ, it would be written:

$$\mu_{\underset{\sim}{A}_j}(x) = \mu \qquad \mu \in [0, 1]$$

Which is to say:

$$\underset{\sim}{A}_j = \underset{\sim}{f}(P_j)$$

In the following a fuzzy subset $\underset{\sim}{U} \subset C$ is defined, called a **threshold subset**.

$$\mu_{\underset{\sim}{u}}(P_j) = \lambda_j \in [0, 1]$$

with which some boolean subsets are found from which and, following the now known process we arrive at the min-terms or atoms and from there the topology called clan.

Before continuing and with the aim of clarifying that which we have just stated, we will turn to an example. Let:

$$E = \{a, b, c, d, e, f, g\}$$
$$C = \{P_1, P_2, P_3\}$$

We assume that, due to the characteristics of the problem studied, we have the following fuzzy subsets:

$\underset{\sim}{A}_1 = \underset{\sim}{f}(P_1) =$

a	b	c	d	e	f	g
.3	1	.8	0	.2	0	.4

$\underset{\sim}{A}_2 = \underset{\sim}{f}(P_2) =$

a	b	c	d	e	f	g
.7	0	.7	.2	.1	.8	1

$\underset{\sim}{A}_3 = \underset{\sim}{f}(P_3) =$

a	b	c	d	e	f	g
1	.8	.7	.7	.3	.9	.2

$\underset{\sim}{U} =$

P_1	P_2	P_3
.6	.3	.8

From here through the $\propto$-cuts of the fuzzy subset of thresholds $\underset{\sim}{U}$, we have:

$A_1^{(0,6)} = f^{0,6}(P_1) =$

a	b	c	d	e	f	g
0	1	1	0	0	0	0

$= \{b, c\}$

$A_2^{(0,3)} = f^{0,3}(P_2) =$

a	b	c	d	e	f	g
1	0	1	0	0	1	1

$= \{a, c, f, g\}$

$A_3^{(0,8)} = f^{0,8}(P_3) =$

a	b	c	d	e	f	g
1	1	0	0	0	1	0

$= \{a, b, f\}$

The type of ∝-cut has been taken:

$$< \propto \;\Rightarrow\; 0$$
$$\geq \propto \;\Rightarrow\; 1$$

We find ourselves in disposition of obtaining the min-terms, in the following way:

$$f(1,2,3) = \varnothing \qquad f(1,2,\overline{3}) = \{c\} \qquad f(1,\overline{2},3) = \{b\}$$
$$f(1,\overline{2},\overline{3}) = \varnothing \qquad f(\overline{1},2,3) = \{a,f\} \qquad f(\overline{1},2,\overline{3}) = \{g\}$$
$$f(\overline{1},\overline{2},3) = \varnothing \qquad f(\overline{1},\overline{2},\overline{3}) = \{d,e\}$$

The obtaining of the corresponding clan now results as immediate.
We assume that we have established the key:

$$P = (P_1 \,\Delta\, P_2) \,\nabla\, (\overline{P}_1 \,\Delta\, \overline{P}_3)$$

corresponding to:

$$\begin{aligned} f(P) &= f((P_1 \,\Delta\, P_2) \,\nabla\, (\overline{P}_1 \,\Delta\, \overline{P}_3)) = \\ &= (A_1^{(0,6)} \cap A_2^{(0,3)}) \cup (\overline{A}_1^{(0,6)} \cap \overline{A}_3^{(0,8)}) = \\ &= (\{b,c\} \cap \{a,c,f,g\}) \cup (\{a,d,e,f,g\} \cap \{c,d,e,g\}) = \\ &= \{c\} \cup \{d,e,g\} = \{c,d,e,g\} \end{aligned}$$

On certain occasions the presentation of this outline takes place by forming a matrix with the fuzzy subsets $\underset{\sim}{A}_j$. This type of matrix results as very convenient to reason from the min-terms and show how these are formed. In our example which follows that which we have just stated can be seen, beginning from the established threshold $\underset{\sim}{U}$. In effect, if we consider:

	a	b	c	d	e	f	g	U
P_1	.3	1	.8	0	.2	0	.4	.6
P_2	.7	0	.7	.2	.1	.8	1	.3
P_3	1	.8	.7	.7	.3	.9	.2	.8

the 8 min-terms will be obtained:

	a	b	c	d	e	f	g
P_1		1	1				
P_2	1		1			1	1
P_3	1	1				1	

	a	b	c	d	e	f	g
P_1		1	1				
P_2	1		1			1	1
$\overline{P}_3$			1	1	1		1

	a	b	c	d	e	f	g
P_1		1	1				
$\overline{P}_2$		1		1	1		
P_3	1	1				1	

	a	b	c	d	e	f	g
P_1		1	1				
$\overline{P}_2$		1		1	1		
$\overline{P}_3$			1	1	1		1

	a	b	c	d	e	f	g
$\overline{P}_1$	1			1	1	1	1
P_2	1		1			1	1
P_3	1	1				1	

	a	b	c	d	e	f	g
$\overline{P}_1$	1			1	1	1	1
P_2	1		1			1	1
$\overline{P}_3$			1	1	1		1

	a	b	c	d	e	f	g
$\overline{P}_1$	1			1	1	1	1
$\overline{P}_2$		1		1	1		
P_3	1	1				1	

	a	b	c	d	e	f	g
$\overline{P}_1$	1			1	1	1	1
$\overline{P}_2$		1		1	1		
$\overline{P}_3$			1	1	1		1

The previous fuzzy relationship is normally called **Emergency Matrix** and the non-fuzzy relationships which which have allowed us to find the min-terms are called **Clan Matrices** for the levels of $\underset{\sim}{U}$. It is obvious that in these matrices the completely identical columns give the elements which belong to the same atom.

With this necessarily theoretical exposition done, we believe it to be illustrative to reproduce an application in the field of finance.

For this we will assume[6] that in the analysis of the economic-financial situation of a company it is considered that certain aspects exist which may give place to difficulties or "illnesses":

M_1 = Lack of liquid assets
M_2 = Difficulty of external financing
M_3 = Problems of capturing self-financing
M_4 = Possibility of arriving at cessation of payments
M_5 = Disequilibrium between the economic and financial structures
M_6 = Continuous losses

The experts estimate that these difficulties can be detected through the value which the magnitudes (symptoms of difficulties) S_1, S_2, S_3, S_4, S_5 and S_6 take.

	M_1	M_2	M_3	M_4	M_5	M_6
For S_1:	[.6, 1]	[.6, 1]	[.8, 1]	[.8, 1]	[.5, 1]	[.6, 1]
For S_2:	< 10.000	< 5.000	< 0	< 1.000	< 20.000 > 30.000	< 8.000
For S_3:	$\leq 1,2$	$\leq 1,3$	$\leq 1,2$	≤ 1	$\leq 1,6$	$\leq 0,8$
For S_4:	$< 0,4$	$< 0,3$	$< 0,2$	$< 0,1$	$< 0,4$	$< 0,2$
For S_5:	$< 0,7$	$< 1,5$	$< 1,4$	$< 0,8$	$< 1,5$	< 1
For S_6:	$\leq 0,6$	$\leq 0,5$	$\leq 0,5$	≤ 0	$\leq 0,6$	≤ 0

It therefore deals with a "table of financial pathologies", valid for a determined type of activity and a determined moment in time.

The "medic" or financial analyst that visits the company or institution should estimate in which situation it will be found in a future moment in time, which is to say which values are expected to be reached for each of the symptoms S_i, $i = 1, 2, .., 6$. Then, the pertinent studies establish the following values, some in precise numerical terms, others in intervals of confidence and one in triplet.

$$S_1 = [0{,}6, 0{,}7] \qquad S_2 = 7.000 \qquad S_3 = [1{,}1,\ 1{,}2]$$

$$S_4 = 0.3 \qquad S_5 = (1,\ 1{,}2,\ 1{,}3) \qquad S_6 = 0.6$$

[6] Kaufmann, A., y Gil Aluja, J.: *Técnicas especiales para la gestión de expertos*. Ed. Milladoiro. Santiago de Compostela, 1993, pages 84-87.

The following boolean matrix is found:

	M_1	M_2	M_3	M_4	M_5	M_6
S_1:	1	1			1	1
S_2:	1				1	1
S_3:	1	1	1		1	
S_4:	1				1	
S_5:		1	1		1	
S_6:	1				1	

A simple look allows the first conclusions to be reached which, despite being superficial, may result as interesting.

This company will have, at the fixed date, a disequilibrium between the economic and financial structures and very probably a lack of treasury (all of the symptoms in M_5 and all except S_5 in M_1).

We now move on to obtain the min-terms or atoms by means of this boolean matrix:

	S_1	S_2	S_3	S_4	S_5	S_6	$\overline{S}_1$	$\overline{S}_2$	$\overline{S}_3$	$\overline{S}_4$	$\overline{S}_5$	$\overline{S}_6$
M_1	1	1	1	1		1					1	
M_2	1		1		1			1		1		1
M_3			1		1		1	1		1		1
M_4							1	1	1	1	1	1
M_5	1	1	1	1	1	1						
M_6	1	1							1	1	1	1

In this case $2^6 = 64$ min-terms exist.

$$f(123456) = \{M_5\} \qquad f(12345\overline{6}) = \varnothing \qquad f(1234\overline{5}6) = \{M_1\}$$

....

$$f(\overline{1234}5\overline{6}) = \varnothing \qquad f(\overline{12345}6) = \varnothing \qquad f(\overline{123456}) = \{M_4\}$$

From non-empty the min-terms we construct the corresponding clan in the now known way.

Finally we assume a key such as:

$$(S_1 \,\Delta\, S_3 \,\Delta\, \overline{S}_4) \,\nabla\, (\overline{S}_2 \,\Delta\, S_3 \,\Delta\, \overline{S}_6)$$

We have:

$$(\{M_1, M_2, M_5, M_6\} \cap \{M_1, M_2, M_3, M_5\} \cap \{M_2, M_3, M_4, M_6\}) \cup$$
$$\cup\, (\{M_2, M_3, M_4\} \cap \{M_2, M_3, M_5\} \cap \{M_2, M_3, M_4, M_6\}) =$$
$$= \{M_2\} \cup \{M_2, M_3\} = \{M_2, M_3\}$$

From this key it is deduced that, by the nature of the established symptoms, the company will have difficulties in obtaining external financing and problems of capturing self-financing.

We believe that what we have shown is enough, be it in a superficial way, to illustrate the possibilities of use of this aspect of the theory of clans.

B) We now move on to the development of the second focus of uncertain topological spaces. As previously done, we will start from a referential of fuzzy subsets. Let us see an example:

$$E = \{\underset{\sim}{P}_1, \underset{\sim}{P}_2, \underset{\sim}{P}_3, \underset{\sim}{P}_4, \underset{\sim}{P}_5, \underset{\sim}{P}_6\}$$

in which:

$\underset{\sim}{P}_1 =$

a	b	c
.3	1	.8

$\underset{\sim}{P}_4 =$

a	b	c
.4	.9	.4

$\underset{\sim}{P}_2 =$

a	b	c
.7	0	.7

$\underset{\sim}{P}_5 =$

a	b	c
.8	.5	.7

$\underset{\sim}{P}_3 =$

a	b	c
1	.8	.7

$\underset{\sim}{P}_6 =$

a	b	c
.5	.8	.2

Now, the fuzzy subset of thresholds has as a referential the same elements as the referential of the fuzzy subsets $\underset{\sim}{P}_j$.

$\underset{\sim}{U} =$

a	b	c
.6	.7	.6

With this information the following matrix whose valuations are all the same or higher than the corresponding threshold may be created.

	a	b	c
P_1		1	.8
P_2	.7		.7
P_3	1	.8	.7
P_4		.9	
P_5	.8		.7
P_6		.8	

From here:

$\underset{\sim}{a} =$

P_1	P_2	P_3	P_4	P_5	P_6
	.7	1		.8	

$= \{P_2, P_3, P_5\}$

$\underset{\sim}{b} =$

P_1	P_2	P_3	P_4	P_5	P_6
1		.8	.9		.8

$= \{P_1, P_3, P_4, P_6\}$

$\underset{\sim}{c} =$

P_1	P_2	P_3	P_4	P_5	P_6
.8	.7	.7		.7	

$= \{P_1, P_2, P_3, P_5\}$

The min-terms are:

$a \cap b \cap c = \{\underset{\sim}{P}_1\}$ $\quad a \cap b \cap \bar{c} = \varnothing$ $\quad a \cap \bar{b} \cap c = \{\underset{\sim}{P}_2, \underset{\sim}{P}_5\}$

$\bar{a} \cap b \cap c = \{\underset{\sim}{P}_1\}$ $\quad a \cap \bar{b} \cap \bar{c} = \varnothing$ $\quad \bar{a} \cap b \cap \bar{c} = \{\underset{\sim}{P}_4, \underset{\sim}{P}_6\}$

$\bar{a} \cap \bar{b} \cap c = \varnothing$ $\quad \bar{a} \cap \bar{b} \cap \bar{c} = \varnothing$

The non-empty min-terms or atoms, all of their unions, together with the empty, form a clan:

$$K(E) = \big\{\varnothing, \{\underset{\sim}{P}_1\}, \{\underset{\sim}{P}_3\}, \{\underset{\sim}{P}_2, \underset{\sim}{P}_5\}, \{\underset{\sim}{P}_4, \underset{\sim}{P}_6\}, \{\underset{\sim}{P}_1, \underset{\sim}{P}_3\}, \{\underset{\sim}{P}_1, \underset{\sim}{P}_2, \underset{\sim}{P}_5\}, \{\underset{\sim}{P}_1, \underset{\sim}{P}_4, \underset{\sim}{P}_6\},$$
$$\{\underset{\sim}{P}_2, \underset{\sim}{P}_3, \underset{\sim}{P}_5\}, \{\underset{\sim}{P}_3, \underset{\sim}{P}_4, \underset{\sim}{P}_6\}, \{\underset{\sim}{P}_1, \underset{\sim}{P}_2, \underset{\sim}{P}_3, \underset{\sim}{P}_5\}, \{\underset{\sim}{P}_1, \underset{\sim}{P}_3, \underset{\sim}{P}_4, \underset{\sim}{P}_6\},$$
$$\{\underset{\sim}{P}_2, \underset{\sim}{P}_4, \underset{\sim}{P}_5, \underset{\sim}{P}_6\}, \{\underset{\sim}{P}_2, \underset{\sim}{P}_3, \underset{\sim}{P}_4, \underset{\sim}{P}_5, \underset{\sim}{P}_6\}, \{\underset{\sim}{P}_1, \underset{\sim}{P}_2, \underset{\sim}{P}_4, \underset{\sim}{P}_5, \underset{\sim}{P}_6\}, E\big\}$$

As it could not be in any other way, the clan forms a Boole lattice which is the following:

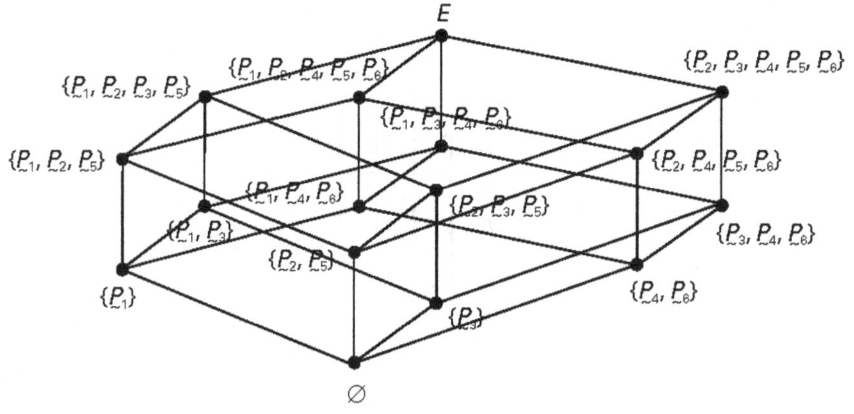

which, at the same time, is a sub-lattice of the corresponding "power set".

The use of this outline does not provoke important difficulties. To confirm this affirmation we believe it enough to recall an approach with which we encountered due to our academical relationships with a great football club in the city in which we live.

In this club a problem was posed as a consequence of an uncomfortable situation in which the manger responsible found himself, upon having to decide the first choice for a position of the team (specifically, that of centre forward) among the four existing players of the squad. We will designate these sportsmen by their initials A, K, O, D. Each of them had certain qualities, characteristics or singularities at a nominally distinct level in a way that a superficial analysis was not enough to adequately choose one specific player among those available to achieve an optimum performance in the position studied for the team.

The first thing to tackle would be the problem of selecting those characteristics which the manager of the club believed to be essential to define a good centre forward under diverse circumstances. The high number of elements used for this objective forces us, in the context of this work, to reduce and summarise them into the following:

a = top speed
b = ability to take advantage of goal scoring opportunities
c = shot strength
d = dribbling ability

The trainers of the club, experts in the characteristics of their sportsmen and in charge of the tracking of their evolution and behaviour, were able to describe each of the sportsmen through the following fuzzy subsets:

$$\underset{\sim}{P}_A = \begin{array}{cccc} a & b & c & d \\ \hline \multicolumn{1}{|c|}{.8} & \multicolumn{1}{c|}{.9} & \multicolumn{1}{c|}{.7} & \multicolumn{1}{c|}{.6} \\ \hline \end{array} \qquad \underset{\sim}{P}_K = \begin{array}{cccc} a & b & c & d \\ \hline \multicolumn{1}{|c|}{.7} & \multicolumn{1}{c|}{.7} & \multicolumn{1}{c|}{.6} & \multicolumn{1}{c|}{.9} \\ \hline \end{array}$$

$$\underset{\sim}{P}_O = \begin{array}{cccc} a & b & c & d \\ \hline \multicolumn{1}{|c|}{.6} & \multicolumn{1}{c|}{.8} & \multicolumn{1}{c|}{.7} & \multicolumn{1}{c|}{.7} \\ \hline \end{array} \qquad \underset{\sim}{P}_D = \begin{array}{cccc} a & b & c & d \\ \hline \multicolumn{1}{|c|}{.9} & \multicolumn{1}{c|}{.5} & \multicolumn{1}{c|}{.6} & \multicolumn{1}{c|}{.8} \\ \hline \end{array}$$

Then, the levels from which it was considered that the sportsmen had the corresponding quality, characteristic or singularity were established. And this through the following fuzzy subsets of thresholds.

$\underset{\sim}{U} =$

a	b	c	d
.7	.8	.7	.8

Upon taking into account these thresholds the following matrix was constructed:

	a	b	c	d
P_A	.8	.9	.7	
P_K	.7			.9
P_O		.8	.7	
P_D	.9			.8

It is easy to find the fuzzy subsets which describe each characteristic, taking the footballers as a referential:

$\underset{\sim}{a} =$

P_A	P_K	P_O	P_D
.8	.7		.9

$= \{\underset{\sim}{P}_A, \underset{\sim}{P}_K, \underset{\sim}{P}_D\}$

$\underset{\sim}{b} =$

P_A	P_K	P_O	P_D
.9		.8	

$= \{\underset{\sim}{P}_A, \underset{\sim}{P}_O\}$

$\underset{\sim}{c} =$

P_A	P_K	P_O	P_D
.7		.7	.8

$= \{\underset{\sim}{P}_A, \underset{\sim}{P}_O, \underset{\sim}{P}_D\}$

$\underset{\sim}{d} =$

P_A	P_K	P_O	P_D
	.9		.8

$= \{\underset{\sim}{P}_K, \underset{\sim}{P}_D\}$

With this information, the obtaining of the corresponding family is immediate:

$$F = \{\{\underset{\sim}{P}_A, \underset{\sim}{P}_K, \underset{\sim}{P}_D\}, \{\underset{\sim}{P}_A, \underset{\sim}{P}_O\}, \{\underset{\sim}{P}_A, \underset{\sim}{P}_O\}, \{\underset{\sim}{P}_K, \underset{\sim}{P}_D\}\}$$

From this family the min-terms or atoms were found. For this the following was taken into account:

$A_1 = \{\underset{\sim}{P}_A, \underset{\sim}{P}_K, \underset{\sim}{P}_D\}$ $\quad A_2 = \{\underset{\sim}{P}_A, \underset{\sim}{P}_O\}$ $\quad A_3 = \{\underset{\sim}{P}_A, \underset{\sim}{P}_O\}$ $\quad A_4 = \{\underset{\sim}{P}_K, \underset{\sim}{P}_D\}$

$\overline{A}_1 = \{\underset{\sim}{P}_O\}$ $\quad \overline{A}_2 = \{\underset{\sim}{P}_K, \underset{\sim}{P}_D\}$ $\quad \overline{A}_3 = \{\underset{\sim}{P}_K, \underset{\sim}{P}_D\}$ $\quad \overline{A}_4 = \{\underset{\sim}{P}_A, \underset{\sim}{P}_O\}$

The min-terms are:

$$A_1 \cap A_2 \cap A_3 \cap A_4 = \varnothing \qquad A_1 \cap A_2 \cap A_3 \cap \overline{A}_4 = \{\underset{\sim}{P}_A\}$$
$$A_1 \cap A_2 \cap \overline{A}_3 \cap A_4 = \varnothing \qquad A_1 \cap \overline{A}_2 \cap A_3 \cap A_4 = \varnothing$$
$$\overline{A}_1 \cap A_2 \cap A_3 \cap A_4 = \varnothing \qquad A_1 \cap A_2 \cap \overline{A}_3 \cap \overline{A}_4 = \varnothing$$
$$A_1 \cap \overline{A}_2 \cap A_3 \cap \overline{A}_4 = \varnothing \qquad \overline{A}_1 \cap A_2 \cap A_3 \cap \overline{A}_4 = \{\underset{\sim}{P}_O\}$$
$$A_1 \cap \overline{A}_2 \cap \overline{A}_3 \cap A_4 = \{\underset{\sim}{P}_K, \underset{\sim}{P}_D\} \qquad \overline{A}_1 \cap A_2 \cap \overline{A}_3 \cap A_4 = \varnothing$$
$$\overline{A}_1 \cap \overline{A}_2 \cap A_3 \cap A_4 = \varnothing \qquad A_1 \cap \overline{A}_2 \cap \overline{A}_3 \cap \overline{A}_4 = \varnothing$$
$$\overline{A}_1 \cap A_2 \cap \overline{A}_3 \cap \overline{A}_4 = \varnothing \qquad \overline{A}_1 \cap \overline{A}_2 \cap A_3 \cap \overline{A}_4 = \varnothing$$
$$\overline{A}_1 \cap \overline{A}_2 \cap \overline{A}_3 \cap A_4 = \varnothing \qquad \overline{A}_1 \cap \overline{A}_2 \cap \overline{A}_3 \cap \overline{A}_4 = \varnothing$$

The non-empty atoms are:

$$\{\underset{\sim}{P}_A\}, \{\underset{\sim}{P}_O\}, \{\underset{\sim}{P}_K, \underset{\sim}{P}_D\}$$

Therefore the clan formed by the family F is:

$$K(E) = \{\varnothing, \{\underset{\sim}{P}_A\}, \{\underset{\sim}{P}_O\}, \{\underset{\sim}{P}_K, \underset{\sim}{P}_D\}, \{\underset{\sim}{P}_A, \underset{\sim}{P}_O\}, \{\underset{\sim}{P}_A, \underset{\sim}{P}_K, \underset{\sim}{P}_D\}, \{\underset{\sim}{P}_O, \underset{\sim}{P}_K, \underset{\sim}{P}_D\}, \{\underset{\sim}{P}_A, \underset{\sim}{P}_O, \underset{\sim}{P}_K, \underset{\sim}{P}_D\}$$

which can be represented by the following Boole lattice:

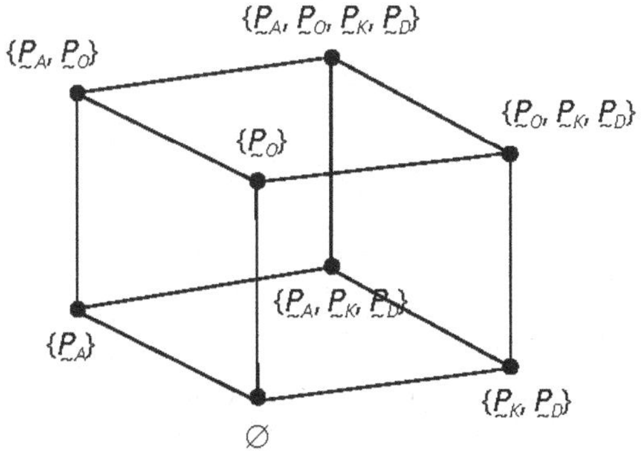

Which is, at the same time, a sub-lattice of that corresponding to the power set, as can be seen in the following figure:

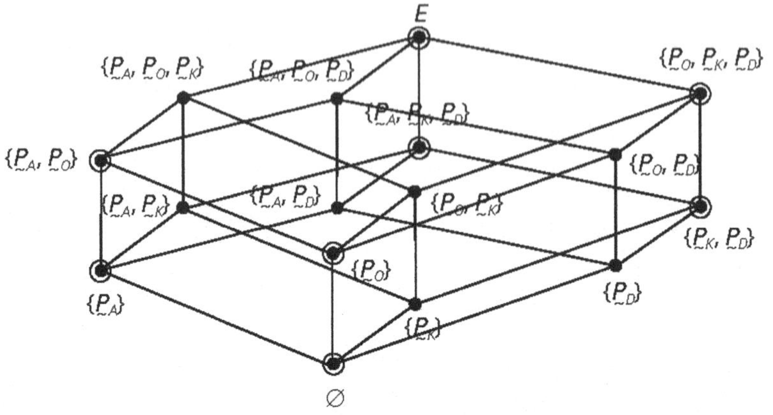

All of this information may be channelled through the keys which can take a high number of forms which, in our case, may be useful for a wide range of circumstances which habitually arise in team sports. In this way, for example, it is efficient in determining which player or players should be on the pitch when the key is:

$$(a \cap b) \cup (c \cap d)$$

which is to say, "top speed", and "ability to take advantage of goal scoring opportunities" or "shot strength" and "dribbling ability". We perform:

$$\begin{aligned}(A_1 \cap A_2) \cup (A_3 \cap A_4) &= (\{\underset{\sim}{P}_A, \underset{\sim}{P}_K, \underset{\sim}{P}_D\} \cap \{\underset{\sim}{P}_A, \underset{\sim}{P}_O\}) \cup (\{\underset{\sim}{P}_A, \underset{\sim}{P}_O\} \cap \{\underset{\sim}{P}_K, \underset{\sim}{P}_D\}) \\ &= \{\underset{\sim}{P}_A\} \cup \varnothing = \{\underset{\sim}{P}_A\}\end{aligned}$$

This key shows us P_A as the single player who fulfils all the established requisites.

Another key which was frequently used is the following:

$$(d \cap b) \cup (a \cap c)$$

which expresses "dribbling ability" and "ability to take advantage of goal scoring opportunities" or "top speed" and "shot strength".

We now perform:

$$\begin{aligned}(A_4 \cap A_2) \cup (A_1 \cap A_3) &= (\{\underset{\sim}{P}_K, \underset{\sim}{P}_D\} \cap \{\underset{\sim}{P}_A, \underset{\sim}{P}_O\}) \cup (\{\underset{\sim}{P}_A, \underset{\sim}{P}_K, \underset{\sim}{P}_D\} \cap \{\underset{\sim}{P}_A, \underset{\sim}{P}_O\} \\ &= \varnothing \cup \{\underset{\sim}{P}_A\} = \{\underset{\sim}{P}_A\}\end{aligned}$$

P_A continues to be our chosen player.

These application, shown in a summarised way, are only a sample of the many possibilities of this instrument of management.

We will now leave these examples to continue with an interesting aspect derived from topological spaces, that of **neighbourhood** in its incorporation to the **theory of clans**.

4 Neighbourhood of a Clan

We believe that it has been sufficiently shown that a clan in E is also a topology in E. As we have been able to see, in all topology a notion of neighbourhood is defined through the following expression:

$$(X \subset E \text{ is a neighbour of } A \subset E)$$
$$\Rightarrow (\exists\, Y \in T(E) / A \subset Y \subset X)$$

In the context of this epigraph we can make $T(E) = K(E)$ with the special feature that $K(E)$ is a Boole lattice.

In the first place we will return to the first example of clan of this block and once again reproduce the corresponding Boole sub-lattice of the "power set".

$$K = \{\emptyset, \{a\}, \{b\}, \{a, b\}, \{c, d\}, \{a, c, d\}, \{b, c, d\}, E\}$$

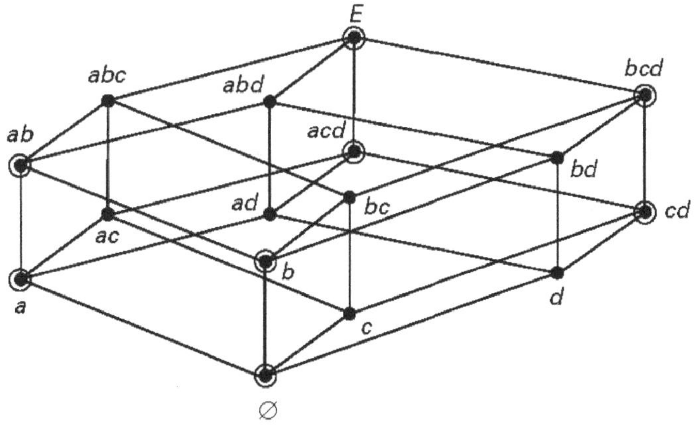

○ Elements of the clan

It has also been possible to confirm that the elements of $P(E)$ which are not open are obtained by the functional application δ. If $A_k \in P(E)$, the interior of A_k is the largest open contained in A_k.

We once again repeat that **in a clan** $K \subset P(E)$, all of the closes are also opens.

To find the neighbourhoods of the previous clan we will use the formula reproduced at the beginning of this epigraph. If, for example, the element $A_j = \{a, c\}$ is considered, there exists an open $Y = \{a, c, d\}$ which is a neighbour of A_j, as $A_j \subset Y$

$\subset K$. In continuation we provide the relationship of all the neighbourhood of the elements of $P(E)$ distinct to those of $K(E)$.

neighbourhood of $\{c\} = \{\{c, d\}, \{a, c, d\}, \{b, c, d\}, E\}$
neighbourhood of $\{d\} = \{\{c, d\}, \{a, c, d\}, \{b, c, d\}, E\}$
neighbourhood of $\{a, c\} = \{\{a, c, d\}, E\}$
neighbourhood of $\{b, c\} = \{\{b, c, d\}, E\}$
neighbourhood of $\{a, d\} = \{\{a, c, d\}, E\}$
neighbourhood of $\{b, d\} = \{\{b, c, d\}, E\}$
neighbourhood of $\{a, b, c\} = \{E\}$
neighbourhood of $\{a, b, d\} = \{E\}$

And in the following for those of $K(E)$:
neighbourhood of $\{a\} = \{\{a\}, \{a, b\}, \{a, c\}, \{a, d\}, \{a, b, c\}, \{a, b, d\}, \{a, c, d\}, E\}$

In effect:

$\{a\} \subset \{a\} \subset \{a\}$ $\qquad$ $\{a\} \subset \{a, b\} \subset \{a, b\}$

$\{a\} \subset \{a\} \subset \{a, c\}$ $\qquad$ $\{a\} \subset \{a\} \subset \{a, d\}$

$\{a\} \subset \{a, b\} \subset \{a, b, c\}$ $\qquad$ $\{a\} \subset \{a, b\} \subset \{a, b, d\}$

$\{a\} \subset \{a, c, d\} \subset \{a, c, d\}$ $\qquad$ $\{a\} \subset E \subset E$

neighbourhood of $\{b\} = \{\{b\}, \{a, b\}, \{b, c\}, \{b, d\}, \{a, b, d\}, \{b, c, d\}, E\}$
neighbourhood of $\{a, b\} = \{\{a, b\}, \{a, b, c\}, \{a, b, d\}, E\}$
neighbourhood of $\{c, d\} = \{\{c, d\}, \{a, b, c\}, \{b, c, d\}, E\}$
neighbourhood of $\{a, c, d\} = \{\{a, c, d\}, E\}$
neighbourhood of $\{b, c, d\} = \{\{b, c, d\}, E\}$
neighbourhood of $\{E\} = \{E\}$

Let us see another supposition, beginning with the same referential. Another clan is considered:

$$K = \{\emptyset, \{c\}, \{d\}, \{c, d\}, E\}$$

which is identified by a circle in the following representative of a Boole lattice of the "power set".

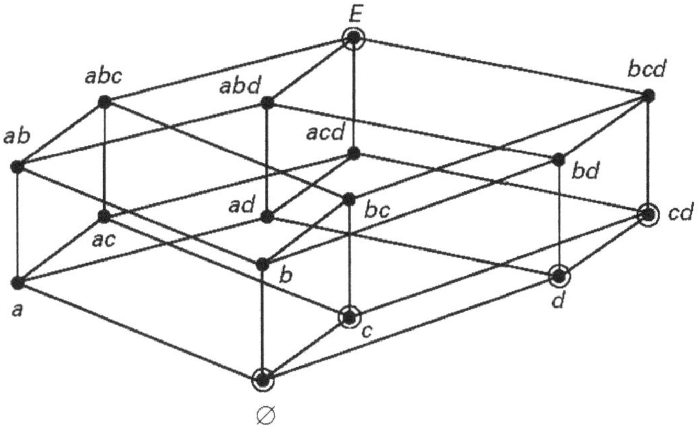

We obtain:

Neighbourhood of $\{a\} = \{E\}$
Neighbourhood of $\{b\} = \{E\}$

All of the neighbourhoods of $A_j \notin K$ give $\{E\}$.

Let us see the $A_j \in K$, $A_j \neq \varnothing$.

neighbourhood of $\{c\}$ = $\{\{c\}, \{a, c\}, \{b, c\}, \{c, d\}, \{a, b, c\}, \{a, c, d\}, \{b, c, d\}, E\}$
neighbourhood of $\{d\}$ = $\{\{d\}, \{a, d\}, \{b, d\}, \{c, d\}, \{a, b, d\}, \{a, c, d\}, \{\mathrm{b, c, d}\}, E\}$
neighbourhood of $\{c, d\} = \{\{c, d\}, \{a, c, d\}, \{b, c, d\}, E\}$
neighbourhood of $\{E\} = \{E\}$

It is easy to see that by construction all of the neighbourhoods of A_j form a boolean sub-lattice in which the lower extreme is A_j and the higher extreme is E. In this way, when $\varnothing$ is incorporated into all of the neighbourhoods of $P(E)$, **sub-clans** are formed.

As has been put forward, if the families of neighbourhoods of all of the elements of $P(E)$ are established as a priority, those which contain their own element form the application δ of the opens or in this case, what leads to the same as $\mathbf{\Gamma} = \delta$, of the closes.

We will now move on to the notion of neighbourhood in the fuzzy area, as much in focus ***A*** as in ***B***. We will begin with the first of these.

A) One of the ways of constructing a **fuzzy clan** consists of considering the existence of a **min-term** from the referential:

$E = \{a, b, c\}$ with:

$\mu_{\underset{\sim}{A}}(a) = 0{,}4 \qquad \mu_{\underset{\sim}{A}}(b) = 0 \qquad \mu_{\underset{\sim}{A}}(c) = 0{,}2$

which gives rise to:

$$\underset{\sim}{A}_1 = \begin{array}{ccc} a & b & c \\ \hline \multicolumn{1}{|c|}{.4} & \multicolumn{1}{c|}{0} & \multicolumn{1}{c|}{.2} \\ \hline \end{array}$$

Let us define $K(E)$ by the fuzzy subsets constructed from $\underset{\sim}{A}_1$.

$$K(E) = \{\varnothing, \overset{\underset{\sim}{A}_1}{\begin{array}{|c|c|c|} \multicolumn{1}{c}{a} & \multicolumn{1}{c}{b} & \multicolumn{1}{c}{c} \\ \hline .4 & 0 & .2 \\ \hline \end{array}}, \overset{\underset{\sim}{A}_2}{\begin{array}{|c|c|c|} \multicolumn{1}{c}{a} & \multicolumn{1}{c}{b} & \multicolumn{1}{c}{c} \\ \hline .4 & 0 & .8 \\ \hline \end{array}}, \overset{\underset{\sim}{A}_3}{\begin{array}{|c|c|c|} \multicolumn{1}{c}{a} & \multicolumn{1}{c}{b} & \multicolumn{1}{c}{c} \\ \hline .4 & 1 & .2 \\ \hline \end{array}}, \overset{\underset{\sim}{A}_4}{\begin{array}{|c|c|c|} \multicolumn{1}{c}{a} & \multicolumn{1}{c}{b} & \multicolumn{1}{c}{c} \\ \hline .4 & 1 & .8 \\ \hline \end{array}},$$

$$\overset{\underset{\sim}{A}_5}{\begin{array}{|c|c|c|} \multicolumn{1}{c}{a} & \multicolumn{1}{c}{b} & \multicolumn{1}{c}{c} \\ \hline .6 & 0 & .2 \\ \hline \end{array}}, \overset{\underset{\sim}{A}_6}{\begin{array}{|c|c|c|} \multicolumn{1}{c}{a} & \multicolumn{1}{c}{b} & \multicolumn{1}{c}{c} \\ \hline .6 & 0 & .8 \\ \hline \end{array}}, \overset{\underset{\sim}{A}_7}{\begin{array}{|c|c|c|} \multicolumn{1}{c}{a} & \multicolumn{1}{c}{b} & \multicolumn{1}{c}{c} \\ \hline .6 & 1 & .2 \\ \hline \end{array}}, \overset{\underset{\sim}{A}_8}{\begin{array}{|c|c|c|} \multicolumn{1}{c}{a} & \multicolumn{1}{c}{b} & \multicolumn{1}{c}{c} \\ \hline .6 & 1 & .8 \\ \hline \end{array}}, E\}$$

In which:

$$\underset{\sim}{A}_8 = \overline{\underset{\sim}{A}_1} \qquad \underset{\sim}{A}_7 = \overline{\underset{\sim}{A}_2} \qquad \underset{\sim}{A}_6, = \overline{\underset{\sim}{A}_3} \qquad \underset{\sim}{A}_5 = \overline{\underset{\sim}{A}_4}$$

It is evident that it satisfies the axiomatic presented at the beginning of this third part and therefore it is a **fuzzy clan**. This clan is presented by way of the following lattice:

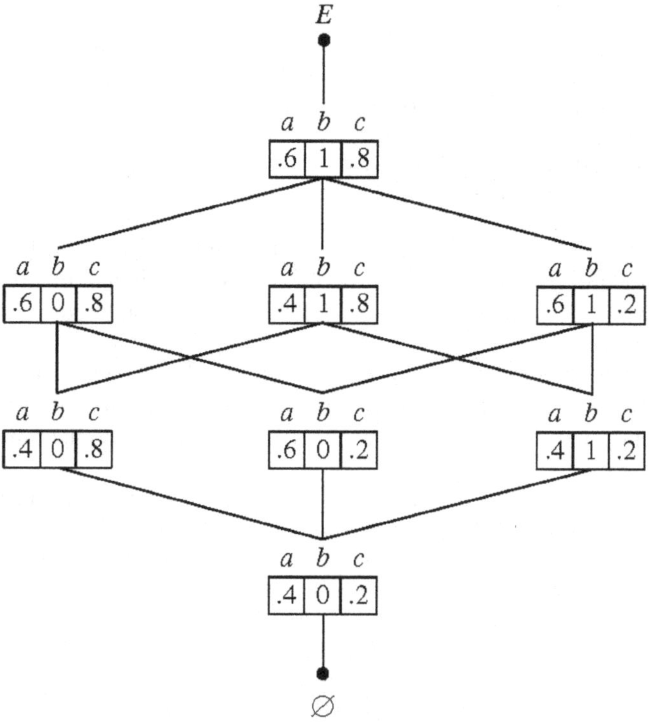

We now assume a fuzzy subset, chosen arbitrarily of the referential $E = \{a, b, c\}$, for example:

a	b	c
.5	0	.1

It can be confirmed in $K(E)$ that there exists a

$\underset{\sim}{Y} =$

a	b	c
.6	0	.2

$\in K(E)$

such as

a	b	c
.5	0	.1

$\subset$

a	b	c
.6	0	.2

and all of the opens of $K(E)$ which have

a	b	c
.6	0	.2

as a lower extreme form a distributive sub-lattice of the distributive lattice of the previous figure.

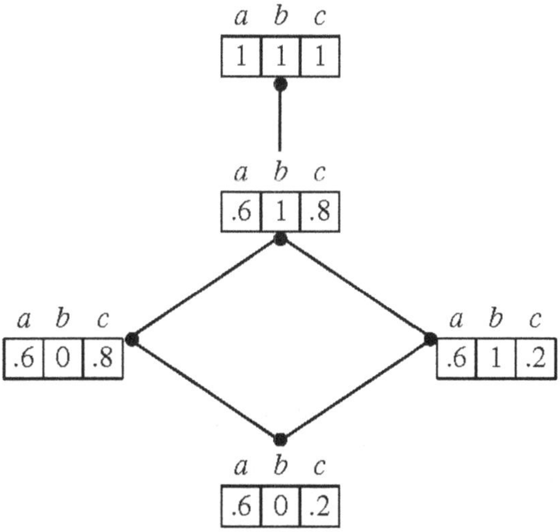

and the neighbourhood of

a	b	c
.5	0	.1

is, therefore:

$$\text{Neighbourhood of } \begin{array}{ccc} a & b & c \\ \hline \multicolumn{1}{|c|}{.5} & \multicolumn{1}{c|}{0} & \multicolumn{1}{c|}{.1} \\ \hline \end{array} = \left\{ \begin{array}{ccc} a & b & c \\ \hline \multicolumn{1}{|c|}{.6} & \multicolumn{1}{c|}{0} & \multicolumn{1}{c|}{.2} \\ \hline \end{array}, \begin{array}{ccc} a & b & c \\ \hline \multicolumn{1}{|c|}{.6} & \multicolumn{1}{c|}{0} & \multicolumn{1}{c|}{.8} \\ \hline \end{array}, \begin{array}{ccc} a & b & c \\ \hline \multicolumn{1}{|c|}{.6} & \multicolumn{1}{c|}{1} & \multicolumn{1}{c|}{.2} \\ \hline \end{array}, \right.$$

$$\left. \begin{array}{ccc} a & b & c \\ \hline \multicolumn{1}{|c|}{.6} & \multicolumn{1}{c|}{1} & \multicolumn{1}{c|}{.8} \\ \hline \end{array}, \; E \right\}$$

The usefulness of the notion of neighbourhood in a fuzzy clan is evident. If we have a fuzzy subset of properties and a clan of economic and business objects characterised by these properties, it may be interesting to know the elements of the clan which dominate the fuzzy subset considered. In the assumption that it is necessary to take into account various fuzzy subsets of reference, their union is taken as is our habit.

B) To show the second way we limit ourselves to reproducing the clan of the sporting example put forward in the previous epigraph, presented in an abstract way:

$$K(E) = \{\emptyset, \{\underset{\sim}{P}_1\}, \{\underset{\sim}{P}_3\}, \{\underset{\sim}{P}_2, \underset{\sim}{P}_4\}, \{\underset{\sim}{P}_1, \underset{\sim}{P}_3\}, \{\underset{\sim}{P}_1, \underset{\sim}{P}_2, \underset{\sim}{P}_4\}, \{\underset{\sim}{P}_2, \underset{\sim}{P}_3, \underset{\sim}{P}_4\}, \{\underset{\sim}{P}_1, \underset{\sim}{P}_2, \underset{\sim}{P}_3, \underset{\sim}{P}_4\}\}$$

In which, we remind ourselves:

$$\underset{\sim}{P}_1 = \begin{array}{cccc} a & b & c & d \\ \hline \multicolumn{1}{|c|}{.8} & \multicolumn{1}{c|}{.9} & \multicolumn{1}{c|}{.7} & \multicolumn{1}{c|}{.6} \\ \hline \end{array} \qquad \underset{\sim}{P}_2 = \begin{array}{cccc} a & b & c & d \\ \hline \multicolumn{1}{|c|}{.7} & \multicolumn{1}{c|}{.7} & \multicolumn{1}{c|}{.6} & \multicolumn{1}{c|}{.9} \\ \hline \end{array}$$

$$\underset{\sim}{P}_3 = \begin{array}{cccc} a & b & c & d \\ \hline \multicolumn{1}{|c|}{.6} & \multicolumn{1}{c|}{.8} & \multicolumn{1}{c|}{.7} & \multicolumn{1}{c|}{.7} \\ \hline \end{array} \qquad \underset{\sim}{P}_4 = \begin{array}{cccc} a & b & c & d \\ \hline \multicolumn{1}{|c|}{.9} & \multicolumn{1}{c|}{.5} & \multicolumn{1}{c|}{.6} & \multicolumn{1}{c|}{.8} \\ \hline \end{array}$$

The elements of the clan within the Boole lattice representative of the "power set" are indicated with a circle.

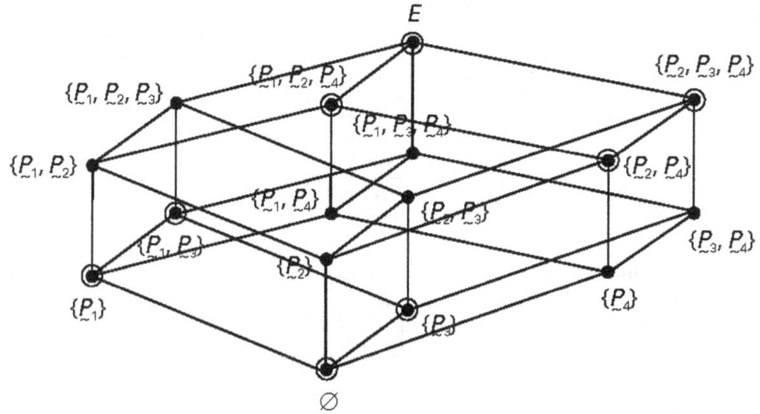

If any element of the "power set" is considered such as $\{\underset{\sim}{P}_4\}$ for example, there exists a:

$$\{\underset{\sim}{P}_2, \underset{\sim}{P}_4\} \in K(E)$$

so that:

$$\{\underset{\sim}{P}_4\} \subset \{\underset{\sim}{P}_2, \underset{\sim}{P}_4\}$$

So, all of the opens of $K(E)$ which contain $\{\underset{\sim}{P}_2, \underset{\sim}{P}_4\}$ form the following Boole sub-lattice:

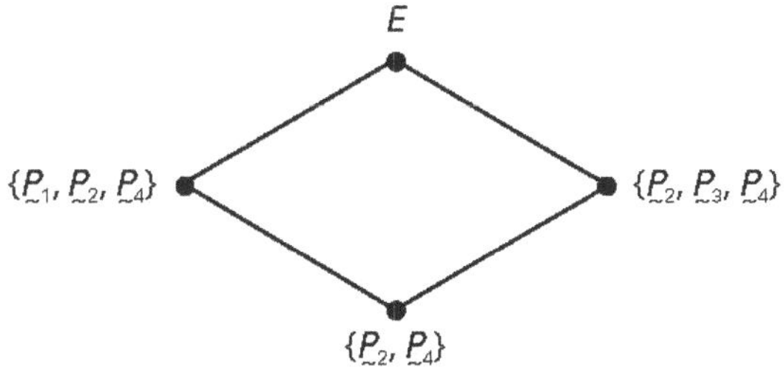

In this way the neighbourhood of $\{\underset{\sim}{P}_4\}$ is:

Neighbourhood $\{\underset{\sim}{P}_4\} = \{\{$

	a	b	c	d
	.7	.7	.6	.9
,	.9	.5	.6	.8

$\}, \{$

a	b	c	d
.8	.9	.7	.6
.7	.7	.6	.9
.9	.5	.6	.8

$\}, \{$

a	b	c	d
.7	.7	.6	.9
.6	.8	.7	.7
.9	.5	.6	.8

$\}, \{$

a	b	c	d
.8	.9	.7	.6
.7	.7	.6	.9
.6	.8	.7	.7
.9	.5	.6	.8

$\}\}$

From this perspective the possible interpretations of the notion of neighbourhood in a fuzzy clan change in relation to the first focus as, now, we begin from a set of physical or mental objects, described by means of fuzzy subsets and the domain should be interpreted around groups (or subsets if prefered) of these objects. Here, the necessary union which we earlier advocated to be able to take into account various fuzzy subsets is no longer necessary, as this comes automatically given by the actual process followed.

Chapter 6
Pretopology, Topology and Affinities

1 A Return to Moore's Fuzzy Pretopology

In the study of isotone pretopologies we have indicated that a certain relationship exists between these and the concept of Moore's closing. What is more, from the common axioms we had defined an **uncertain Moore pretopology.** In continuation we will recall these aspects of pretopological spaces to try to advance in the elaboration of outlines apt for the treatment of economics and management problems.

The following table may be a good starting point.

Moore Closing	Isotone Pretopology
	$\Gamma\varnothing = \varnothing$
$\underset{\sim}{A}_j \subset M\underset{\sim}{A}_j$	$\underset{\sim}{A}_j \subset \Gamma\, \underset{\sim}{A}_j$
$M(M\underset{\sim}{A}_j) = M\, \underset{\sim}{A}_j$	$\Gamma(\Gamma\underset{\sim}{A}_j) \supset \Gamma\, \underset{\sim}{A}_j$
$(\underset{\sim}{A}_j \subset \underset{\sim}{A}_k) \Rightarrow (M\, \underset{\sim}{A}_j \subset M\, \underset{\sim}{A}_k)$	$(\underset{\sim}{A}_j \subset \underset{\sim}{A}_k) \Rightarrow (\Gamma\, \underset{\sim}{A}_j \subset \Gamma\, \underset{\sim}{A}_k)$

If in an isotone pretopology the idempotency $\mathbf{\Gamma}(\mathbf{\Gamma}\underset{\sim}{A}_j) = \mathbf{\Gamma}\underset{\sim}{A}_j$ is established, one has a **Moore closing** in which $\mathbf{\Gamma}\varnothing = \varnothing$ is also fulfilled.

As known, the idempotency may be obtained by performing:

$$\hat{\Gamma}\, \underset{\sim}{A}_j = \Gamma^n \underset{\sim}{A}_j = \Gamma^{n+1} \underset{\sim}{A}_j$$

In effect:

$$\begin{aligned}\hat{\Gamma}\,(\hat{\Gamma}\, \underset{\sim}{A}_j) &= \Gamma^n (\Gamma^n \underset{\sim}{A}_j)\\ &= \Gamma^{2n} \underset{\sim}{A}_j\\ &= \Gamma^n \underset{\sim}{A}_j\\ &= \hat{\Gamma}\, \underset{\sim}{A}_j\end{aligned}$$

In this way, a pretopology $\hat{\Gamma}\, \underset{\sim}{A}_j$ obtained from the isotone pretopology is also a Moore closing.

J. Gil Aluja & A.M. Gil Lafuente: Towards an Advanced Modelling, STUDFUZZ 276, pp. 217–266.
springerlink.com

Therefore, upon considering $\hat{\Gamma}\,\underset{\sim}{A}_j$ idempotency is always fulfilled and so, in this case, the axiomatic Moore closing, **when it is an isotone pretopology**, may be expressed in the following way:

$$\hat{\Gamma}\,\varnothing = \varnothing$$
$$\underset{\sim}{A}_j \subset \hat{\Gamma}\,\underset{\sim}{A}_j$$
$$(\underset{\sim}{A}_j \subset \underset{\sim}{A}_k) \Rightarrow (\hat{\Gamma}\,\underset{\sim}{A}_j \subset \hat{\Gamma}\,\underset{\sim}{A}_k)$$

given that, upon fulfilling:

$$\underset{\sim}{A}_j \subset \hat{\Gamma}\,\underset{\sim}{A}_j$$

it also fulfils:

$$\hat{\Gamma}\,(\hat{\Gamma}\underset{\sim}{A}_j) = \hat{\Gamma}\,\underset{\sim}{A}_j$$

and also upon being:

$$\underset{\sim}{A}_j \subset \hat{\Gamma}\,\underset{\sim}{A}_j$$

it gives:

$$\hat{\Gamma}\,E = E$$

However, although this Moore closing is a pretopology, specifically isotone, that which fulfils the four following axioms is given to be called an uncertain Moore pretopology:

$$\Gamma\,\varnothing = \varnothing$$
$$\underset{\sim}{A}_j \subset \Gamma\,\underset{\sim}{A}_j$$
$$\Gamma\,(\Gamma\,\underset{\sim}{A}_j) = \Gamma\,\underset{\sim}{A}_j$$
$$\Gamma\,(\underset{\sim}{A}_j \cup \underset{\sim}{A}_k) = \Gamma\,\underset{\sim}{A}_j \cup \Gamma\,\underset{\sim}{A}_k$$

We can see that in this case the **distributivity** axiom is established in place of the **isotone** axiom as the fourth axiom. We remind ourselves that a distributive pretopological space is isotone, even though the reciprocal proposal is not certain. We will reproduce its confirmation which has already been shown in the epigraph on distributive pretopology.

In effect, in $\underset{\sim}{A}_j \cup \underset{\sim}{A}_k$ we assume:

$$\underset{\sim}{A}_j \subset \underset{\sim}{A}_k$$

so:

$$\underset{\sim}{A}_j \cup \underset{\sim}{A}_k = \underset{\sim}{A}_k$$

and, therefore:

$$\Gamma\,(\underset{\sim}{A}_j \cup \underset{\sim}{A}_k) = \Gamma\,\underset{\sim}{A}_k$$

If:

$$\Gamma\,\underset{\sim}{A}_j \subset \Gamma\,\underset{\sim}{A}_k$$

it can be written:

$$\Gamma\,\underset{\sim}{A}_j \cup \Gamma\,\underset{\sim}{A}_k = \Gamma\,\underset{\sim}{A}_k$$

It this way it results that:

$$\Gamma\,(\underset{\sim}{A}_j \cup \underset{\sim}{A}_k) = \Gamma\,\underset{\sim}{A}_k \cup \Gamma\,\underset{\sim}{A}_k$$

Our proposal has once again been confirmed.

Now that these basic concepts have been recalled, we will present the process through which a functional application **Γ** fulfills the axioms of the isotone pretopology which is a Moore closing. For this we will start from the following referential of the **B** focus:

$$E = \{\underset{\sim}{P}_1, \underset{\sim}{P}_2, \underset{\sim}{P}_3, \underset{\sim}{P}_4\}$$

In continuation we consider the following functional application **Γ** among the elements of the set of its parts $P(E)$:

$$\Gamma\varnothing = \varnothing \qquad \Gamma\,\{\underset{\sim}{P}_1\} = \{\underset{\sim}{P}_1, \underset{\sim}{P}_4\} \qquad \Gamma\,\{\underset{\sim}{P}_2\} = \{\underset{\sim}{P}_2\} \qquad \Gamma\,\{\underset{\sim}{P}_3\} = \{\underset{\sim}{P}_3\}$$

$$\Gamma\,\{\underset{\sim}{P}_4\} = \{\underset{\sim}{P}_1, \underset{\sim}{P}_4\} \qquad \Gamma\,\{\underset{\sim}{P}_1, \underset{\sim}{P}_2\} = \{\underset{\sim}{P}_1, \underset{\sim}{P}_2, \underset{\sim}{P}_4\} \qquad \Gamma\,\{\underset{\sim}{P}_1, \underset{\sim}{P}_3\} = \{\underset{\sim}{P}_1, \underset{\sim}{P}_3, \underset{\sim}{P}_4\}$$

$$\Gamma\,\{\underset{\sim}{P}_1, \underset{\sim}{P}_4\} = \{\underset{\sim}{P}_1, \underset{\sim}{P}_4\} \qquad \Gamma\,\{\underset{\sim}{P}_2, \underset{\sim}{P}_3\} = E \qquad \Gamma\,\{\underset{\sim}{P}_2, \underset{\sim}{P}_4\} = \{\underset{\sim}{P}_1, \underset{\sim}{P}_2, \underset{\sim}{P}_4\}$$

$$\Gamma\,\{\underset{\sim}{P}_3, \underset{\sim}{P}_4\} = \{\underset{\sim}{P}_1, \underset{\sim}{P}_3, \underset{\sim}{P}_4\} \qquad \Gamma\,\{\underset{\sim}{P}_1, \underset{\sim}{P}_2, \underset{\sim}{P}_3\} = E \qquad \Gamma\,\{\underset{\sim}{P}_1, \underset{\sim}{P}_2, \underset{\sim}{P}_4\} = \{\underset{\sim}{P}_1, \underset{\sim}{P}_2, \underset{\sim}{P}_4\}$$

$$\Gamma\,\{\underset{\sim}{P}_1, \underset{\sim}{P}_3, \underset{\sim}{P}_4\} = \{\underset{\sim}{P}_1, \underset{\sim}{P}_3, \underset{\sim}{P}_4\} \qquad \Gamma\,\{\underset{\sim}{P}_2, \underset{\sim}{P}_3, \underset{\sim}{P}_4\} = E \qquad \Gamma\,E = E$$

In this application it can be seen at first glance that the following axioms are fulfilled:

$$\Gamma\varnothing = \varnothing$$

$$\underset{\sim}{A}_j \subset \Gamma\,\underset{\sim}{A}_j$$

Let us see if the idempotency is fulfilled:

a) In a sagittate representation we have:

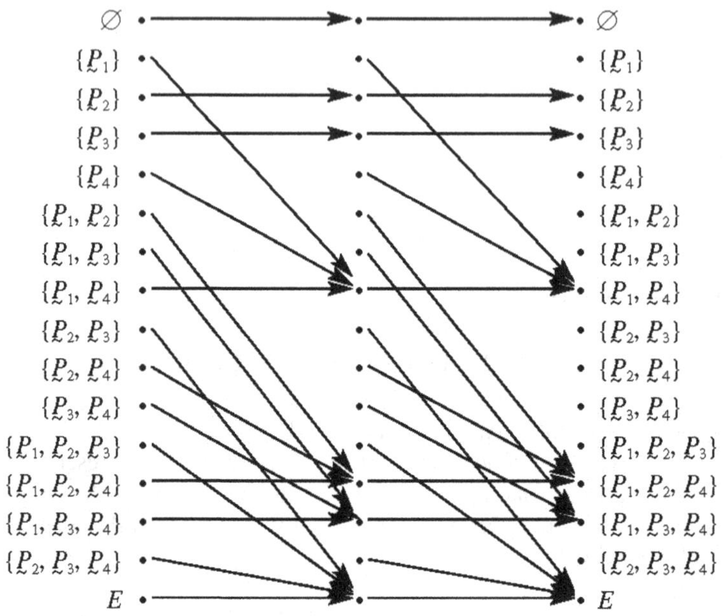

b) In the matrix representation it is enough to perform the maxmin composition. We have:

	$\varnothing$	$\{P_1\}$	$\{P_2\}$	$\{P_3\}$	$\{P_4\}$	$\{P_1, P_2\}$	$\{P_1, P_3\}$	$\{P_1, P_4\}$	$\{P_2, P_3\}$	$\{P_2, P_4\}$	$\{P_3, P_4\}$	$\{P_1, P_2, P_3\}$	$\{P_1, P_2, P_4\}$	$\{P_1, P_3, P_4\}$	$\{P_2, P_3, P_4\}$	E	
$\varnothing$	1																
$\{P_1\}$								1									
$\{P_2\}$			1														
$\{P_3\}$				1													
$\{P_4\}$								1									
$\{P_1, P_2\}$													1				
$\{P_1, P_2\}$														1			
$\{P_1, P_4\}$								1									
$\{P_2, P_3\}$																1	0
$\{P_2, P_4\}$													1				
$\{P_3, P_4\}$														1			
$\{P_1, P_2, P_3\}$																1	
$\{P_1, P_2, P_4\}$													1				
$\{P_1, P_3, P_4\}$														1			
$\{P_2, P_3, P_4\}$																1	
E																1	

	$\varnothing$	$\{\underset{\sim}{P}_1\}$	$\{\underset{\sim}{P}_2\}$	$\{\underset{\sim}{P}_3\}$	$\{\underset{\sim}{P}_4\}$	$\{\underset{\sim}{P}_1, \underset{\sim}{P}_2\}$	$\{\underset{\sim}{P}_1, \underset{\sim}{P}_3\}$	$\{\underset{\sim}{P}_1, \underset{\sim}{P}_4\}$	$\{\underset{\sim}{P}_2, \underset{\sim}{P}_3\}$	$\{\underset{\sim}{P}_2, \underset{\sim}{P}_4\}$	$\{\underset{\sim}{P}_3, \underset{\sim}{P}_4\}$	$\{\underset{\sim}{P}_1, \underset{\sim}{P}_2, \underset{\sim}{P}_3\}$	$\{\underset{\sim}{P}_1, \underset{\sim}{P}_2, \underset{\sim}{P}_4\}$	$\{\underset{\sim}{P}_1, \underset{\sim}{P}_3, \underset{\sim}{P}_4\}$	$\{\underset{\sim}{P}_2, \underset{\sim}{P}_3, \underset{\sim}{P}_4\}$	E
$\varnothing$	1															
$\{\underset{\sim}{P}_1\}$								1								
$\{\underset{\sim}{P}_2\}$			1													
$\{\underset{\sim}{P}_3\}$				1												
$\{\underset{\sim}{P}_4\}$								1								
$\{\underset{\sim}{P}_1, \underset{\sim}{P}_2\}$													1			
$\{\underset{\sim}{P}_1, \underset{\sim}{P}_2\}$														1		
$\{\underset{\sim}{P}_1, \underset{\sim}{P}_4\}$								1								
$\{\underset{\sim}{P}_2, \underset{\sim}{P}_3\}$																1
$\{\underset{\sim}{P}_2, \underset{\sim}{P}_4\}$													1			
$\{\underset{\sim}{P}_3, \underset{\sim}{P}_4\}$														1		
$\{\underset{\sim}{P}_1, \underset{\sim}{P}_2, \underset{\sim}{P}_3\}$																1
$\{\underset{\sim}{P}_1, \underset{\sim}{P}_2, \underset{\sim}{P}_4\}$													1			
$\{\underset{\sim}{P}_1, \underset{\sim}{P}_3, \underset{\sim}{P}_4\}$														1		
$\{\underset{\sim}{P}_2, \underset{\sim}{P}_3, \underset{\sim}{P}_4\}$																1
E																1

$=$

	$\varnothing$	$\{\underset{\sim}{P}_1\}$	$\{\underset{\sim}{P}_2\}$	$\{\underset{\sim}{P}_3\}$	$\{\underset{\sim}{P}_4\}$	$\{\underset{\sim}{P}_1, \underset{\sim}{P}_2\}$	$\{\underset{\sim}{P}_1, \underset{\sim}{P}_3\}$	$\{\underset{\sim}{P}_1, \underset{\sim}{P}_4\}$	$\{\underset{\sim}{P}_2, \underset{\sim}{P}_3\}$	$\{\underset{\sim}{P}_2, \underset{\sim}{P}_4\}$	$\{\underset{\sim}{P}_3, \underset{\sim}{P}_4\}$	$\{\underset{\sim}{P}_1, \underset{\sim}{P}_2, \underset{\sim}{P}_3\}$	$\{\underset{\sim}{P}_1, \underset{\sim}{P}_2, \underset{\sim}{P}_4\}$	$\{\underset{\sim}{P}_1, \underset{\sim}{P}_3, \underset{\sim}{P}_4\}$	$\{\underset{\sim}{P}_2, \underset{\sim}{P}_3, \underset{\sim}{P}_4\}$	E
$\varnothing$	1															
$\{\underset{\sim}{P}_1\}$								1								
$\{\underset{\sim}{P}_2\}$			1													
$\{\underset{\sim}{P}_3\}$				1												
$\{\underset{\sim}{P}_4\}$								1								
$\{\underset{\sim}{P}_1, \underset{\sim}{P}_2\}$													1			
$\{\underset{\sim}{P}_1, \underset{\sim}{P}_2\}$														1		
$\{\underset{\sim}{P}_1, \underset{\sim}{P}_4\}$								1								
$\{\underset{\sim}{P}_2, \underset{\sim}{P}_3\}$																1
$\{\underset{\sim}{P}_2, \underset{\sim}{P}_4\}$													1			
$\{\underset{\sim}{P}_3, \underset{\sim}{P}_4\}$														1		
$\{\underset{\sim}{P}_1, \underset{\sim}{P}_2, \underset{\sim}{P}_3\}$																1
$\{\underset{\sim}{P}_1, \underset{\sim}{P}_2, \underset{\sim}{P}_4\}$													1			
$\{\underset{\sim}{P}_1, \underset{\sim}{P}_3, \underset{\sim}{P}_4\}$														1		
$\{\underset{\sim}{P}_2, \underset{\sim}{P}_3, \underset{\sim}{P}_4\}$																1
E																1

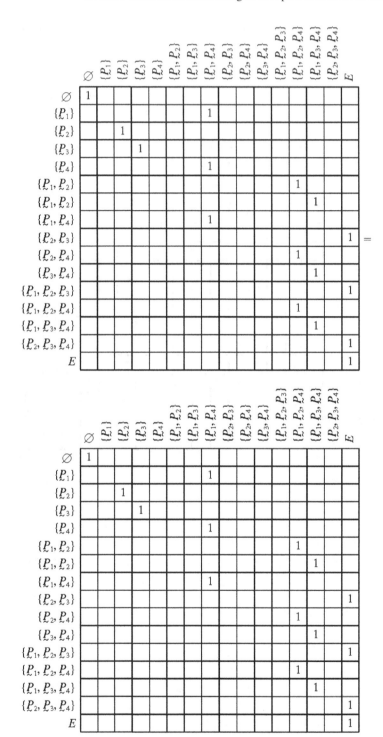

Finally, we confirm if the condition of isotone is verified:

$\{\underset{\sim}{P}_1\} \subset \{\underset{\sim}{P}_1, \underset{\sim}{P}_2\}$	$(\Gamma\,\{\underset{\sim}{P}_1\} = \{\underset{\sim}{P}_1, \underset{\sim}{P}_4\}) \subset (\Gamma\,\{\underset{\sim}{P}_1, \underset{\sim}{P}_2\} = \{\underset{\sim}{P}_1, \underset{\sim}{P}_2, \underset{\sim}{P}_4\})$
$\{\underset{\sim}{P}_1\} \subset \{\underset{\sim}{P}_1, \underset{\sim}{P}_3\}$	$\Gamma\,\{\underset{\sim}{P}_1\} \subset (\Gamma\,\{\underset{\sim}{P}_1, \underset{\sim}{P}_3\} = \{\underset{\sim}{P}_1, \underset{\sim}{P}_3, \underset{\sim}{P}_4\})$
$\{\underset{\sim}{P}_1\} \subset \{\underset{\sim}{P}_1, \underset{\sim}{P}_4\}$	$\Gamma\,\{\underset{\sim}{P}_1\} \subset (\Gamma\,\{\underset{\sim}{P}_1, \underset{\sim}{P}_4\} = \{\underset{\sim}{P}_1, \underset{\sim}{P}_4\})$
$\{\underset{\sim}{P}_1\} \subset \{\underset{\sim}{P}_1, \underset{\sim}{P}_2, \underset{\sim}{P}_3\}$	$\Gamma\,\{\underset{\sim}{P}_1\} \subset (\Gamma\,\{\underset{\sim}{P}_1, \underset{\sim}{P}_2, \underset{\sim}{P}_3\} = E)$
$\{\underset{\sim}{P}_1, \underset{\sim}{P}_2\} \subset \{\underset{\sim}{P}_1, \underset{\sim}{P}_2, \underset{\sim}{P}_3\}$	$(\Gamma\,\{\underset{\sim}{P}_1, \underset{\sim}{P}_2\} = \{\underset{\sim}{P}_1, \underset{\sim}{P}_2, \underset{\sim}{P}_4\}) \subset (\Gamma\,\{\underset{\sim}{P}_1, \underset{\sim}{P}_2, \underset{\sim}{P}_3\} = E)$
$\{\underset{\sim}{P}_1, \underset{\sim}{P}_2\} \subset \{\underset{\sim}{P}_1, \underset{\sim}{P}_2, \underset{\sim}{P}_4\}$	$\Gamma\,\{\underset{\sim}{P}_1, \underset{\sim}{P}_2\} \subset (\Gamma\,\{\underset{\sim}{P}_1, \underset{\sim}{P}_2, \underset{\sim}{P}_4\}\} = \{\underset{\sim}{P}_1, \underset{\sim}{P}_2, \underset{\sim}{P}_4\})$
$\{\underset{\sim}{P}_1, \underset{\sim}{P}_2\} \subset E$	$\Gamma\,\{\underset{\sim}{P}_1, \underset{\sim}{P}_2\} \subset (\Gamma\,E = E)$
$\{\underset{\sim}{P}_1, \underset{\sim}{P}_3\} \subset \{\underset{\sim}{P}_1, \underset{\sim}{P}_2, \underset{\sim}{P}_3\}$	$(\Gamma\,\{\underset{\sim}{P}_1, \underset{\sim}{P}_3\} = \{\underset{\sim}{P}_1, \underset{\sim}{P}_3, \underset{\sim}{P}_4\}) \subset (\Gamma\,\{\underset{\sim}{P}_1, \underset{\sim}{P}_2, \underset{\sim}{P}_3\} = E)$
$\{\underset{\sim}{P}_1, \underset{\sim}{P}_3\} \subset \{\underset{\sim}{P}_1, \underset{\sim}{P}_3, \underset{\sim}{P}_4\}$	$\Gamma\,\{\underset{\sim}{P}_1, \underset{\sim}{P}_3\} \subset (\Gamma\,\{\underset{\sim}{P}_1, \underset{\sim}{P}_3, \underset{\sim}{P}_4\}\} = \{\underset{\sim}{P}_1, \underset{\sim}{P}_3, \underset{\sim}{P}_4\})$
$\{\underset{\sim}{P}_1, \underset{\sim}{P}_3\} \subset E$	$\Gamma\,\{\underset{\sim}{P}_1, \underset{\sim}{P}_3\} \subset (\Gamma\,E = E)$
$\{\underset{\sim}{P}_1, \underset{\sim}{P}_4\} \subset \{\underset{\sim}{P}_1, \underset{\sim}{P}_2, \underset{\sim}{P}_4\}$	$(\Gamma\,\{\underset{\sim}{P}_1, \underset{\sim}{P}_4\} = \{\underset{\sim}{P}_1, \underset{\sim}{P}_4\}) \subset (\Gamma\,\{\underset{\sim}{P}_1, \underset{\sim}{P}_2, \underset{\sim}{P}_4\} = \{\underset{\sim}{P}_1, \underset{\sim}{P}_2, \underset{\sim}{P}_4\})$
$\{\underset{\sim}{P}_1, \underset{\sim}{P}_4\} \subset \{\underset{\sim}{P}_1, \underset{\sim}{P}_3, \underset{\sim}{P}_4\}$	$\Gamma\,\{\underset{\sim}{P}_1, \underset{\sim}{P}_4\} \subset (\Gamma\,\{\underset{\sim}{P}_1, \underset{\sim}{P}_3, \underset{\sim}{P}_4\} = \{\underset{\sim}{P}_1, \underset{\sim}{P}_3, \underset{\sim}{P}_4\})$
$\{\underset{\sim}{P}_1, \underset{\sim}{P}_4\} \subset E$	$\Gamma\,\{\underset{\sim}{P}_1, \underset{\sim}{P}_4\} \subset (\Gamma\,E = E)$
$\{\underset{\sim}{P}_2, \underset{\sim}{P}_3\} \subset \{\underset{\sim}{P}_1, \underset{\sim}{P}_2, \underset{\sim}{P}_3\}$	$(\Gamma\,\{\underset{\sim}{P}_2, \underset{\sim}{P}_3\} = E) \subset (\Gamma\,\{\underset{\sim}{P}_1, \underset{\sim}{P}_2, \underset{\sim}{P}_3\} = E)$
$\{\underset{\sim}{P}_2, \underset{\sim}{P}_3\} \subset \{\underset{\sim}{P}_2, \underset{\sim}{P}_3, \underset{\sim}{P}_4\}$	$\Gamma\,\{\underset{\sim}{P}_2, \underset{\sim}{P}_3\} \subset (\Gamma\,\{\underset{\sim}{P}_2, \underset{\sim}{P}_3, \underset{\sim}{P}_4\} = E)$
$\{\underset{\sim}{P}_2, \underset{\sim}{P}_3\} \subset E$	$\Gamma\,\{\underset{\sim}{P}_2, \underset{\sim}{P}_3\} \subset (\Gamma\,E = E)$
$\{\underset{\sim}{P}_2, \underset{\sim}{P}_4\} \subset \{\underset{\sim}{P}_1, \underset{\sim}{P}_2, \underset{\sim}{P}_4\}$	$(\Gamma\,\{\underset{\sim}{P}_2, \underset{\sim}{P}_4\} = \{\underset{\sim}{P}_1, \underset{\sim}{P}_2, \underset{\sim}{P}_4\}) \subset (\Gamma\,\{\underset{\sim}{P}_1, \underset{\sim}{P}_2, \underset{\sim}{P}_4\} = \{\underset{\sim}{P}_1, \underset{\sim}{P}_2, \underset{\sim}{P}_4\})$
$\{\underset{\sim}{P}_2, \underset{\sim}{P}_4\} \subset \{\underset{\sim}{P}_2, \underset{\sim}{P}_3, \underset{\sim}{P}_4\}$	$\Gamma\,\{\underset{\sim}{P}_2, \underset{\sim}{P}_4\} \subset (\Gamma\,\{\underset{\sim}{P}_2, \underset{\sim}{P}_3, \underset{\sim}{P}_4\} = E)$
$\{\underset{\sim}{P}_2, \underset{\sim}{P}_4\} \subset E$	$\Gamma\,\{\underset{\sim}{P}_2, \underset{\sim}{P}_4\} \subset (\Gamma\,E = E)$
$\{\underset{\sim}{P}_3, \underset{\sim}{P}_4\} \subset \{\underset{\sim}{P}_1, \underset{\sim}{P}_3, \underset{\sim}{P}_4\}$	$(\Gamma\,\{\underset{\sim}{P}_3, \underset{\sim}{P}_4\} = \{\underset{\sim}{P}_1, \underset{\sim}{P}_3, \underset{\sim}{P}_4\}) \subset (\Gamma\,\{\underset{\sim}{P}_1, \underset{\sim}{P}_3, \underset{\sim}{P}_4\} = \{\underset{\sim}{P}_1, \underset{\sim}{P}_3, \underset{\sim}{P}_4\})$
$\{\underset{\sim}{P}_3, \underset{\sim}{P}_4\} \subset \{\underset{\sim}{P}_2, \underset{\sim}{P}_3, \underset{\sim}{P}_4\}$	$\Gamma\,\{\underset{\sim}{P}_3, \underset{\sim}{P}_4\} \subset (\Gamma\,\{\underset{\sim}{P}_2, \underset{\sim}{P}_3, \underset{\sim}{P}_4\} = E)$
$\{\underset{\sim}{P}_3, \underset{\sim}{P}_4\} \subset E$	$\Gamma\,\{\underset{\sim}{P}_3, \underset{\sim}{P}_4\} \subset (\Gamma\,E = E)$
$\{\underset{\sim}{P}_1\} \subset \{\underset{\sim}{P}_1, \underset{\sim}{P}_2, \underset{\sim}{P}_4\}$	$\Gamma\,\{\underset{\sim}{P}_1\} \subset (\Gamma\,\{\underset{\sim}{P}_1, \underset{\sim}{P}_2, \underset{\sim}{P}_4\} = \{\underset{\sim}{P}_1, \underset{\sim}{P}_2, \underset{\sim}{P}_4\})$
$\{\underset{\sim}{P}_1\} \subset \{\underset{\sim}{P}_1, \underset{\sim}{P}_3, \underset{\sim}{P}_4\}$	$\Gamma\,\{\underset{\sim}{P}_1\} \subset (\Gamma\,\{\underset{\sim}{P}_1, \underset{\sim}{P}_3, \underset{\sim}{P}_4\} = \{\underset{\sim}{P}_1, \underset{\sim}{P}_3, \underset{\sim}{P}_4\})$
$\{\underset{\sim}{P}_1\} \subset E$	$\Gamma\,\{\underset{\sim}{P}_1\} \subset (\Gamma\,E = E)$
$\{\underset{\sim}{P}_2\} \subset \{\underset{\sim}{P}_1, \underset{\sim}{P}_2\}$	$(\Gamma\,\{\underset{\sim}{P}_2\} = \{\underset{\sim}{P}_2\}) \subset (\Gamma\,\{\underset{\sim}{P}_1, \underset{\sim}{P}_2\} = \{\underset{\sim}{P}_1, \underset{\sim}{P}_2, \underset{\sim}{P}_4\})$
$\{\underset{\sim}{P}_2\} \subset \{\underset{\sim}{P}_2, \underset{\sim}{P}_3\}$	$\Gamma\,\{\underset{\sim}{P}_2\} \subset (\Gamma\,\{\underset{\sim}{P}_2, \underset{\sim}{P}_3\} = E)$
$\{\underset{\sim}{P}_2\} \subset \{\underset{\sim}{P}_2, \underset{\sim}{P}_4\}$	$\Gamma\,\{\underset{\sim}{P}_2\} \subset (\Gamma\,\{\underset{\sim}{P}_2, \underset{\sim}{P}_4\} = \{\underset{\sim}{P}_1, \underset{\sim}{P}_2, \underset{\sim}{P}_4\})$

$\{\underset{\sim}{P}_2\} \subset \{\underset{\sim}{P}_1, \underset{\sim}{P}_2, \underset{\sim}{P}_3\}$	$\Gamma\,\{\underset{\sim}{P}_2\} \subset (\Gamma\,\{\underset{\sim}{P}_1, \underset{\sim}{P}_2, \underset{\sim}{P}_3\} = E)$
$\{\underset{\sim}{P}_2\} \subset \{\underset{\sim}{P}_1, \underset{\sim}{P}_2, \underset{\sim}{P}_4\}$	$\Gamma\,\{\underset{\sim}{P}_2\} \subset (\Gamma\,\{\underset{\sim}{P}_1, \underset{\sim}{P}_2, \underset{\sim}{P}_4\} = \{\underset{\sim}{P}_1, \underset{\sim}{P}_2, \underset{\sim}{P}_4\})$
$\{\underset{\sim}{P}_2\} \subset \{\underset{\sim}{P}_2, \underset{\sim}{P}_3, \underset{\sim}{P}_4\}$	$\Gamma\,\{\underset{\sim}{P}_2\} \subset (\Gamma\,\{\underset{\sim}{P}_2, \underset{\sim}{P}_3, \underset{\sim}{P}_4\} = E)$
$\{\underset{\sim}{P}_2\} \subset E$	$\Gamma\,\{\underset{\sim}{P}_2\} \subset (\Gamma\,E = E)$
$\{\underset{\sim}{P}_3\} \subset \{\underset{\sim}{P}_1, \underset{\sim}{P}_3\}$	$(\Gamma\,\{\underset{\sim}{P}_3\} = \{\underset{\sim}{P}_3\}) \subset (\Gamma\,\{\underset{\sim}{P}_1, \underset{\sim}{P}_3\} = \{\underset{\sim}{P}_1, \underset{\sim}{P}_3, \underset{\sim}{P}_4\})$
$\{\underset{\sim}{P}_3\} \subset \{\underset{\sim}{P}_2, \underset{\sim}{P}_3\}$	$\Gamma\,\{\underset{\sim}{P}_3\} \subset (\Gamma\,\{\underset{\sim}{P}_2, \underset{\sim}{P}_3\} = E)$
$\{\underset{\sim}{P}_3\} \subset \{\underset{\sim}{P}_3, \underset{\sim}{P}_4\}$	$\Gamma\,\{\underset{\sim}{P}_3\} \subset (\Gamma\,\{\underset{\sim}{P}_3, \underset{\sim}{P}_4\} = \{\underset{\sim}{P}_1, \underset{\sim}{P}_3, \underset{\sim}{P}_4\})$
$\{\underset{\sim}{P}_3\} \subset \{\underset{\sim}{P}_1, \underset{\sim}{P}_2, \underset{\sim}{P}_3\}$	$\Gamma\,\{\underset{\sim}{P}_3\} \subset (\Gamma\,\{\underset{\sim}{P}_1, \underset{\sim}{P}_2, \underset{\sim}{P}_3\} = E)$
$\{\underset{\sim}{P}_3\} \subset \{\underset{\sim}{P}_1, \underset{\sim}{P}_3, \underset{\sim}{P}_4\}$	$\Gamma\,\{\underset{\sim}{P}_3\} \subset (\Gamma\,\{\underset{\sim}{P}_1, \underset{\sim}{P}_3, \underset{\sim}{P}_4\} = \{\underset{\sim}{P}_1, \underset{\sim}{P}_3, \underset{\sim}{P}_4\})$
$\{\underset{\sim}{P}_3\} \subset \{\underset{\sim}{P}_2, \underset{\sim}{P}_3, \underset{\sim}{P}_4\}$	$\Gamma\,\{\underset{\sim}{P}_3\} \subset (\Gamma\,\{\underset{\sim}{P}_2, \underset{\sim}{P}_3, \underset{\sim}{P}_4\} = E)$
$\{\underset{\sim}{P}_3\} \subset E$	$\Gamma\,\{\underset{\sim}{P}_3\} \subset (\Gamma\,E = E)$
$\{\underset{\sim}{P}_4\} \subset \{\underset{\sim}{P}_1, \underset{\sim}{P}_4\}$	$(\Gamma\,\{\underset{\sim}{P}_4\} = \{\underset{\sim}{P}_1, \underset{\sim}{P}_4\}) \subset (\Gamma\,\{\underset{\sim}{P}_1, \underset{\sim}{P}_4\} = \{\underset{\sim}{P}_1, \underset{\sim}{P}_4\})$
$\{\underset{\sim}{P}_4\} \subset \{\underset{\sim}{P}_2, \underset{\sim}{P}_4\}$	$\Gamma\,\{\underset{\sim}{P}_4\} \subset (\Gamma\,\{\underset{\sim}{P}_2, \underset{\sim}{P}_4\} = \{\underset{\sim}{P}_1, \underset{\sim}{P}_2, \underset{\sim}{P}_4\})$
$\{\underset{\sim}{P}_4\} \subset \{\underset{\sim}{P}_3, \underset{\sim}{P}_4\}$	$\Gamma\,\{\underset{\sim}{P}_4\} \subset (\Gamma\,\{\underset{\sim}{P}_3, \underset{\sim}{P}_4\} = \{\underset{\sim}{P}_1, \underset{\sim}{P}_3, \underset{\sim}{P}_4\})$
$\{\underset{\sim}{P}_4\} \subset \{\underset{\sim}{P}_1, \underset{\sim}{P}_2, \underset{\sim}{P}_4\}$	$\Gamma\,\{\underset{\sim}{P}_4\} \subset (\Gamma\,\{\underset{\sim}{P}_1, \underset{\sim}{P}_2, \underset{\sim}{P}_3\} = \{\underset{\sim}{P}_1, \underset{\sim}{P}_2, \underset{\sim}{P}_4\})$
$\{\underset{\sim}{P}_4\} \subset \{\underset{\sim}{P}_1, \underset{\sim}{P}_3, \underset{\sim}{P}_4\}$	$\Gamma\,\{\underset{\sim}{P}_4\} \subset (\Gamma\,\{\underset{\sim}{P}_1, \underset{\sim}{P}_3, \underset{\sim}{P}_4\} = \{\underset{\sim}{P}_1, \underset{\sim}{P}_3, \underset{\sim}{P}_4\})$
$\{\underset{\sim}{P}_4\} \subset \{\underset{\sim}{P}_2, \underset{\sim}{P}_3, \underset{\sim}{P}_4\}$	$\Gamma\,\{\underset{\sim}{P}_4\} \subset (\Gamma\,\{\underset{\sim}{P}_2, \underset{\sim}{P}_3, \underset{\sim}{P}_4\} = E)$
$\{\underset{\sim}{P}_4\} \subset E$	$\Gamma\,\{\underset{\sim}{P}_4\} \subset (\Gamma\,E = E)$
$\{\underset{\sim}{P}_1, \underset{\sim}{P}_2, \underset{\sim}{P}_3\} \subset E$	$(\Gamma\,\{\underset{\sim}{P}_1, \underset{\sim}{P}_2, \underset{\sim}{P}_3\} = E) \subset (\Gamma\,E = E)$
$\{\underset{\sim}{P}_1, \underset{\sim}{P}_2, \underset{\sim}{P}_4\} \subset E$	$(\Gamma\,\{\underset{\sim}{P}_1, \underset{\sim}{P}_2, \underset{\sim}{P}_4\} = \{\underset{\sim}{P}_1, \underset{\sim}{P}_2, \underset{\sim}{P}_4\}) \subset (\Gamma\,E = E)$
$\{\underset{\sim}{P}_1, \underset{\sim}{P}_3, \underset{\sim}{P}_4\} \subset E$	$(\Gamma\,\{\underset{\sim}{P}_1, \underset{\sim}{P}_3, \underset{\sim}{P}_4\} = \{\underset{\sim}{P}_1, \underset{\sim}{P}_3, \underset{\sim}{P}_4\}) \subset (\Gamma\,E = E)$
$\{\underset{\sim}{P}_2, \underset{\sim}{P}_3, \underset{\sim}{P}_4\} \subset E$	$(\Gamma\,\{\underset{\sim}{P}_2, \underset{\sim}{P}_3, \underset{\sim}{P}_4\} = E) \subset (\Gamma\,E = E)$

In this way it has been confirmed that we are dealing with an isotone pretopology which is also a Moore closing, in which it is also given that:

$$\boldsymbol{\Gamma}\,\emptyset = \emptyset$$

2 Obtaining of Moore Closing from a Family

In the example which we have just developed we started from a functional application $\boldsymbol{\Gamma}$, which we knew combined the requisites of a **Moore closing** “ex-ante”.

It may result as interesting, and most of all useful, to establish a way through which a Moore closing can be found when beginning from a very elemental axiomatic. The notion of **Moore family** may be adequate in this sense. In effect, if we designate $F(E) \subset P(E)$ to a family when it fulfils the two following properties:

$$E \in F(E)$$
$$(\underset{\sim}{A}_j, \underset{\sim}{A}_k \in F(E)) \Rightarrow (\underset{\sim}{A}_j \cap \underset{\sim}{A}_k \in F(E))$$

it is said that $F(E)$ is a **Moore family**.

Let us see an example, beginning with the referential:

$$E = \{\underset{\sim}{P}_1, \underset{\sim}{P}_2, \underset{\sim}{P}_3, \underset{\sim}{P}_4\}$$

We create a family choosing some elements of the set of the parts $P(E)$ which fulfil the stated axioms.

$$F(E) = \{\varnothing, \{\underset{\sim}{P}_1\}, \{\underset{\sim}{P}_3\}, \{\underset{\sim}{P}_1, \underset{\sim}{P}_3\}, \{\underset{\sim}{P}_2, \underset{\sim}{P}_3\}, E\}$$

The subset of elements of $F(E)$ which contain $\underset{\sim}{A}_j$ is designated by $F_{\underset{\sim}{A}_j}(E)$.

The Moore closing $M\underset{\sim}{A}_j$, which is the only one for a family is constructed by:

$$M\underset{\sim}{A}_j = \bigcap_{F \in F_{\underset{\sim}{A}_j}(E)} F$$

In continuation we will use this expression in the family $F(E)$ proposed. For this we will obtain $F_{\underset{\sim}{A}_j}(E)$ for all of the elements of the "power set".

$$F_{\varnothing}(E) = \{\varnothing, \{\underset{\sim}{P}_1\}, \{\underset{\sim}{P}_3\}, \{\underset{\sim}{P}_1, \underset{\sim}{P}_3\}, \{\underset{\sim}{P}_2, \underset{\sim}{P}_3\}, E\} \qquad \bigcap_{F \in F_{\varnothing}(E)} F = \varnothing$$

$$F_{\underset{\sim}{P}_1}(E) = \{\{\underset{\sim}{P}_1\}, \{\underset{\sim}{P}_1, \underset{\sim}{P}_3\}, E\} \qquad \bigcap_{F \in F_{\underset{\sim}{P}_1}(E)} F = \{\underset{\sim}{P}_1\}$$

$$F_{\underset{\sim}{P}_2}(E) = \{\{\underset{\sim}{P}_2, \underset{\sim}{P}_3\}, E\} \qquad \bigcap_{F \in F_{\underset{\sim}{P}_2}(E)} F = \{\underset{\sim}{P}_2, \underset{\sim}{P}_3\}$$

$$F_{\underset{\sim}{P}_3}(E) = \{\{\underset{\sim}{P}_3\}, \{\underset{\sim}{P}_1, \underset{\sim}{P}_3\}, \{\underset{\sim}{P}_2, \underset{\sim}{P}_3\}, E\} \qquad \bigcap_{F \in F_{\underset{\sim}{P}_3}(E)} F = \{\underset{\sim}{P}_3\}$$

$$F_{\underset{\sim}{P}_4}(E) = E \qquad \bigcap_{F \in F_{\underset{\sim}{P}_4}(E)} F = E$$

$$F_{\underset{\sim}{P}_1, \underset{\sim}{P}_2}(E) = E \qquad \bigcap_{F \in F_{\underset{\sim}{P}_1, \underset{\sim}{P}_2}(E)} F = E$$

$$F_{\underset{\sim}{P}_1, \underset{\sim}{P}_3}(E) = \{\{\underset{\sim}{P}_1, \underset{\sim}{P}_3\}, E\} \qquad \bigcap_{F \in F_{\underset{\sim}{P}_1, \underset{\sim}{P}_3}(E)} F = \{\underset{\sim}{P}_1, \underset{\sim}{P}_3\}$$

$$F_{\underset{\sim}{P}_1, \underset{\sim}{P}_4}(E) = E \qquad \bigcap_{F \in F_{\underset{\sim}{P}_1, \underset{\sim}{P}_4}(E)} F = E$$

$$F_{P_2,P_3}(E) = \{\{P_2, P_3\}, E\} \qquad \bigcap_{F \in F_{P_2,P_3}(E)} F = \{P_2, P_3\}$$

$$F_{P_2,P_4}(E) = E \qquad \bigcap_{F \in F_{P_2,P_4}(E)} F = E$$

$$F_{P_3,P_4}(E) = E \qquad \bigcap_{F \in F_{P_3,P_4}(E)} F = E$$

$$F_{P_1,P_2,P_3}(E) = E \qquad \bigcap_{F \in F_{P_1,P_2,P_3}(E)} F = E$$

If we continue with this process it can be seen that from here all of the elements have the referential E as an intersection.

With these results the corresponding Moore closing can be constructed for the family F(E), which is also an isotone pretopology with idempotency. We also remind ourselves that a **Moore family** can give rise to one, and only one, **Moore closing**. In our case it is the following, expressed in sagittate form:

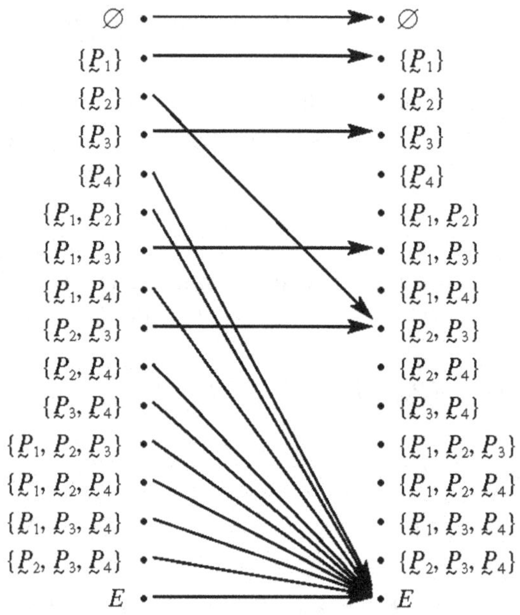

It is easy to confirm, even visually, that this application **Γ**, or M if preferred, fulfils the axioms established for a Moore closing and, also, those of isotone pre-topology (upon having initially demanded $\Gamma\,\varnothing=\varnothing$). The idempotency gives that the elements of P(E) which are the image of another element of P(E) are here in the same way their own image. They are those called "closes". Therefore, **a Moore family is a set of closes**. In effect, it can be seen that, in our case, the closes are:

$$C_e = \{\varnothing, \{P_1\}, \{P_3\}, \{P_1, P_3\}, \{P_2, P_3\}, E\}$$

which coincide with the Moore family initially established.

These closes, or in the same way, the Moore family, form the following lattice:

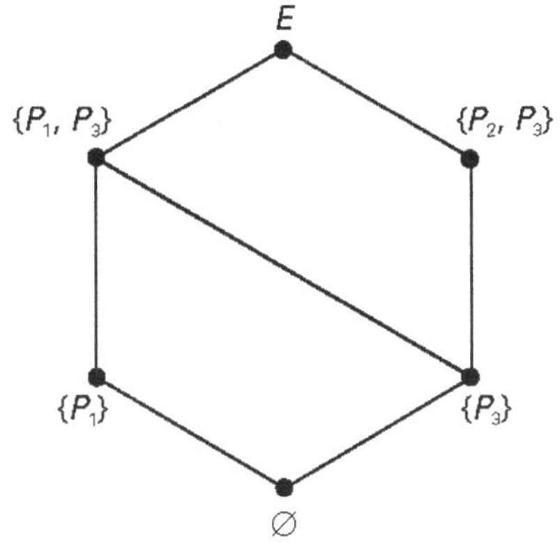

We indicate with a large circle ● the Moore family within the Boole lattice corresponding to the "power set":

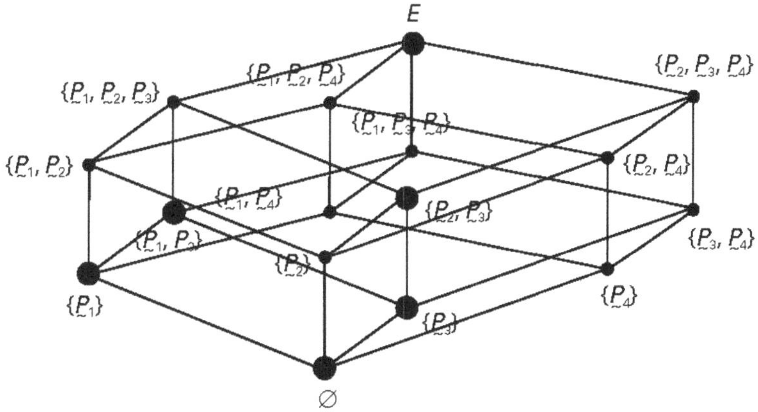

Up to now we have been able to see how it is possible to find **Moore closings** from a **Moore family**. But, it is also possible to arrive at this isotone pretopology by taking a **relationship** between the elements of a set as a starting point.

3 Obtaining of Moore Closing through a Relationship

With the aim of finding a Moore closing when starting from a relationship, it will be necessary to express this by means of a graph, which may equally be in matrix or sagittate form.

In effect, let us take the following referential:

$$E = \{\underset{\sim}{P}_1, \underset{\sim}{P}_2, \underset{\sim}{P}_3, \underset{\sim}{P}_4\}$$

and a fuzzy relationship among these elements:

$$\underset{\sim}{R} \subset E \cdot E$$

A boolean graph is defined from these thresholds or level $\propto$, which we designate in the following way:

$$R_\alpha = \{(\underset{\sim}{P}_j, \underset{\sim}{P}_k) \subset E \cdot E / \mu_{\underset{\sim}{R}}(\underset{\sim}{P}_j, \underset{\sim}{P}_k) \geq \alpha\}, \alpha \in [0, 1]$$

We arbitrarily assign values in [0, 1] to the fuzzy relationship $\underset{\sim}{R}$:

$\underset{\sim}{R} =$

	$\underset{\sim}{P}_1$	$\underset{\sim}{P}_2$	$\underset{\sim}{P}_3$	$\underset{\sim}{P}_4$
$\underset{\sim}{P}_1$	.5	1	.7	.8
$\underset{\sim}{P}_2$	.9	.8	.2	0
$\underset{\sim}{P}_3$	.3	.9	.8	1
$\underset{\sim}{P}_4$	0	.1	.3	.8

If we take $\propto \; \geq 0.7$ as a threshold, the following boolean matrix is obtained:

$\underset{\sim}{R}_{0,7} =$

	$\underset{\sim}{P}_1$	$\underset{\sim}{P}_2$	$\underset{\sim}{P}_3$	$\underset{\sim}{P}_4$
$\underset{\sim}{P}_1$		1	1	1
$\underset{\sim}{P}_2$	1	1		
$\underset{\sim}{P}_3$		1	1	1
$\underset{\sim}{P}_4$				1

The associated sagittate graph is the following:

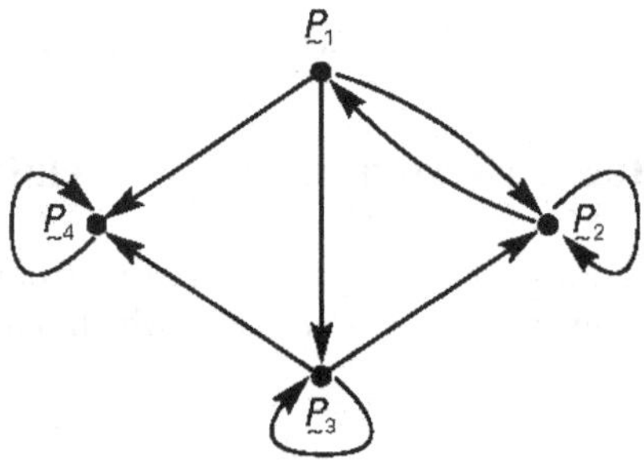

Now that a fuzzy relationship and a threshold which allows the movement into the boolean field are known we will define two important concepts, **predecessor level $\propto$** and **successor level $\propto$**.

Given a boolean relationship R_α it is said that $\underset{\sim}{P}_j$ is a **predecessor of** $\underset{\sim}{P}_k$ at level $\propto$ or also that $\underset{\sim}{P}_k$ is a **successor of** $\underset{\sim}{P}_j$ at level $\propto$ if $(\underset{\sim}{P}_j, \underset{\sim}{P}_k) \in R_\propto$. When $(\underset{\sim}{P}_j, \underset{\sim}{P}_k) \in R_\propto$ and $(\underset{\sim}{P}_j, \underset{\sim}{P}_k) \in R_\alpha$, $\underset{\sim}{P}_j$ and $\underset{\sim}{P}_k$ are both predecessor and successor at the same time. In our example $\underset{\sim}{P}_1$ is **one** predecessor of $\underset{\sim}{P}_2$, **that** of $\underset{\sim}{P}_3$ and **one** of $\underset{\sim}{P}_4$: $\underset{\sim}{P}_2$ is predecessor of $\underset{\sim}{P}_1$ and of itself, etc.

These two notions allow the definition of the **connection to the right** and **connection to the left** at level $\propto$[1].

Let a fuzzy relationship $\underset{\sim}{R} \subset E \times E$, the level $\propto$ of $\underset{\sim}{R}$ is called **connection to the right** of the functional application R^+_α of $P(E)$ in $P(E)$ so that for all $\underset{\sim}{A}_j \in P(E)$, R^+_α is the subset of elements of E which are predecessors of the level $\propto$ of all elements which belong to $\underset{\sim}{A}_j$. This may be expressed in the following way:

$$R^+_\alpha \underset{\sim}{A}_j = \{\underset{\sim}{P}_k \in E \,/\, (\underset{\sim}{P}_j, \underset{\sim}{P}_k) \in R_\alpha, \forall\, \underset{\sim}{P}_j \in \underset{\sim}{A}_j\}, \quad \text{with: } R^+_\alpha \varnothing = E$$

Once again, let $\underset{\sim}{R} \subset E \times E$, the level $\propto$ of $\underset{\sim}{R}$ is called **connection to the left** of the functional application R^+_α of $P(E)$ in $P(E)$ so that for all $\underset{\sim}{A}_j \in P(E)$, R^+_α is the subset of elements of E which are predecessors of the level $\propto$ of all elements which belong to $\underset{\sim}{A}_j$. This is written in this way:

$$R^-_\alpha \underset{\sim}{A}_j = \{\underset{\sim}{P}_k \in E \,/\, (\underset{\sim}{P}_k, \underset{\sim}{P}_j) \in R_\alpha, \forall\, \underset{\sim}{P}_k \in \underset{\sim}{A}_j\}, \quad \text{with: } R^-_\alpha \varnothing = E$$

In our example we would have:

$R^+_{0,7} \varnothing = E$	$R^+_{0,7}\{\underset{\sim}{P}_1\} = \{\underset{\sim}{P}_2, \underset{\sim}{P}_3, \underset{\sim}{P}_4\}$	$R^+_{0,7}\{\underset{\sim}{P}_2\} = \{\underset{\sim}{P}_1, \underset{\sim}{P}_2\}$
$R^+_{0,7}\{\underset{\sim}{P}_3\} = \{\underset{\sim}{P}_2, \underset{\sim}{P}_3, \underset{\sim}{P}_4\}$	$R^+_{0,7}\{\underset{\sim}{P}_4\} = \{\underset{\sim}{P}_4\}$	$R^+_{0,7}\{\underset{\sim}{P}_1, \underset{\sim}{P}_2\} = \{\underset{\sim}{P}_2\}$
$R^+_{0,7}\{\underset{\sim}{P}_1, \underset{\sim}{P}_3\} = \{\underset{\sim}{P}_2, \underset{\sim}{P}_3, \underset{\sim}{P}_4\}$	$R^+_{0,7}\{\underset{\sim}{P}_1, \underset{\sim}{P}_4\} = \{\underset{\sim}{P}_4\}$	$R^+_{0,7}\{\underset{\sim}{P}_2, \underset{\sim}{P}_3\} = \{\underset{\sim}{P}_2\}$
$R^+_{0,7}\{\underset{\sim}{P}_2, \underset{\sim}{P}_4\} = \varnothing$	$R^+_{0,7}\{\underset{\sim}{P}_3, \underset{\sim}{P}_4\} = \{\underset{\sim}{P}_4\}$	$R^+_{0,7}\{\underset{\sim}{P}_1, \underset{\sim}{P}_2, \underset{\sim}{P}_3\} = \{\underset{\sim}{P}_2\}$
$R^+_{0,7}\{\underset{\sim}{P}_1, \underset{\sim}{P}_2, \underset{\sim}{P}_4\} = \varnothing$	$R^+_{0,7}\{\underset{\sim}{P}_1, \underset{\sim}{P}_3, \underset{\sim}{P}_4\} = \{\underset{\sim}{P}_4\}$	$R^+_{0,7}\{\underset{\sim}{P}_2, \underset{\sim}{P}_3, \underset{\sim}{P}_4\} = \varnothing$
$R^+_{0,7} E = \varnothing$		

[1] Kaufmann, A., y Gil Aluja, J.: Técnicas de gestión de empresas. Previsiones, decisiones y estrategias. Ed. Pirámide, Madrid, 1992, page 353.

With the following sagittate representation:

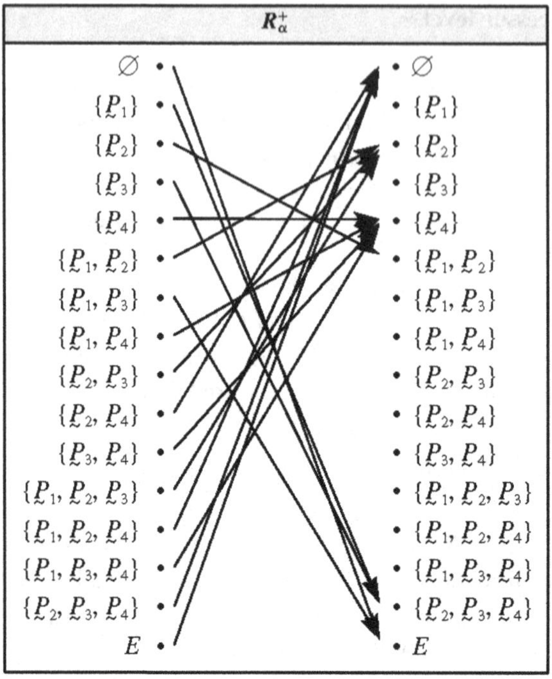

For the connections to that left it results that:

$R^-_{0,7}\varnothing = E$ $\quad R^-_{0,7}\{\underset{\sim}{P}_1\} = \{\underset{\sim}{P}_2\}$ $\quad R^-_{0,7}\{\underset{\sim}{P}_2\} = \{\underset{\sim}{P}_1, \underset{\sim}{P}_2, \underset{\sim}{P}_3\}$

$R^-_{0,7}\{\underset{\sim}{P}_3\} = \{\underset{\sim}{P}_1, \underset{\sim}{P}_3\}$ $\quad R^-_{0,7}\{\underset{\sim}{P}_4\} = \{\underset{\sim}{P}_1, \underset{\sim}{P}_3, \underset{\sim}{P}_4\}$ $\quad R^-_{0,7}\{\underset{\sim}{P}_1, \underset{\sim}{P}_2\} = \{\underset{\sim}{P}_2\}$

$R^-_{0,7}\{\underset{\sim}{P}_1, \underset{\sim}{P}_3\} = \varnothing$ $\quad R^-_{0,7}\{\underset{\sim}{P}_1, \underset{\sim}{P}_4\} = \varnothing$ $\quad R^-_{0,7}\{\underset{\sim}{P}_2, \underset{\sim}{P}_3\} = \{\underset{\sim}{P}_1, \underset{\sim}{P}_3\}$

$R^-_{0,7}\{\underset{\sim}{P}_2, \underset{\sim}{P}_4\} = \{\underset{\sim}{P}_1, \underset{\sim}{P}_3\}$ $\quad R^-_{0,7}\{\underset{\sim}{P}_3, \underset{\sim}{P}_4\} = \{\underset{\sim}{P}_1, \underset{\sim}{P}_3\}$ $\quad R^-_{0,7}\{\underset{\sim}{P}_1, \underset{\sim}{P}_2, \underset{\sim}{P}_3\} = \varnothing$

$R^-_{0,7}\{\underset{\sim}{P}_1, \underset{\sim}{P}_2, \underset{\sim}{P}_4\} = \varnothing$ $\quad R^-_{0,7}\{\underset{\sim}{P}_1, \underset{\sim}{P}_3, \underset{\sim}{P}_4\} = \varnothing$ $\quad R^-_{0,7}\{\underset{\sim}{P}_2, \underset{\sim}{P}_3, \underset{\sim}{P}_4\} = \{\underset{\sim}{P}_1, \underset{\sim}{P}_3\}$

$R^-_{0,7}E = \varnothing$

and the sagittate representation:

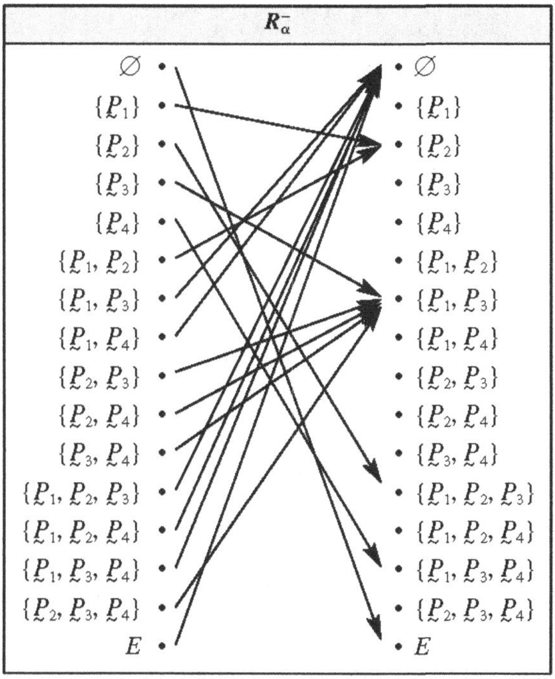

The previously stated definitions of connection to the right $R_\alpha^+ \underset{\sim}{A}_j$ and the to the left $R_\alpha^- \underset{\sim}{A}_j$ allow us to immediately confirm the two following properties:

$\forall \underset{\sim}{P}_j \subset \underset{\sim}{A}_j \in P(E)$:

$$R_\alpha^+ \underset{\sim}{A}_j = \bigcap_{\underset{\sim}{P}_j \in A_j} R_\alpha^+ \{\underset{\sim}{P}_j\}$$

$$R_\alpha^- \underset{\sim}{A}_j = \bigcap_{\underset{\sim}{P}_j \in A_j} R_\alpha^- \{\underset{\sim}{P}_j\}$$

which allow the direct finding of $R_\alpha^+ \underset{\sim}{A}_j$ and $R_\alpha^- \underset{\sim}{A}_j$ by means of direct reading and intersections in the relationship R_α.

As we have reached this point we should enunciate the following theorems:

1. The connection to the right (and, respectively, that to the left) is antitone in relation to that included in the elements of $P(E)$. We express this in the following way:

$\forall \underset{\sim}{A}_j, \underset{\sim}{A}_k \in P(E)$:

$$(\underset{\sim}{A}_j \subset \underset{\sim}{A}_k) \Rightarrow (R_\alpha^+ \underset{\sim}{A}_j \supset R_\alpha^+ \underset{\sim}{A}_k)$$

$$(\underset{\sim}{A}_j \subset \underset{\sim}{A}_k) \Rightarrow (R_\alpha^- \underset{\sim}{A}_j \supset R_\alpha^- \underset{\sim}{A}_k)$$

The demonstration is immediate, we have:

$$\bigcap_{\underset{\sim}{P}_j \in A_j} R_\alpha^+ \{\underset{\sim}{P}_j\} \supset \bigcap_{\underset{\sim}{P}_j \in A_k} R_\alpha^+ \{\underset{\sim}{P}_j\}$$

But, as we have seen that:

$$R_\alpha^+ \underset{\sim}{A}_j = \bigcap_{\underset{\sim}{P}_j \in A_j} R_\alpha^+ \{\underset{\sim}{P}_j\}$$

We have, definitively:

$$R_\alpha^+ \underset{\sim}{A}_j \supset R_\alpha^+ \underset{\sim}{A}_k$$

The same can be done for the connection to the left.

That which we have just stated is shown in our example. So:

$$(\{\underset{\sim}{P}_1, \underset{\sim}{P}_2\} \subset \{\underset{\sim}{P}_1, \underset{\sim}{P}_2, \underset{\sim}{P}_3\}) \Rightarrow (\{\underset{\sim}{P}_2\} \supset \{\underset{\sim}{P}_2\})$$
$$(\{\underset{\sim}{P}_1, \underset{\sim}{P}_2\} \subset \{\underset{\sim}{P}_1, \underset{\sim}{P}_2, \underset{\sim}{P}_4\}) \Rightarrow (\{\underset{\sim}{P}_2\} \supset \varnothing)$$
$$(\{\underset{\sim}{P}_1, \underset{\sim}{P}_2\} \subset E) \Rightarrow (\{\underset{\sim}{P}_2\} \supset \varnothing)$$

We would continue in the same way with the successive elements of the "power set".

If R_α^+ and R_α^- are, respectively, the connections to the right and to the left of a fuzzy relationship $\underset{\sim}{R} \subset E \times E$ at the level $\propto$, the maxmin compositions of R_α^- with R_α^+ and of R_α^+ with R_α^- give the Moore closings, which is to say isotone pretopologies with idempotency. This is written as:

$$M_\alpha^{(1)} = R_\alpha^- \bullet R_\alpha^+$$
$$M_\alpha^{(2)} = R_\alpha^+ \bullet R_\alpha^-$$

in which $M_\alpha^{(1)}$, $M_\alpha^{(2)}$ are Moore closings.

We move on to the confirmation, taking $M_\alpha^{(2)}$ as a reference.

a) In first place we should demonstrate that $M_\alpha^{(2)}$ is extensive.

If $\underset{\sim}{A}_j \in P(E)$, $M_\alpha^{(2)} \underset{\sim}{A}_j = (R_\alpha^+ \circ R_\alpha^-)$ is the subset of elements of E which are captured by each element of E, which is to say:

$$M_\alpha^{(2)} \underset{\sim}{A}_j = \{\underset{\sim}{P}_j \in E \,/\, (\underset{\sim}{P}_j, \underset{\sim}{P}_k) \in R_\alpha^-, \forall\, \underset{\sim}{P}_k \in R_\alpha^+ \underset{\sim}{A}_j\}$$

in which all element $\underset{\sim}{P}_j \in \underset{\sim}{A}_j$ belongs to $M_\alpha^{(2)} \underset{\sim}{A}_j$, and this as $(\underset{\sim}{P}_j\,,\, \underset{\sim}{P}_k) \in R_\alpha^-$ when $\underset{\sim}{P}_k \in R_\alpha^{+-} \underset{\sim}{A}_j$. Therefore:

$$\underset{\sim}{A}_j \subset (R_\alpha^+ \circ R_\alpha^-)\, \underset{\sim}{A}_j$$

or, which is the same:

$$\underset{\sim}{A}_j \subset M_\alpha^{(2)} \underset{\sim}{A}_j$$

b) We verify that $M_\alpha^{(2)}$ is idempotent.
$\underset{\sim}{A}_j$ is any subset of E. We may also write that:

$$R_\alpha^- \underset{\sim}{A}_j \subset M_\alpha^{(2)} (R_\alpha^- \underset{\sim}{A}_j)$$

which is to say that:

$$R_\alpha^- \underset{\sim}{A}_j \subset (R_\alpha^- \circ R_\alpha^+) \circ R_\alpha^- \underset{\sim}{A}_j$$

On the other hand, we have seen that R_α^- is antitone upon:

$$(\underset{\sim}{A}_j \subset \underset{\sim}{A}_k) \Rightarrow (R_\alpha^- \underset{\sim}{A}_j \supset R_\alpha^- \underset{\sim}{A}_k)$$

As well as the proposal $M_\alpha^{(2)} = R_\alpha^{++} \circ R_\alpha^-$ and as a consequence of $\underset{\sim}{A}_j \subset M_\alpha^{(2)} \underset{\sim}{A}_j$, we have:

$$R_\alpha^- \underset{\sim}{A}_j \supset (R_\alpha^- \circ R_\alpha^+) \circ R_\alpha^- \underset{\sim}{A}_j$$

But with:

$$R_\alpha^- \underset{\sim}{A}_j \supseteq\!\!\!\!\subseteq (R_\alpha^- \circ R_\alpha^+) \circ R_\alpha^- \underset{\sim}{A}_j$$

It is concluded that:

$$R_\alpha^- \underset{\sim}{A}_j = (R_\alpha^- \circ R_\alpha^+) \circ R_\alpha^- \underset{\sim}{A}_j$$

The same occurs for R_α^+. So:

$$\begin{aligned} M_\alpha^{(2)} \circ M_\alpha^{(2)} &= M_\alpha^{(2)} \circ (R_\alpha^+ \circ R_\alpha^-) \underset{\sim}{A}_j \\ &= (R_\alpha^+ \circ R_\alpha^- \circ R_\alpha^+ \circ R_\alpha^-) \underset{\sim}{A}_j \\ &= (R_\alpha^+ \circ R_\alpha^-) \underset{\sim}{A}_j \\ &= M_\alpha^{(2)} \underset{\sim}{A}_j \end{aligned}$$

c) We state that $M_\alpha^{(2)}$ is isotone.

If

$$\underset{\sim}{A}_j \in P(E) \text{ y } \underset{\sim}{A}_k \in P(E)$$

with

$$\underset{\sim}{A}_j \subset \underset{\sim}{A}_k$$

Upon being antitone in relation with the inclusion, we first have that:

$$R_\alpha^+ \underset{\sim}{A}_j \supset R_\alpha^+ \underset{\sim}{A}_k$$

and then that:

$$(R_\alpha^+ \circ R_\alpha^-)\, \underset{\sim}{A}_j \subset (R_\alpha^+ \circ R_\alpha^-)\, \underset{\sim}{A}_k$$

which is to say:

$$M_\alpha^{(2)}\, \underset{\sim}{A}_j \subset M_\alpha^{(2)}\, \underset{\sim}{A}_k$$

The same reasoning is valid for $M_\alpha^{(1)}$.

To illustrate this important theorem we will return to our previous example.

We reproduce the boolean matrix which resulted from the establishment of $\propto\ \geq$ 0.7 as a threshold in the fuzzy matrix $\underset{\sim}{R}$:

$R_{0,7} =$

	$\underset{\sim}{P}_1$	$\underset{\sim}{P}_2$	$\underset{\sim}{P}_3$	$\underset{\sim}{P}_4$
$\underset{\sim}{P}_1$		1	1	1
$\underset{\sim}{P}_2$	1	1		
$\underset{\sim}{P}_3$		1	1	1
$\underset{\sim}{P}_4$				1

From the connection to the right R_α^- and the connection to the left R_α^+, $M_\alpha^{(1)}$ is found by maxmin composition:

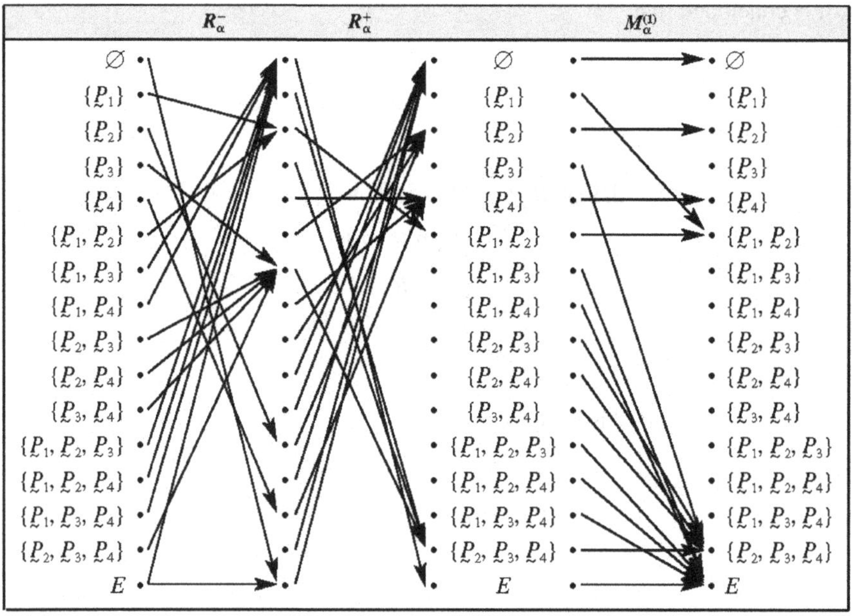

Which has as closes:

$$C_e^{(1)} = \{\varnothing, \{\underset{\sim}{P}_2\}, \{\underset{\sim}{P}_4\}, \{\underset{\sim}{P}_1, \underset{\sim}{P}_2\}, \{\underset{\sim}{P}_2, \underset{\sim}{P}_3, \underset{\sim}{P}_4\}, E\}$$

as they fulfil the demand:

$$\forall\, \underset{\sim}{A}_j \in P(E):$$

$$\underset{\sim}{A}_j = M\underset{\sim}{A}_j$$

In the same way, by means of the maxmin composition between the connection to the right $R^{+}_{\alpha\,\infty}{}^{+}$ and the connection to the left R^{-}_{α}, $M^{(2)}_{\alpha}$ is obtained:

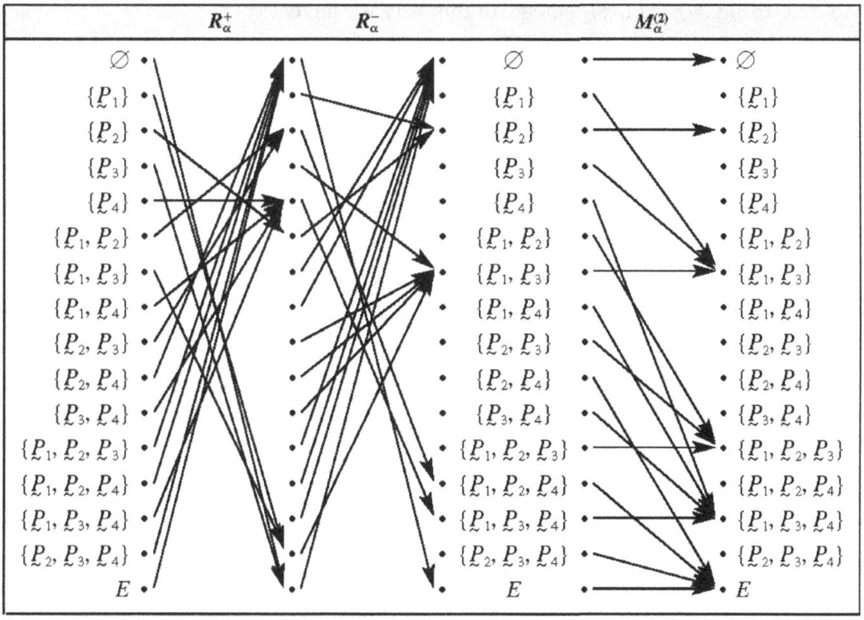

The set of closes is now:

$$C^{(2)}_{e} = \{\varnothing, \{\underset{\sim}{P}_2\}, \{\underset{\sim}{P}_1, \underset{\sim}{P}_3\}, \{\underset{\sim}{P}_1, \underset{\sim}{P}_2, \underset{\sim}{P}_3\}, \{\underset{\sim}{P}_1, \underset{\sim}{P}_3, \underset{\sim}{P}_4\}, E\}$$

We now move on to invoke some properties of the **closes**.

1. We have seen that, as a consequence of idempotency we have:

$$\begin{aligned} R^{-}_{\alpha}\underset{\sim}{A}_j &= (R^{-}_{\alpha} \circ R^{+}_{\alpha}) \circ R^{-}_{\alpha}\underset{\sim}{A}_j \\ &= M^{(2)}_{\alpha} \circ R^{-}_{\alpha}\underset{\sim}{A}_j \qquad \text{(by definition)} \\ &= M^{(2)}_{\alpha}(R^{-}_{\alpha}\underset{\sim}{A}_j) \qquad \text{(using another notation)} \end{aligned}$$

Therefore, $R^{-}_{\alpha}\underset{\sim}{A}_j$ is a close of $P(E)$ for $M^{(2)}_{\alpha}$.

In this way, for $C(E, M^{(2)}_{\alpha})$ the subset of closes of $P(E)$ corresponding to the Moore closing $M^{(2)}_{\alpha}$ is designated. We may therefore write:

$$C(E, M^{(2)}_{\alpha}) = \bigcup_{\underset{\sim}{A}_j \in P(E)} R^{-}_{\alpha}\underset{\sim}{A}_j$$

In the same way we obtain:

$$\begin{aligned} R_\alpha^+ \underset{\sim}{A}_j &= (R_\alpha^+ \circ R_\alpha^-) \circ R_\alpha^+ \underset{\sim}{A}_j \\ &= M_\alpha^{(1)} \circ R_\alpha^+ \underset{\sim}{A}_j \\ &= M_\alpha^{(1)} (R_\alpha^+ \underset{\sim}{A}_j) \end{aligned}$$

Therefore, $R_\alpha^+ \underset{\sim}{A}_j$ is a close of $P(E)$ for $M_\alpha^{(1)}$.

In this way, for $C(E, M_\alpha^{(1)})$ the subset of closes of $P(E)$ corresponding to the Moore closing $M_\alpha^{(1)}$ is designated. In this way we have:

$$C(E, M_\alpha^{(1)}) = \bigcup_{\underset{\sim}{A}_j \in P(E)} R_\alpha^+ \underset{\sim}{A}_j$$

We move on to another property.

The two families $C(E, M_\alpha^{(1)})$ and $C(E, M_\alpha^{(2)})$ are isomorphic between each other and the relations between these are dual. This may be expressed in the following way:

$$\underset{\sim}{A}_j \in C(E, M_\alpha^{(1)}) \Rightarrow (\underset{\sim}{A}_k = R_\alpha^- \underset{\sim}{A}_j \in C(E, M_\alpha^{(2)}) \text{ y } R_\alpha^+ \underset{\sim}{A}_k = \underset{\sim}{A}_j)$$
$$\underset{\sim}{A}_k \in C(E, M_\alpha^{(2)}) \Rightarrow (\underset{\sim}{A}_j = R_\alpha^+ \underset{\sim}{A}_k \in C(E, M_\alpha^{(1)}) \text{ y } R_\alpha^- \underset{\sim}{A}_j = \underset{\sim}{A}_k)$$

We move on to the confirmation. For a Moore closing the set of the closed subsets for $M_\propto^{(1)}$ on one hand and for $M_\propto^{(2)}$ on the other constitute finite lattices. We point out that:

R_α^- is a functional application of $C(E, M_\alpha^{(1)})$ in $C(E, M_\alpha^{(2)})$
R_α^+ is a functional application of $C(E, M_\alpha^{(2)})$ in $C(E, M_\alpha^{(1)})$

For greater commodity we will designate as H_α^- the functional application R_α^- reduced to $C(E, M_\alpha^{(1)})$ as a source and $C(E, M_\alpha^{(2)})$ as an image. In the same way, H_α^+ for $R_{\alpha\,\propto}^{+\;\;+}$ in relation to $C(E, M_\alpha^{(2)})$ and $C(E, M_\alpha^{(1)})$.

For the lattices $C(E, M_\propto^{(2)})$ and $C(E, M_\propto^{(1)})$ to be isomorphic and dual it is necessary that they fulfil the three following properties:

a) The functional applications should be bijective and inverse of one another.
b) $H_\propto^+$ should be antitone of $C(E, M_\alpha^{(2)})$ in $C(E, M_\alpha^{(1)})$.
c) $H_\propto^-$ should be antitone of $C(E, M_\alpha^{(1)})$ in $C(E, M_\alpha^{(2)})$.

In that referred to in b) and c), the antitone is shown by R_α^+ and R_α^- and this property is evidently kept by H_α^+ and H_α^-. Condition a) is the consequence of the idempotency and, therefore:

$$R_\alpha^- \underset{\sim}{A}_j = (R_\alpha^- \circ R_\alpha^+) \circ R_\alpha^- \underset{\sim}{A}_j$$

We assume that $\underset{\sim}{A}_j \in C(E, M_\alpha^{(1)})$. It can be written that $(H_\alpha^- \circ R_\alpha^+)\, \underset{\sim}{A}_k = \underset{\sim}{A}_k$, which means that $H_\alpha^- \circ H_\alpha^+$ is the same as the identical bijective functional application of $C(E, M_\alpha^{(1)})$ in $C(E, M_\alpha^{(1)})$. In the same way, that $H_\alpha^+ \circ H_\alpha^-$ is the same as the functional application of $C(E, M_\alpha^{(2)})$ in $C(E, M_\alpha^{(2)})$. In this way it is concluded that $H_\alpha^+ \circ H_\alpha^-$ are bijective functional applications which are the inverse of each other.

$$H_\alpha^- = (H_\alpha^+)^{-1}$$

and also:

$$H_\alpha^+ = (H_\alpha^-)^{-1}$$

We return to our example. We have seen that the closes were, respectively:

$$C(E, M_\alpha^{(1)}) = \{\varnothing, \{\underset{\sim}{P}_2\}, \{\underset{\sim}{P}_4\}, \{\underset{\sim}{P}_1, \underset{\sim}{P}_2\}, \{\underset{\sim}{P}_2, \underset{\sim}{P}_3, \underset{\sim}{P}_4\}, E\}$$
$$C(E, M_\alpha^{(2)}) = \{\varnothing, \{\underset{\sim}{P}_2\}, \{\underset{\sim}{P}_1, \underset{\sim}{P}_3\}, \{\underset{\sim}{P}_1, \underset{\sim}{P}_2, \underset{\sim}{P}_3\}, \{\underset{\sim}{P}_1, \underset{\sim}{P}_3, \underset{\sim}{P}_4\}, E\}$$

Between the connection to the left $R_\propto^-$ and the connection to the right R_α^- we will choose those which refer to the set of closes $C(E, M_\alpha^{(1)})$ and $C(E, M_\alpha^{(2)})$, which we have designated by H_α^- and H_α^+. In this way we have:

a) $H_\alpha^- \varnothing = E$ $\quad H_\alpha^- \{\underset{\sim}{P}_2\} = \{\underset{\sim}{P}_1, \underset{\sim}{P}_2, \underset{\sim}{P}_3\}$ $\quad H_\alpha^- \{\underset{\sim}{P}_4\} = \{\underset{\sim}{P}_1, \underset{\sim}{P}_3, \underset{\sim}{P}_4\}$
$H_\alpha^- \{\underset{\sim}{P}_1, \underset{\sim}{P}_2\} = \{\underset{\sim}{P}_2\}$ $\quad H_\alpha^- \{\underset{\sim}{P}_2, \underset{\sim}{P}_3, \underset{\sim}{P}_4\} = \{\underset{\sim}{P}_1, \underset{\sim}{P}_3\}$ $\quad H_\alpha^- E = \varnothing$

b) $H_\alpha^+ \varnothing = E$ $\quad H_\alpha^+ \{\underset{\sim}{P}_2\} = \{\underset{\sim}{P}_1, \underset{\sim}{P}_2\}$ $\quad H_\alpha^+ \{\underset{\sim}{P}_1, \underset{\sim}{P}_3\} = \{\underset{\sim}{P}_2, \underset{\sim}{P}_3, \underset{\sim}{P}_4\}$
$H_\alpha^+ \{\underset{\sim}{P}_1, \underset{\sim}{P}_2, \underset{\sim}{P}_3\} = \{\underset{\sim}{P}_2\}$ $\quad H_\alpha^+ \{\underset{\sim}{P}_1, \underset{\sim}{P}_3, \underset{\sim}{P}_4\} = \{\underset{\sim}{P}_4\}$ $\quad H_\alpha^+ E = \varnothing$

We construct the corresponding lattices in which the applications are bijective, inverse and antitone.

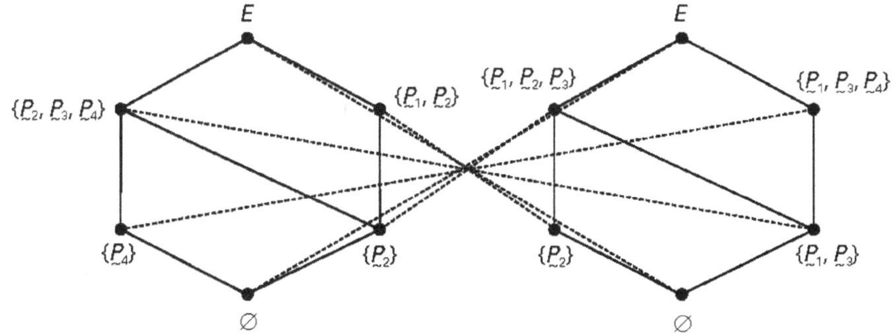

We believe that with this example the duality is evident, also reflected in the following figure:

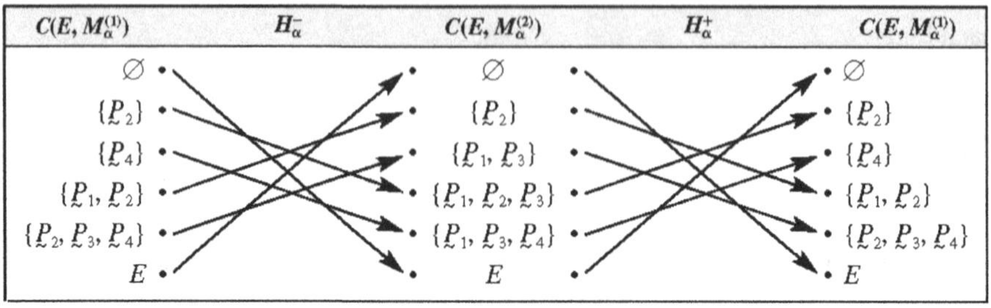

The previous development took the transformation of a fuzzy relationship in a boolean relationship upon establishing as a threshold $\propto\ \geq 0.7$ as a starting point. This takes us to the possibility of considering various groups of connections to the right and connections to the left as levels of $\propto$ which are established.

4 Establishment of Connections from ∝-Cuts

For each level of ∝, the maxmin composition of the connection to the left with the connection to the right and the maxmin composition of the connection to the right with the connection to the left gives the two Moore closings.

For a greater insight into that which we have just stated, we will turn to a example[2] which is simple, but which we believe to be sufficient to state the most significant aspects of the variation of the lattices when they move from one level of $\propto$ to another. For this we will begin with the following fuzzy matrix:

$\underset{\sim}{R} =$

	$\underset{\sim}{P}_1$	$\underset{\sim}{P}_2$	$\underset{\sim}{P}_3$
$\underset{\sim}{P}_1$	.7	.5	1
$\underset{\sim}{P}_2$	1	0	.2
$\underset{\sim}{P}_3$	.8	.2	1

The successive $\propto$-cuts are considered and the corresponding Moore closings are obtained.

[2] Kaufmann, A., y Gil Aluja, J.: *Técnicas para la gestión de empresa. Previsiones, decisiones y estrategias.* Ed. Pirámide. Madrid, 1992, pages 359-362; and also Gil Aluja, J.: *La pretopología en la gestión de la incertidumbre.* Speech for the purpose of entering as Doctor «Honoris Causa» of the Universidad de León. Publ. Universidad de León 2001, pages 11-18.

For $\propto = 1$:

$R_1 =$

	P_1	P_2	P_3
P_1			1
P_2	1		
P_3			1

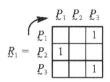

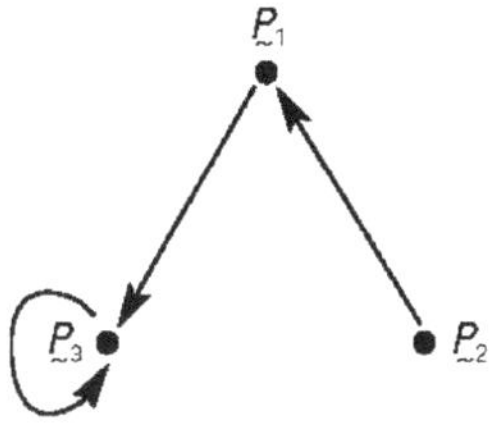

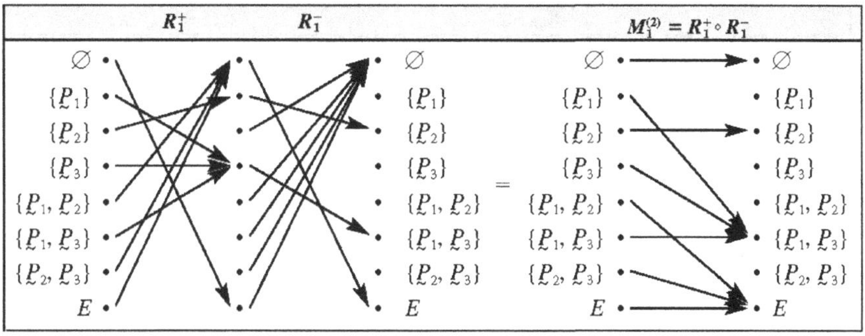

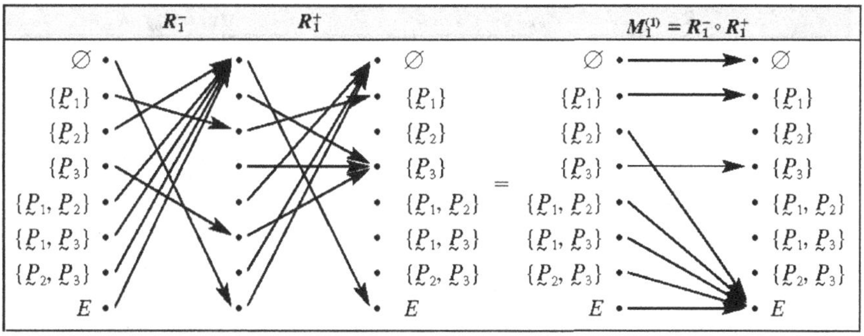

For ∝ = 0.8:

$R_{0,8} =$

	P_1	P_2	P_3
P_1			1
P_2	1		
P_3	1		1

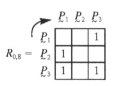

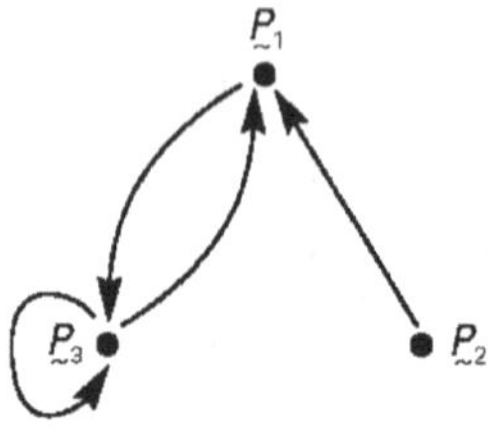

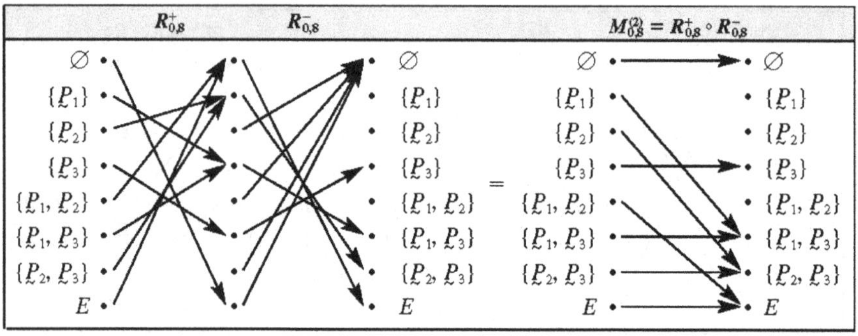

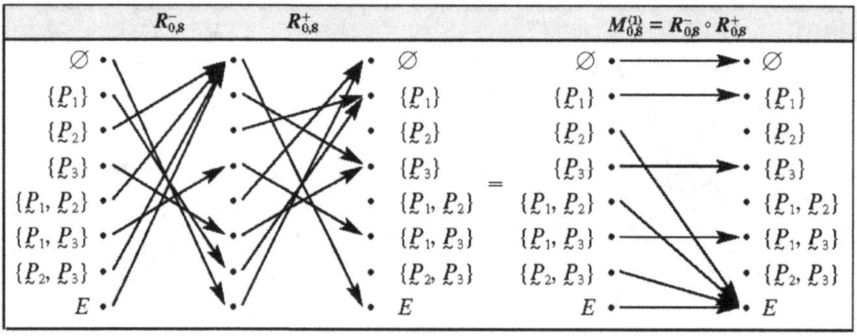

For $\propto = 0.7$:

$R_{0,7} =$

	$\underset{\sim}{P}_1$	$\underset{\sim}{P}_2$	$\underset{\sim}{P}_3$
$\underset{\sim}{P}_1$	1		1
$\underset{\sim}{P}_2$	1		
$\underset{\sim}{P}_3$	1		1

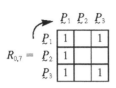

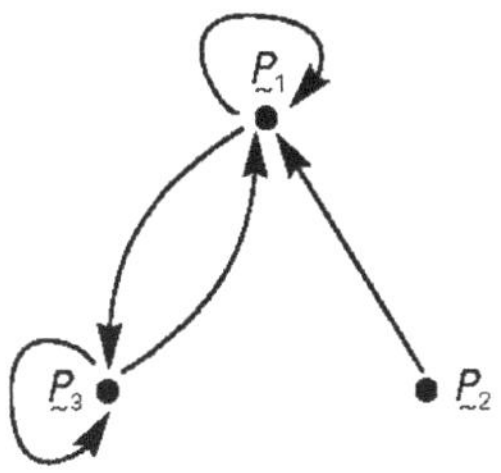

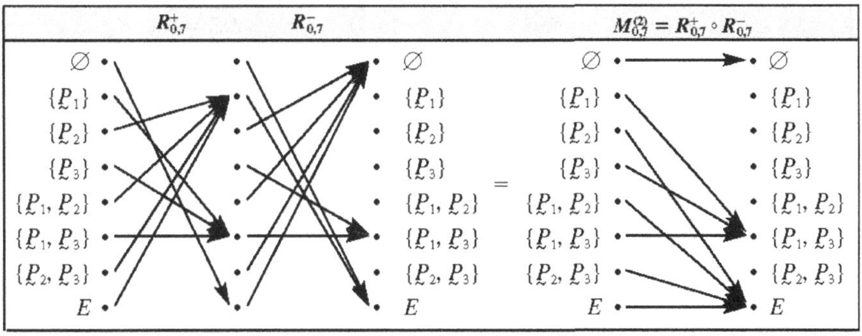

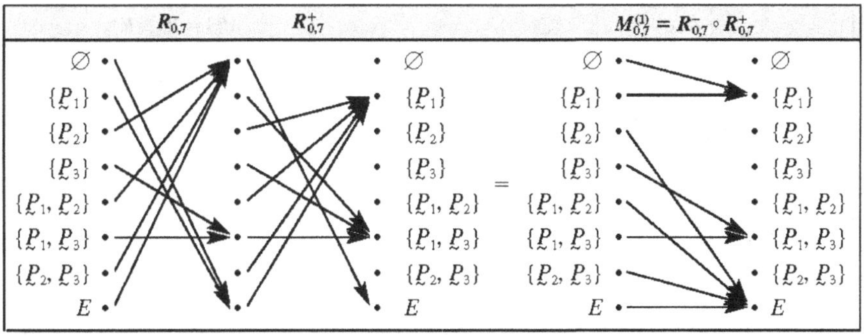

For ∝ = 0.5:

$$R_{0,5} = \begin{array}{c|ccc} & \underset{\sim}{P}_1 & \underset{\sim}{P}_2 & \underset{\sim}{P}_3 \\ \hline \underset{\sim}{P}_1 & 1 & 1 & 1 \\ \underset{\sim}{P}_2 & 1 & 0 & 0 \\ \underset{\sim}{P}_3 & 1 & 0 & 1 \end{array}$$

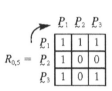

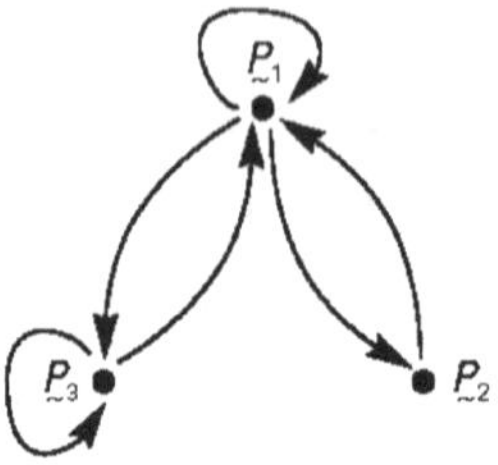

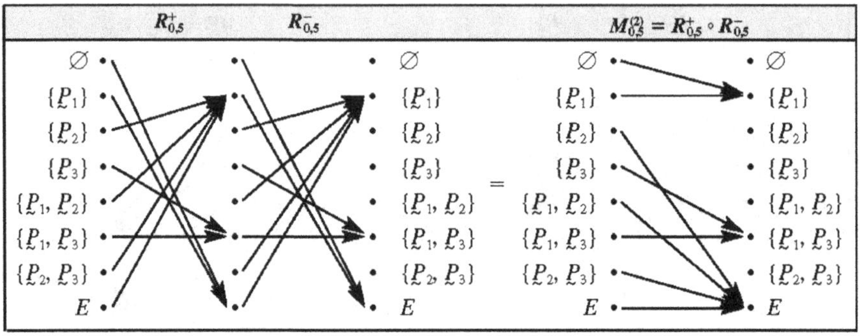

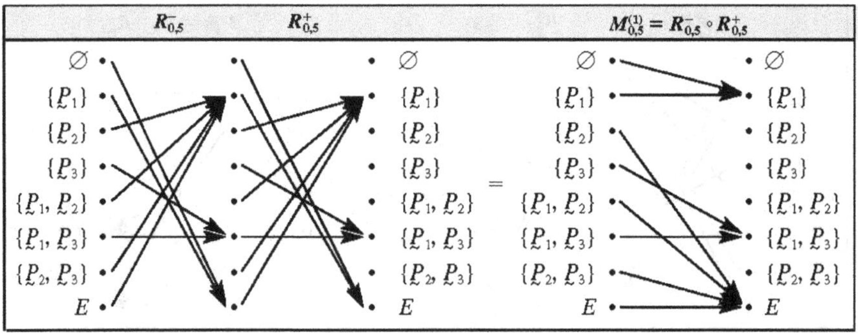

For $\propto = 0.2$:

$R_{0,2} =$

	P_1	P_2	P_3
P_1	1	1	1
P_2	1	0	1
P_3	1	1	1

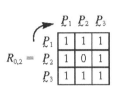

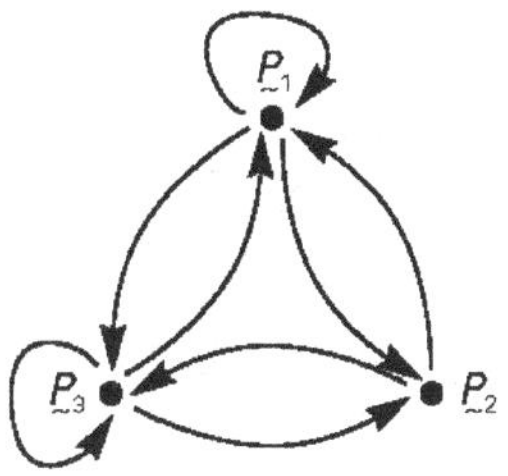

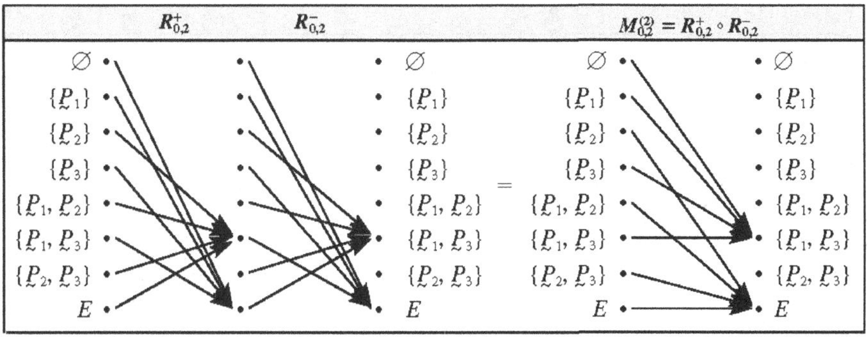

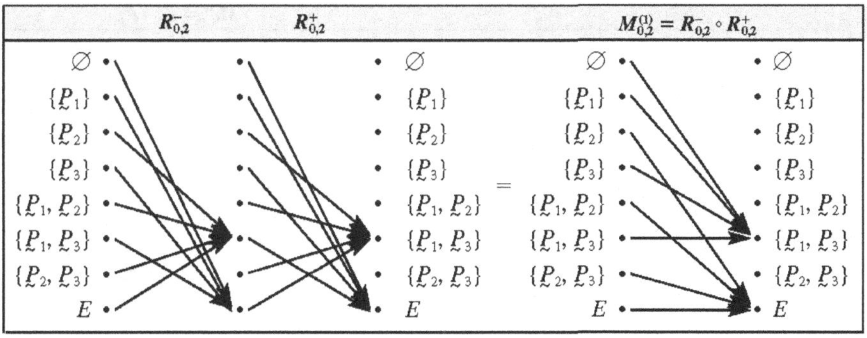

For $\propto = 0$:

$$R_0 = \begin{array}{c|ccc} & \underset{\sim}{P}_1 & \underset{\sim}{P}_2 & \underset{\sim}{P}_3 \\ \hline \underset{\sim}{P}_1 & 1 & 1 & 1 \\ \underset{\sim}{P}_2 & 1 & 1 & 1 \\ \underset{\sim}{P}_3 & 1 & 1 & 1 \end{array}$$

R_0^+		R_0^-		$M_0^{(2)} = R_0^+ \circ R_0^-$	
$\varnothing$		$\varnothing$	$\varnothing$		$\varnothing$
$\{P_1\}$		$\{P_1\}$	$\{P_1\}$		$\{P_1\}$
$\{P_2\}$		$\{P_2\}$	$\{P_2\}$		$\{P_2\}$
$\{P_3\}$		$\{P_3\}$	$\{P_3\}$		$\{P_3\}$
$\{P_1, P_2\}$		$\{P_1, P_2\}$	$\{P_1, P_2\}$		$\{P_1, P_2\}$
$\{P_1, P_3\}$		$\{P_1, P_3\}$	$\{P_1, P_3\}$		$\{P_1, P_3\}$
$\{P_2, P_3\}$		$\{P_2, P_3\}$	$\{P_2, P_3\}$		$\{P_2, P_3\}$
E		E	E		E

R_0^-		R_0^+		$M_0^{(1)} = R_0^- \circ R_0^+$	
$\varnothing$		$\varnothing$	$\varnothing$		$\varnothing$
$\{P_1\}$		$\{P_1\}$	$\{P_1\}$		$\{P_1\}$
$\{P_2\}$		$\{P_2\}$	$\{P_2\}$		$\{P_2\}$
$\{P_3\}$		$\{P_3\}$	$\{P_3\}$		$\{P_3\}$
$\{P_1, P_2\}$		$\{P_1, P_2\}$	$\{P_1, P_2\}$		$\{P_1, P_2\}$
$\{P_1, P_3\}$		$\{P_1, P_3\}$	$\{P_1, P_3\}$		$\{P_1, P_3\}$
$\{P_2, P_3\}$		$\{P_2, P_3\}$	$\{P_2, P_3\}$		$\{P_2, P_3\}$
E		E	E		E

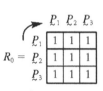

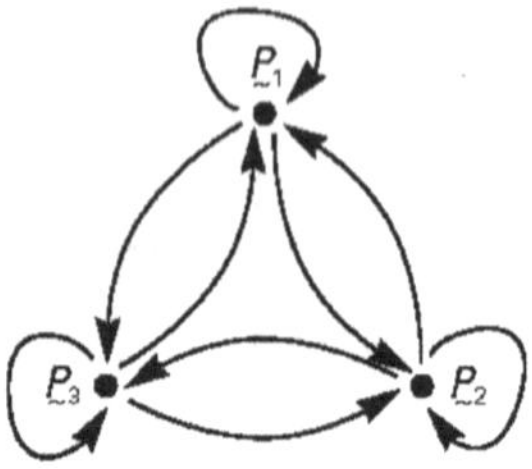

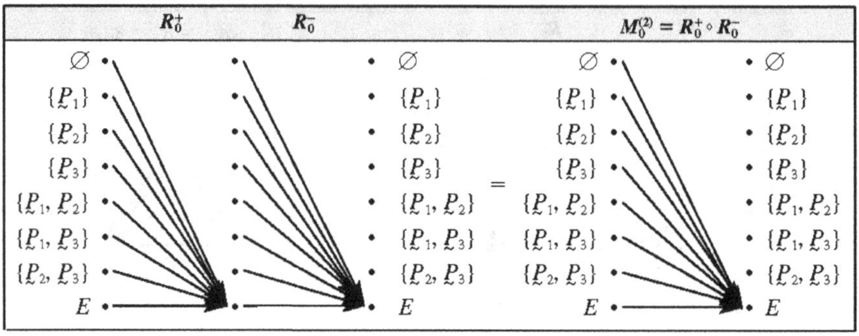

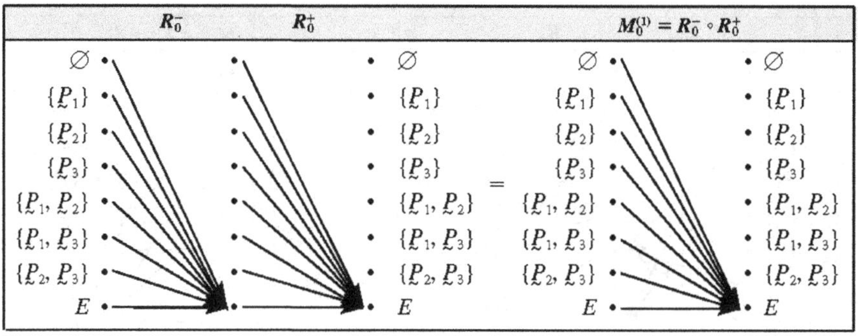

We have seen that, on the other hand, for a Moore closing the subsets of closes corresponding to $M_\alpha^{(1)}$ on one side and $M_\alpha^{(2)}$ on the other constitute complete lattices[3] in the sense that all vertices A_j of the lattice have an upper extreme and a lower extreme in the lattice (a finite lattice is always complete).

---

[3] Dubreil, P.: Leçons d'algébre moderne. Ed. Dunod. París, 1964, page 177.

For our example, in the distinct levels of $\propto$ we have the following closes, to which we associate the isomorphic and dual lattices presented in continuation:

$$M_1^{(2)} = \{\varnothing, \{\underset{\sim}{P}_2\}, \{\underset{\sim}{P}_1, \underset{\sim}{P}_3\}, E\} \qquad M_1^{(1)} = \{\varnothing, \{\underset{\sim}{P}_1\}, \{\underset{\sim}{P}_3\}, E\}$$

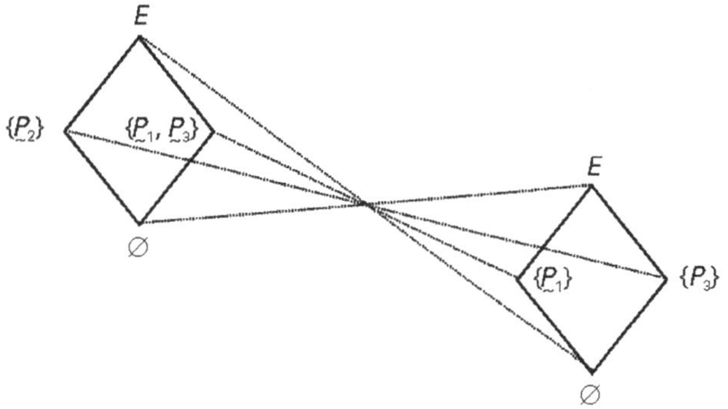

$$M_{0,8}^{(2)} = \{\varnothing, \{\underset{\sim}{P}_3\}, \{\underset{\sim}{P}_1, \underset{\sim}{P}_3\}, \{\underset{\sim}{P}_2, \underset{\sim}{P}_3\}, E\} \qquad M_{0,8}^{(1)} = \{\varnothing, \{\underset{\sim}{P}_1\}, \{\underset{\sim}{P}_3\}, \{\underset{\sim}{P}_1, \underset{\sim}{P}_3\}, E\}$$

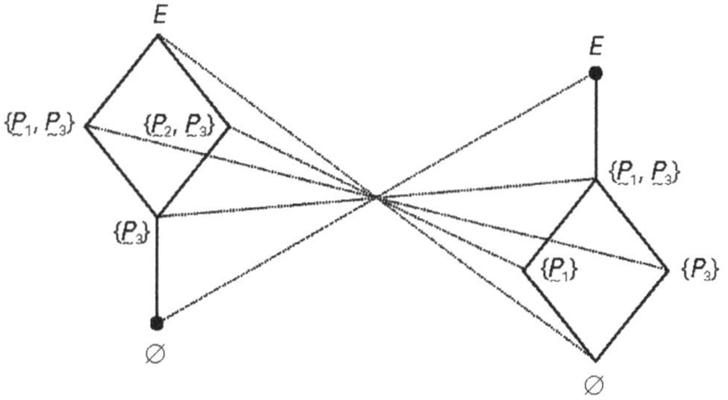

$$M_{0,7}^{(2)} = \{\varnothing, \{\underset{\sim}{P}_1, \underset{\sim}{P}_3\}, E\} \qquad M_{0,7}^{(1)} = \{\{\underset{\sim}{P}_1\}, \{\underset{\sim}{P}_1, \underset{\sim}{P}_3\}, E\}$$

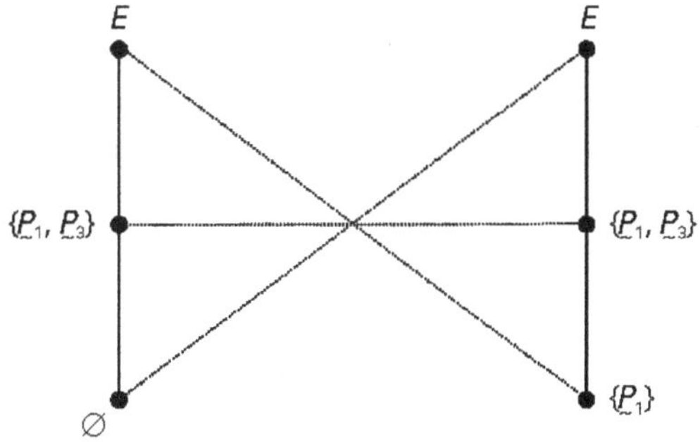

$M^{(2)}_{0,5} = \{\{\underset{\sim}{P}_1\}, \{\underset{\sim}{P}_1, \underset{\sim}{P}_3\}, E\}$ $\quad$ $M^{(1)}_{0,5} = \{\{\underset{\sim}{P}_1\}, \{\underset{\sim}{P}_1, \underset{\sim}{P}_3\}, E\}$

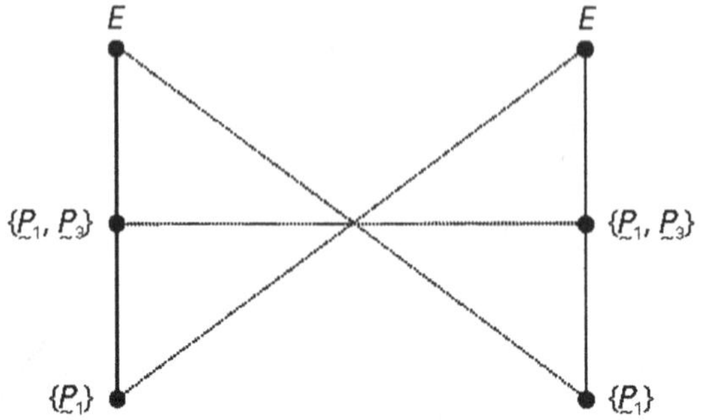

$M^{(2)}_{0,2} = \{\{\underset{\sim}{P}_1, \underset{\sim}{P}_3\}, E\}$ $\quad$ $M^{(1)}_{0,2} = \{\{\underset{\sim}{P}_1, \underset{\sim}{P}_3\}, E\}$

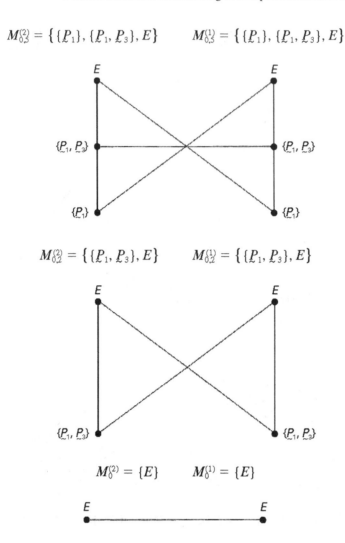

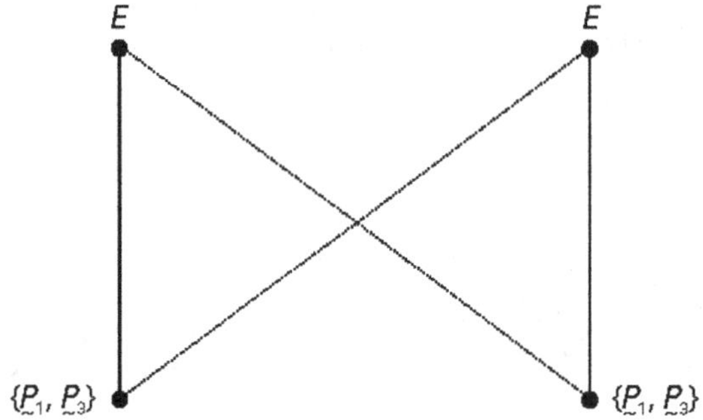

$M^{(2)}_{0} = \{E\}$ $\quad$ $M^{(1)}_{0} = \{E\}$

We believe that the simple example proposed will be enough to show how the lattice and, therefore, the elements of the "power set" which make up its vertices, change when we move from from level to another in the same fuzzy relationship $\underset{\sim}{R} \subset E \times E$.

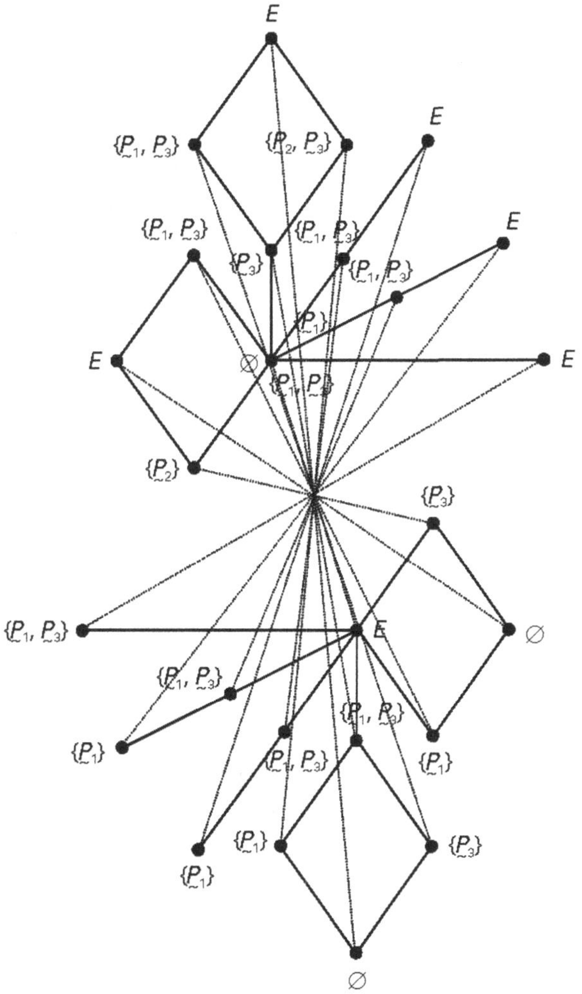

With this figure we have wished to create a summary of the existing mutations in the morphology of a lattice as the level of $\propto$ changes.

5 Generalisation by Means of a Rectangular Fuzzy Relationship

As we have reached this point, it seems justifiable to ask ourselves about the possibility of generalising that which we have just stated on the supposition of a rectangular fuzzy relationship $\underset{\sim}{R} \subset E_1 \times E_2$ in which, therefore, the cardinals E_1 and E_2 are normally distinct. This fact should allow us to relate elements of the two distinct "power sets".

So, the meaning of the the connections to the right R^+_α and to the left R^-_α may be reasoned out in the following way[4]:

"The obtaining of the connection to the right R^+_α shows the elements of the "power set" of the set E_1, in function of one or various elements of the "power set" of E_2. In this phase of the process it is given as fact that the same element of the "power set" of E_1 may join together various elements of the "power set" of E_2 which include elements of the rest. It is therefore necessary to continue with the obtaining of R^-_α. The connection to the left R^-_α shows the elements of the "power set" of E_2 in function of one or various elements of the "power set" of E_1. In R^-_α the phenomenon of the existence of various elements of the "power set" of E_1 for a same element of the "power set" of E_2 is also produced, such as the fact of the existence of an element of the "power set" of E_1 which includes the rest. It is due to this that the Moore closings are obtained by means of the corresponding max-min compositions and, from there, the "closes"."

We propose the visualisation of that which we have just stated with a simple example. We begin from two sets:

$$E_1 = \{\underset{\sim}{P}_1, \underset{\sim}{P}_2, \underset{\sim}{P}_3\}$$
$$E_2 = \{\underset{\sim}{Q}_1, \underset{\sim}{Q}_2, \underset{\sim}{Q}_3, \underset{\sim}{Q}_4\}$$

and we assume a relationship between these, expressed by means of the the following fuzzy matrix:

$R_0 =$

	$\underset{\sim}{Q}_1$	$\underset{\sim}{Q}_2$	$\underset{\sim}{Q}_3$	$\underset{\sim}{Q}_4$
$\underset{\sim}{P}_1$	.9	.6	.8	.2
$\underset{\sim}{P}_2$	.4	.7	.9	.5
$\underset{\sim}{P}_3$	.2	1	.3	.8

As a threshold we establish $\propto \geq 0.6$ and we obtain the boolean matrix:

$R_{0,6} =$

	$\underset{\sim}{Q}_1$	$\underset{\sim}{Q}_2$	$\underset{\sim}{Q}_3$	$\underset{\sim}{Q}_4$
$\underset{\sim}{P}_1$	1	1	1	
$\underset{\sim}{P}_2$		1	1	
$\underset{\sim}{P}_3$		1		1

The connection to the right $R^+_{0,6}$ is:

$$R^+_{0,6}\,\varnothing = E_2 \qquad R^+_{0,6}\,\{\underset{\sim}{P}_1\} = \{\underset{\sim}{Q}_1, \underset{\sim}{Q}_2, \underset{\sim}{Q}_3\} \qquad R^+_{0,6}\,\{\underset{\sim}{P}_2\} = \{\underset{\sim}{Q}_2, \underset{\sim}{Q}_3\}$$
$$R^+_{0,6}\,\{\underset{\sim}{P}_3\} = \{\underset{\sim}{Q}_2, \underset{\sim}{Q}_4\} \qquad R^+_{0,6}\,\{\underset{\sim}{P}_1, \underset{\sim}{P}_2\} = \{\underset{\sim}{Q}_2, \underset{\sim}{Q}_3\}$$
$$R^+_{0,6}\,\{\underset{\sim}{P}_1, \underset{\sim}{P}_3\} = \{\underset{\sim}{Q}_2\} \qquad R^+_{0,6}\,\{\underset{\sim}{P}_2, \underset{\sim}{P}_3\} = \{\underset{\sim}{Q}_2\} \qquad R^+_{0,6}\,E_1 = \{\underset{\sim}{Q}_2\}$$

4 Gil Aluja, J.: *Elements for a theory of decision in incertainty*. Kluwer Academic Publishers. Londres, Boston, Dordrecht, 1999, pages 205-206.

The connection to the right $R^{+}_{0,6}$ is:

$R^{-}_{0,6}\,\emptyset = E_1$ $\quad R^{-}_{0,6}\,\{\underset{\sim}{Q}_1\} = \{\underset{\sim}{P}_1\}$ $\quad R^{-}_{0,6}\,\{\underset{\sim}{Q}_2\} = E_1$

$R^{-}_{0,6}\,\{\underset{\sim}{Q}_3\} = \{\underset{\sim}{P}_1, \underset{\sim}{P}_2\}$ $\quad R^{-}_{0,6}\,\{\underset{\sim}{Q}_4\} = \{\underset{\sim}{P}_3\}$ $\quad R^{-}_{0,6}\,\{\underset{\sim}{Q}_1, \underset{\sim}{Q}_2\} = \{\underset{\sim}{P}_1\}$

$R^{-}_{0,6}\,\{\underset{\sim}{Q}_1, \underset{\sim}{Q}_3\} = \{\underset{\sim}{P}_1\}$ $\quad R^{-}_{0,6}\,\{\underset{\sim}{Q}_1, \underset{\sim}{Q}_4\} = \emptyset$ $\quad R^{-}_{0,6}\,\{\underset{\sim}{Q}_2, \underset{\sim}{Q}_3\} = \{\underset{\sim}{P}_1, \underset{\sim}{P}_2\}$

$R^{-}_{0,6}\,\{\underset{\sim}{Q}_2, \underset{\sim}{Q}_4\} = \{\underset{\sim}{P}_3\}$ $\quad R^{-}_{0,6}\,\{\underset{\sim}{Q}_3, \underset{\sim}{Q}_4\} = \emptyset$ $\quad R^{-}_{0,6}\,\{\underset{\sim}{Q}_1, \underset{\sim}{Q}_2, \underset{\sim}{Q}_3\} = \{\underset{\sim}{P}_1\}$

$R^{-}_{0,6}\,\{\underset{\sim}{Q}_1, \underset{\sim}{Q}_2, \underset{\sim}{Q}_4\} = \emptyset$ $\quad R^{-}_{0,6}\,\{\underset{\sim}{Q}_1, \underset{\sim}{Q}_3, \underset{\sim}{Q}_4\} = \emptyset$ $\quad R^{-}_{0,6}\,\{\underset{\sim}{Q}_2, \underset{\sim}{Q}_3, \underset{\sim}{Q}_4\} = \emptyset$

$R^{-}_{0,6}\,E_2 = \emptyset$

in the same way as in the case of a single referential, the Moore closings are obtained with the compositions of the connections:

$$M^{(1)}_{0,6} = R^{-}_{0,6} \circ R^{+}_{0,6} \qquad M^{(2)}_{0,6} = R^{+}_{0,6} \circ R^{-}_{0,6}$$

Which we see from a sagittate representation. For $M^{(1)}_{0,6}$:

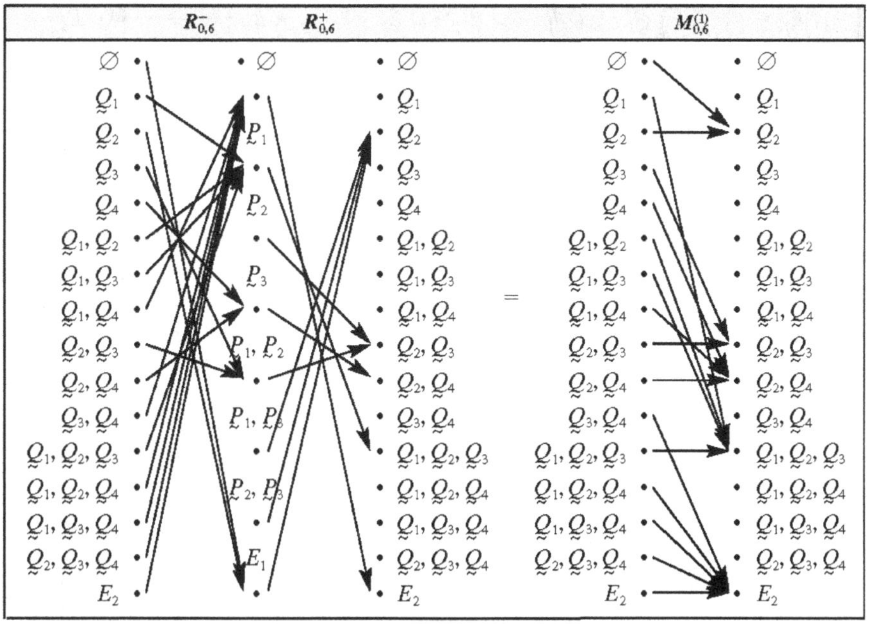

We observe that the closes form the following set:

$$C(E_2, M^{(1)}_{0,6}) = \left\{ \{\underset{\sim}{Q}_2\}, \{\underset{\sim}{Q}_2, \underset{\sim}{Q}_3\}, \{\underset{\sim}{Q}_2, \underset{\sim}{Q}_4\}, \{\underset{\sim}{Q}_1, \underset{\sim}{Q}_2, \underset{\sim}{Q}_3\}, E_2 \right\}$$

For $M_{0,6}^{(2)}$:

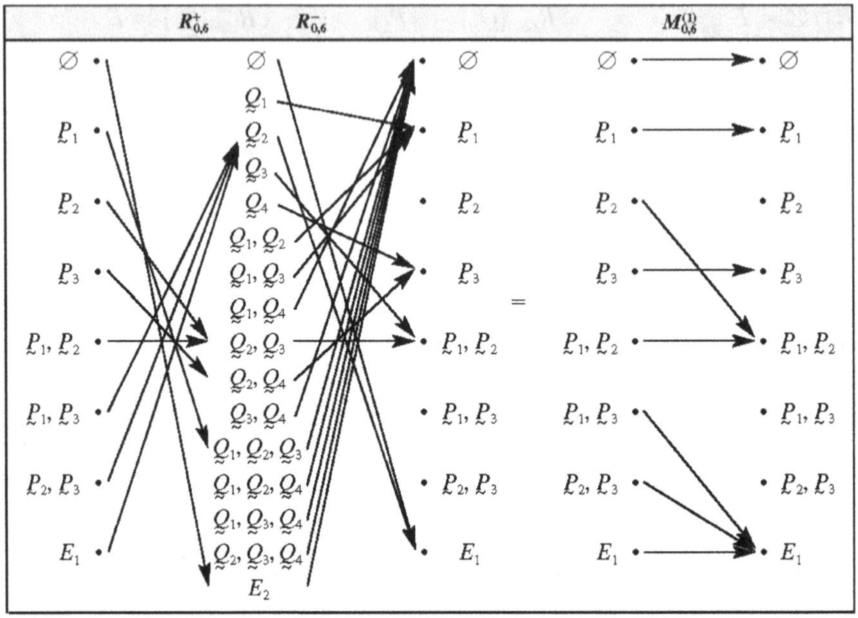

As a set of closes we have:

$$C(E_1, M_{0,6}^{(2)}) = \{\emptyset, \{\underset{\sim}{P}_1\}, \{\underset{\sim}{P}_3\}, \{\underset{\sim}{P}_1, \underset{\sim}{P}_2\}, E_1\}$$

As in that which occurred in the supposition of a single referential E, on this occasion the family of closes corresponding to the Moore closing $M_\alpha^{(1)}$, $C(E, M_\alpha^{(1)})$ gives the groupings with the greatest possible number of elements of E_2, and the family of closes relative to the Moore closing $M_\alpha^{(2)}$, $C(E, M_\alpha^{(2)})$ is made up of the greatest possible grouping of elements of E_1. These families of closes may combine with each other, and so:

1- They have the same number of elements, which is to say the same cardinal:

$$\text{Car. } C(E_2, M_\alpha^{(1)}) = \text{Card. } C(E_1, M_\alpha^{(2)})$$

2- To all elements of the family of closes $C(E\ M_\alpha^{(1)})$ one and only one element of the family of closes $C(E, M_\alpha^{(2)})$ corresponds to which, in the same way, one and only one element of $C(E, M_\alpha^{(1)})$ corresponds.

In effect:

$$R_{0,6}^{-}\{\underset{\sim}{Q}_2\} = E_1 \qquad R_{0,6}^{-}\{\underset{\sim}{Q}_2, \underset{\sim}{Q}_3\} = \{\underset{\sim}{P}_1, \underset{\sim}{P}_2\} \qquad R_{0,6}^{-}\{\underset{\sim}{Q}_2, \underset{\sim}{Q}_4\} = \{\underset{\sim}{P}_3\}$$

$$R_{0,6}^{-}\{\underset{\sim}{Q}_1, \underset{\sim}{Q}_2, \underset{\sim}{Q}_3\} = \{\underset{\sim}{P}_1\} \qquad R_{0,6}^{-} E_2 = \emptyset$$

But also:

$$R^{+}_{0,6}\, E_1 = \{\underset{\sim}{Q}_2\} \qquad R^{+}_{0,6}\, \{\underset{\sim}{P}_1, \underset{\sim}{P}_2\} = \{\underset{\sim}{Q}_2, \underset{\sim}{Q}_3\} \qquad R^{+}_{0,6}\, \{\underset{\sim}{P}_3\} = \{\underset{\sim}{Q}_2, \underset{\sim}{Q}_4\}$$

$$R^{+}_{0,6}\, \{\underset{\sim}{P}_1\} = \{\underset{\sim}{Q}_1, \underset{\sim}{Q}_2, \underset{\sim}{Q}_3\} \qquad R^{+}_{0,6}\, \varnothing = E_2$$

Therefore, these families are dual and make up isomorphic lattices[5] of an antitone nature. In this way, in the formation of groups, the groupings of elements of E_1 of a family of closes are accompanied by the groupings of elements of E_2 of the other family of closes with these related.

In effect, if it is taken into account that each family of closes forms a lattice it can be seen that both lattices are isomorphic:

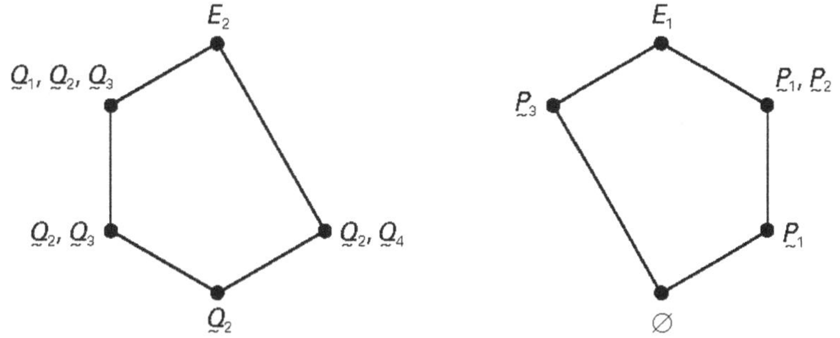

But we have also seen that the functional applications were:

And so, if we continue with our example and graphically represent these applications, in the two isomorphic lattices we have:

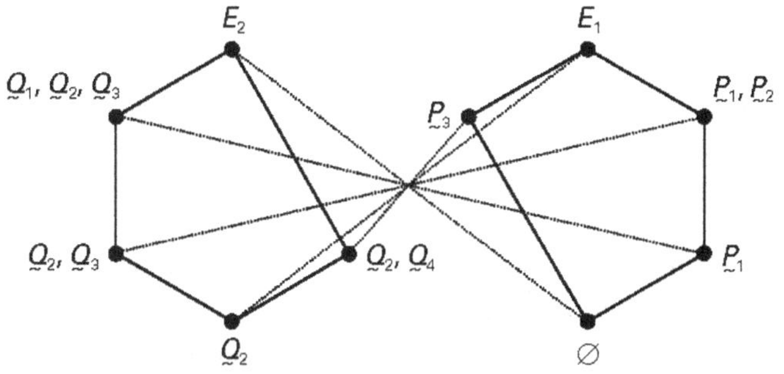

[5] Dubreil, P.: Leçons d'algèbre moderne. Ed. Dunod. París, 1964, page 177 et esq.

Now, upon superimposing the two lattices we find:

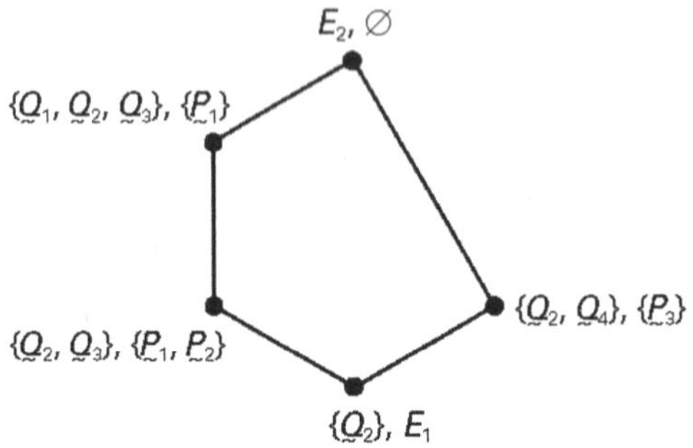

This single lattice collects in each of its vertices the relationship of the groupings of elements of a set E_1 with those of the other set E_2, at the previously established level, which in our case is $\propto \geq 0.6$. When this occurs it is said that an **affinity** at the corresponding level exists between them. In the presented supposition the affinities are:

$$\{\{\underset{\sim}{Q}_2\}, E_1\} = \{\{\underset{\sim}{Q}_2, \underset{\sim}{Q}_3\}, \{\underset{\sim}{P}_1, \underset{\sim}{P}_2\}\}, \{\{\underset{\sim}{Q}_2, \underset{\sim}{Q}_4\}, \{\underset{\sim}{P}_3\}\}, \{\{\underset{\sim}{Q}_1, \underset{\sim}{Q}_2, \underset{\sim}{Q}_3\}, \{\underset{\sim}{P}_1\}\}, \{E_2, \emptyset\}$$

In the same way, the lattice establishes an ordering and structuring of these relationships given that as a displacement takes place from the vertex (Q_2, E_1) to the vertex (E_2, $\emptyset$), or vice versa, the number of grouped elements of one referential grows and the number of grouped elements of the other decreases.

6 The Concept of Affinity

An elemental study of this lattice shows that if two sets E_1 and E_2 are taken into account, and also the respective "power sets" $P(E_1)$ and $Q(E_2)$, the following relationships of order are fulfilled in all cases:

$$\rightarrow \forall \underset{\sim}{A}_j, \underset{\sim}{A}_k \in P(E_1) \quad , \quad \forall \underset{\sim}{B}_l, \underset{\sim}{B}_m \in Q(E_2):$$

$$((\underset{\sim}{A}_j, \underset{\sim}{B}_l) \leq (\underset{\sim}{A}_k, \underset{\sim}{B}_m)) \Leftrightarrow (\underset{\sim}{A}_j \subset \underset{\sim}{A}_k, \underset{\sim}{B}_l \supset \underset{\sim}{B}_m)$$

$$\rightarrow \forall \underset{\sim}{A}_j, \underset{\sim}{A}_k \in P(E_1) \quad , \quad \forall \underset{\sim}{B}_l, \underset{\sim}{B}_m \in Q(E_2):$$

$$((\underset{\sim}{A}_j, \underset{\sim}{B}_l) \geq (\underset{\sim}{A}_k, \underset{\sim}{B}_m)) \Leftrightarrow (\underset{\sim}{A}_j \supset \underset{\sim}{A}_k, \underset{\sim}{B}_l \subset \underset{\sim}{B}_m)$$

If the first of these relationships is incorporated into an **upper extreme** to which we will designate as $(\underset{\sim}{A}_j, \underset{\sim}{B}_l) \nabla (\underset{\sim}{A}_k, \underset{\sim}{B}_m)$ and to the second we introduce a **lower extreme** represented by $(\underset{\sim}{A}_j, \underset{\sim}{B}_l) \Delta (\underset{\sim}{A}_k, \underset{\sim}{B}_m)$ and (E_2, $\emptyset$) is considered as an upper extreme of the first relationship and a lower of the second, as ($\emptyset$, E_1) as upper extreme of the second relationship and lower of the first, the following properties are verified:

1. $((\underset{\sim}{X}, \underset{\sim}{Y}) = (\underset{\sim}{A}_j, \underset{\sim}{B}_l) \nabla (\underset{\sim}{A}_k, \underset{\sim}{B}_m)) \Rightarrow (\underset{\sim}{X} \supset \underset{\sim}{A}_j \cup \underset{\sim}{A}_k, \underset{\sim}{Y} \subset \underset{\sim}{B}_l \cap \underset{\sim}{B}_m)$
2. $((\underset{\sim}{Z}, \underset{\sim}{V}) = (\underset{\sim}{A}_j, \underset{\sim}{B}_l) \Delta (\underset{\sim}{A}_k, \underset{\sim}{B}_m)) \Rightarrow (\underset{\sim}{Z} \subset \underset{\sim}{A}_j \cap \underset{\sim}{A}_k, \underset{\sim}{V} \supset \underset{\sim}{B}_l \cup \underset{\sim}{B}_m)$

When a lattice fulfils these two requisites we find ourselves before a **Galois lattice**. So, when the lattice which the two sets of Moore closings forms has as upper and lower extremes the relationships (E_2, ∅), (∅, E_1) it is a Galois lattice. In the supposition that one or both of these are lacked it is enough to formally add them to reach a Galois lattice. This is the case in our example. In effect, if (∅, E_1) is added to the affinities found then the lattice fulfils the properties necessary to be able to speak of a **Galois Lattice**. We present this in continuation:

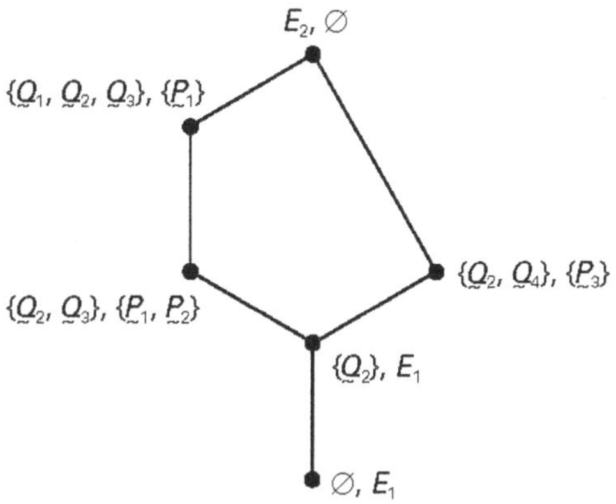

As we have just seen, the Galois lattice visually shows the relationships between the elements of the "power set" of one set with those of the other set, keeping homogeneity to or at the demanded levels. At the same time, lattice presentation of the affinities establishes a perfect structuring of the groupings obtained, such as organisation from less to more elements of the referential which contain each element of the "power set", linked with an order of less to more elements of the referential contained in each element of the other "power set". These properties have propelled us to establish the **definition of affinities** in the following terms[6]:

We define the affinities as those homogenous groupings at determined levels, orderly structured, which join elements of the two sets of a distinct nature, related by the self essence of the phenomena which they represent.

The importance which the theory of affinities has had and is acquiring for the economics and management of companies is unquestionable. As a significant

6 Gil Aluja, J.: Elements for a theory of decision in uncertainty. Kluwer Academic Publishers. London, Boston, Dordrecht, 1999, pages 205-206.

example it is enough to consider the revolution that has been produced in the processes of "segmentation", in which the possibility of presenting all of the homogenous groupings at the same level structured by means of a Galois lattice allows global but at the same time individually ordered information which has not been available up to now. There have been many and significantly decisive applications performed from this finding. Some of these have opened new channels for the solution of problems which were clumsily dealt with in the past: personnel management[7], financial analysis[8], organisational management, commercial management[9] and sports management[10], to only cite a few. But the possibilities of these formal structures, sheltered under the cover of pretopology, have not run out with these findings, but, on the contrary, these possibilities grow as time passes[11].

Including from a merely intuitive perspective, the ample practical possibilities of the previous development cannot escape the work of economists and agencies, with so many properties, interesting relationships and optimal groupings. To aid this task some **algorithms** have been elaborated which make the concept of affinities operative, giving it the adequate form to be immediately used for the solution of social, economic and management problems. Among these algorithms we will describe two of them[12] as an example, that of the **maximum inverse correspondence** and that of **maximum complete sub-matrices**.

a) Maximum inverse correspondence algorithm

- When starting from a fuzzy relationship $\underset{\sim}{R}$, this becomes a boolean R_α, establishing the level of acceptation of homogeneity.
- The set which has the lowest number of elements is chosen between E_1 and E_2.
- We construct the "power set" $P(E_1)$ where E_1 is the set with the least elements, chosen following that established in the previous phase.
- The "connection to the right" R_α^+ is obtained, which is to say, for all $\underset{\sim}{A}_j \in P(E_1)$, $R_\alpha^+ \underset{\sim}{A}_j$ collects the successors of all of the elements which belong to $\underset{\sim}{A}_j$.
- We choose that corresponding to $\underset{\sim}{A}_j$ which has the highest number of elements of the referential E for each of the non-empty sets of $R_\alpha^+ \underset{\sim}{A}_j$.

---

7 Gil Aluja, J.: Interactive management of human resources in uncertainty. Kluwer Academie Publishers. London, Boston, Dordrecht, 1998, pages 130-138.

8 Gil Lafuente, A. M.: Nuevas estrategias para el análisis financiero en la incertidumbre. Ed. Ariel. Barcelona, 2001, pages 332-335.

9 Gil Lafuente, J.: Model for the homogeneous grouping of the sales force. Proceedings of the M. S Congress. 2001. Changsha (Hunan, R. P. China), 25-27 September 2001, pages 332-335.

10 Gil Lafuente, J.: Algoritmos para la excelencia. Claves para el éxito en la gestión deportiva. Ed. Milladorio. Vigo, 2002, pages 237-248.

11 Gil Aluja, J.: La pretopología en la gestión de la incertidumbre. Speech for the purpose of entering as Doctor «Honoris Causa» of the Universidad de León. Publicaciones de la Universidad de León. León, 2002, page 21.

12 For more detail, Gil Aluja, J.: Elements for a theory of decision in uncertainty. Kluwer Academic Publishers. Boston, London, Dordretch, 1999, pages 215-222 may be consulted.

- The relationships found form a lattice which allows a perfect structuring and ordering of all of the possible affinities. If we add the empty-referential and referential-empty vertices in its occurrence we find ourselves before a Galois lattice.

We use this algorithm starting from the same boolean matrix as the previous development. In this way we confirm the coincidence of results.

$R_{0,6} =$

	$\underset{\sim}{Q}_1$	$\underset{\sim}{Q}_2$	$\underset{\sim}{Q}_3$	$\underset{\sim}{Q}_4$
$\underset{\sim}{P}_1$	1	1	1	
$\underset{\sim}{P}_2$		1	1	
$\underset{\sim}{P}_3$		1		1

- The set which has the least elements is E_1:

$$E_1 = \{\underset{\sim}{P}_1, \underset{\sim}{P}_2, \underset{\sim}{P}_3\}$$

- The "power set" $P(E_1)$ is:

$$\{\emptyset, \{\underset{\sim}{P}_1\}, \{\underset{\sim}{P}_2\}, \{\underset{\sim}{P}_3\}, \{\underset{\sim}{P}_1, \underset{\sim}{P}_2\}, \{\underset{\sim}{P}_1, \underset{\sim}{P}_3\}, \{\underset{\sim}{P}_2, \underset{\sim}{P}_3\}, E_1\}$$

- The connection to the right $R^{+}_{0,6}$ is:

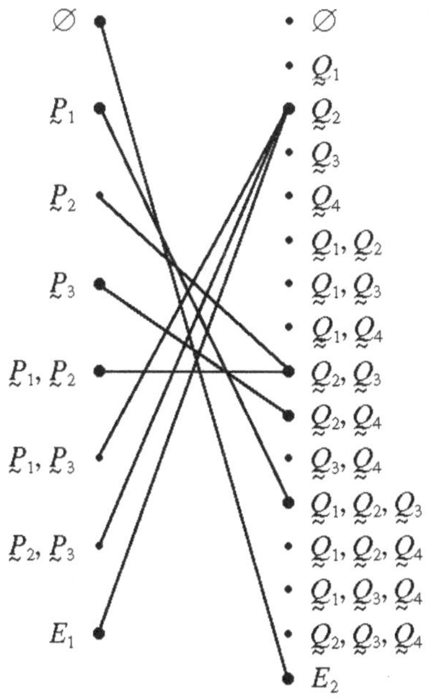

- The highest number of elements of the referential E_I corresponding to each non-empty set of $R^{+}_{0,6}$ is:

$$\{\underset{\sim}{Q}_2\} \rightarrow E_1 \qquad \{\underset{\sim}{Q}_2, \underset{\sim}{Q}_3\} \rightarrow \{\underset{\sim}{P}_1, \underset{\sim}{P}_2\} \qquad \{\underset{\sim}{Q}_2, \underset{\sim}{Q}_4\} \rightarrow \{\underset{\sim}{P}_3\}$$

$$\{\underset{\sim}{Q}_1, \underset{\sim}{Q}_2, \underset{\sim}{Q}_3\} \rightarrow \{\underset{\sim}{P}_1\} \qquad E_2 \rightarrow \varnothing$$

- The result coincides with that previously found, and so we will also have the same Galois lattice.

b) Maximum complete sub-matrices algorithm

- We start from a boolean matrix R_α.
- The connections to the right R^{++}_α and to the right R^{-}_α are found.
- We form the families $C(E_I, M^{(2)}_\alpha)$ and $C(E_2, M^{(1)}_\alpha)$ with the elements of the images in R^{++}_α and R^{-}_α respectively.
- We obtain the cartesian product of all of the elements of $C(E_I, M^{(2)}_\alpha)$ with all of those of $C(E_2, M^{(1)}_\alpha)$ in this way finding the complex and incomplete matrices.
- The incomplete matrices are eliminated.
- The matrices contained in others are eliminated.
- The matrices which remain are complete and maximum.
- Upon adding the empty-referential, referential-empty pairs a Galois lattice is obtained.

We check this algorithm by using the same boolean matrix as in the previous example.

- The connections to the right $R^{+}_{0,6}$ and to the left $R^{-}_{0,6}$ allow us to find the following families:

$$C(E_1, M^{(2)}_{0,6}) = \{\varnothing, \{\underset{\sim}{P}_1\}, \{\underset{\sim}{P}_3\}, \{\underset{\sim}{P}_1, \underset{\sim}{P}_2\}, E_1\}$$

$$C(E_2, M^{(1)}_{0,6}) = \{\{\underset{\sim}{Q}_2\}, \{\underset{\sim}{Q}_2, \underset{\sim}{Q}_3\}, \{\underset{\sim}{Q}_2, \underset{\sim}{Q}_4\}, \{\underset{\sim}{Q}_1, \underset{\sim}{Q}_2, \underset{\sim}{Q}_3\}, E_2\}$$

- The cartesian product is:

$$\{\underset{\sim}{P}_1\} \cdot \{\underset{\sim}{Q}_2\},\ \{\underset{\sim}{P}_1\} \cdot \{\underset{\sim}{Q}_2, \underset{\sim}{Q}_3\},\ \{\underset{\sim}{P}_1\} \cdot \{\underset{\sim}{Q}_2, \underset{\sim}{Q}_4\},\ \{\underset{\sim}{P}_1\} \cdot \{\underset{\sim}{Q}_1, \underset{\sim}{Q}_2, \underset{\sim}{Q}_3\},\ \{\underset{\sim}{P}_1\} \cdot E_2$$

$$\{\underset{\sim}{P}_3\} \cdot \{\underset{\sim}{Q}_2\},\ \{\underset{\sim}{P}_3\} \cdot \{\underset{\sim}{Q}_2, \underset{\sim}{Q}_3\},\ \{\underset{\sim}{P}_3\} \cdot \{\underset{\sim}{Q}_2, \underset{\sim}{Q}_4\},\ \{\underset{\sim}{P}_3\} \cdot \{\underset{\sim}{Q}_1, \underset{\sim}{Q}_2, \underset{\sim}{Q}_3\},\ \{\underset{\sim}{P}_3\} \cdot E_2$$

$$\{\underset{\sim}{P}_1, \underset{\sim}{P}_2\} \cdot \{\underset{\sim}{Q}_2\},\ \{\underset{\sim}{P}_1, \underset{\sim}{P}_2\} \cdot \{\underset{\sim}{Q}_2, \underset{\sim}{Q}_3\},\ \{\underset{\sim}{P}_1, \underset{\sim}{P}_2\} \cdot \{\underset{\sim}{Q}_2, \underset{\sim}{Q}_4\},$$
$$\{\underset{\sim}{P}_1, \underset{\sim}{P}_2\} \cdot \{\underset{\sim}{Q}_1, \underset{\sim}{Q}_2, \underset{\sim}{Q}_3\},\ \{\underset{\sim}{P}_1, \underset{\sim}{P}_2\} \cdot E_2$$

$$E_1 \cdot \{\underset{\sim}{Q}_2\},\ E_1 \cdot \{\underset{\sim}{Q}_2, \underset{\sim}{Q}_3\},\ E_1 \cdot \{\underset{\sim}{Q}_2, \underset{\sim}{Q}_4\},\ E_1 \cdot \{\underset{\sim}{Q}_1, \underset{\sim}{Q}_2, \underset{\sim}{Q}_3\},\ E_1 \cdot E_2$$

– We express each of the elements of the cartesian product in sub-matrix form:

	$\underset{\sim}{Q}_2$
$\underset{\sim}{P}_1$	1

	$\underset{\sim}{Q}_2$	$\underset{\sim}{Q}_3$
$\underset{\sim}{P}_1$	1	1

	$\underset{\sim}{Q}_2$	$\underset{\sim}{Q}_4$
$\underset{\sim}{P}_1$	1	

	$\underset{\sim}{Q}_1$	$\underset{\sim}{Q}_2$	$\underset{\sim}{Q}_3$
$\underset{\sim}{P}_1$	1	1	1

	$\underset{\sim}{Q}_1$	$\underset{\sim}{Q}_2$	$\underset{\sim}{Q}_3$	$\underset{\sim}{Q}_4$
$\underset{\sim}{P}_1$	1	1	1	

	$\underset{\sim}{Q}_2$
$\underset{\sim}{P}_3$	1

	$\underset{\sim}{Q}_2$	$\underset{\sim}{Q}_3$
$\underset{\sim}{P}_3$	1	

	$\underset{\sim}{Q}_2$	$\underset{\sim}{Q}_4$
$\underset{\sim}{P}_3$	1	1

	$\underset{\sim}{Q}_1$	$\underset{\sim}{Q}_2$	$\underset{\sim}{Q}_3$
$\underset{\sim}{P}_3$		1	

	$\underset{\sim}{Q}_1$	$\underset{\sim}{Q}_2$	$\underset{\sim}{Q}_3$	$\underset{\sim}{Q}_4$
$\underset{\sim}{P}_3$		1		1

	$\underset{\sim}{Q}_2$
$\underset{\sim}{P}_1$	1
$\underset{\sim}{P}_2$	1

	$\underset{\sim}{Q}_2$	$\underset{\sim}{Q}_3$
$\underset{\sim}{P}_1$	1	1
$\underset{\sim}{P}_2$	1	1

	$\underset{\sim}{Q}_2$	$\underset{\sim}{Q}_4$
$\underset{\sim}{P}_1$	1	
$\underset{\sim}{P}_2$	1	

	$\underset{\sim}{Q}_1$	$\underset{\sim}{Q}_2$	$\underset{\sim}{Q}_3$
$\underset{\sim}{P}_1$	1	1	1
$\underset{\sim}{P}_2$		1	1

	$\underset{\sim}{Q}_1$	$\underset{\sim}{Q}_2$	$\underset{\sim}{Q}_3$	$\underset{\sim}{Q}_4$
$\underset{\sim}{P}_1$	1	1	1	
$\underset{\sim}{P}_2$		1	1	

	$\underset{\sim}{Q}_2$
$\underset{\sim}{P}_1$	1
$\underset{\sim}{P}_2$	1
$\underset{\sim}{P}_3$	1

	$\underset{\sim}{Q}_2$	$\underset{\sim}{Q}_3$
$\underset{\sim}{P}_1$	1	1
$\underset{\sim}{P}_2$	1	1
$\underset{\sim}{P}_3$	1	

	$\underset{\sim}{Q}_2$	$\underset{\sim}{Q}_4$
$\underset{\sim}{P}_1$	1	
$\underset{\sim}{P}_2$	1	
$\underset{\sim}{P}_3$	1	1

	$\underset{\sim}{Q}_1$	$\underset{\sim}{Q}_2$	$\underset{\sim}{Q}_3$
$\underset{\sim}{P}_1$	1	1	1
$\underset{\sim}{P}_2$		1	1
$\underset{\sim}{P}_3$		1	

	$\underset{\sim}{Q}_1$	$\underset{\sim}{Q}_2$	$\underset{\sim}{Q}_3$	$\underset{\sim}{Q}_4$
$\underset{\sim}{P}_1$	1	1	1	
$\underset{\sim}{P}_2$		1	1	
$\underset{\sim}{P}_3$		1		1

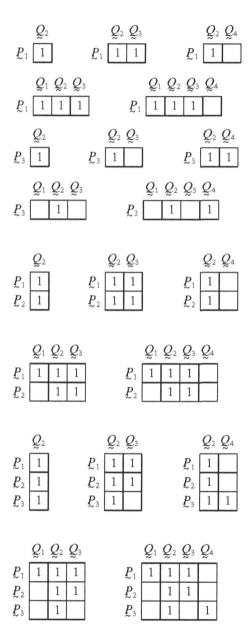

Upon eliminating the incomplete matrices we have:

$$\begin{array}{c|c|} & \underset{\sim}{Q}_2 \\ \hline \underset{\sim}{P}_1 & 1 \\ \hline \end{array} \qquad \begin{array}{c|c|c|} & \underset{\sim}{Q}_2 & \underset{\sim}{Q}_3 \\ \hline \underset{\sim}{P}_1 & 1 & 1 \\ \hline \end{array} \qquad \begin{array}{c|c|c|c|} & \underset{\sim}{Q}_1 & \underset{\sim}{Q}_2 & \underset{\sim}{Q}_3 \\ \hline \underset{\sim}{P}_1 & 1 & 1 & 1 \\ \hline \end{array}$$

$$\begin{array}{c|c|} & \underset{\sim}{Q}_2 \\ \hline \underset{\sim}{P}_3 & 1 \\ \hline \end{array} \qquad \begin{array}{c|c|c|} & \underset{\sim}{Q}_2 & \underset{\sim}{Q}_4 \\ \hline \underset{\sim}{P}_3 & 1 & 1 \\ \hline \end{array}$$

$$\begin{array}{c|c|} & \underset{\sim}{Q}_2 \\ \hline \underset{\sim}{P}_1 & 1 \\ \hline \underset{\sim}{P}_2 & 1 \\ \hline \end{array} \qquad \begin{array}{c|c|c|} & \underset{\sim}{Q}_2 & \underset{\sim}{Q}_3 \\ \hline \underset{\sim}{P}_1 & 1 & 1 \\ \hline \underset{\sim}{P}_3 & 1 & 1 \\ \hline \end{array}$$

$$\begin{array}{c|c|} & \underset{\sim}{Q}_2 \\ \hline \underset{\sim}{P}_1 & 1 \\ \hline \underset{\sim}{P}_2 & 1 \\ \hline \underset{\sim}{P}_3 & 1 \\ \hline \end{array}$$

Upon eliminating the matrices contained in others we obtain:

$$\begin{array}{c|c|c|c|} & \underset{\sim}{Q}_1 & \underset{\sim}{Q}_2 & \underset{\sim}{Q}_3 \\ \hline \underset{\sim}{P}_1 & 1 & 1 & 1 \\ \hline \end{array} \qquad \begin{array}{c|c|c|} & \underset{\sim}{Q}_2 & \underset{\sim}{Q}_4 \\ \hline \underset{\sim}{P}_3 & 1 & 1 \\ \hline \end{array} \qquad \begin{array}{c|c|c|} & \underset{\sim}{Q}_2 & \underset{\sim}{Q}_3 \\ \hline \underset{\sim}{P}_1 & 1 & 1 \\ \hline \underset{\sim}{P}_2 & 1 & 1 \\ \hline \end{array} \qquad \begin{array}{c|c|} & \underset{\sim}{Q}_2 \\ \hline \underset{\sim}{P}_1 & 1 \\ \hline \underset{\sim}{P}_2 & 1 \\ \hline \underset{\sim}{P}_3 & 1 \\ \hline \end{array}$$

All of which gives the following affinities:

$$\{\{\underset{\sim}{P}_1\}, \{\underset{\sim}{Q}_1, \underset{\sim}{Q}_2, \underset{\sim}{Q}_3\}\} \quad \{\{\underset{\sim}{P}_3\}, \{\underset{\sim}{Q}_2, \underset{\sim}{Q}_4\}\} \quad \{\{\underset{\sim}{P}_1, \underset{\sim}{P}_2\}, \{\underset{\sim}{Q}_2, \underset{\sim}{Q}_3\}\} \quad \{E_1, \{\underset{\sim}{Q}_2\}\}$$

which, as it could be in no other way, coincides with that found upon using the previous algorithm.

If we add the pairs $\{E_1, \varnothing\}$, $\{\varnothing, E_2\}$ to these affinities the corresponding Galois lattice can be constructed, which is that known and that we reproduce:

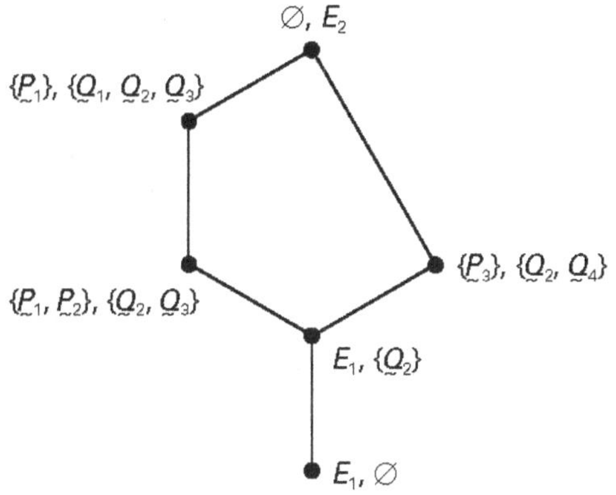

All that we have just stated constitutes a theoretical body susceptible to being used in the areas of social, economic and management life. We ourselves have directed or participated in some of the applications carried out up to now. Only as an example we will present one case which has merited the attention of specialists on this matter. We evidently limit ourselves to a summary, directed to a greater formal understanding of the outlines employed at the time, expounded in a didactical way.

7 An Application of the Theory of Affinities

When faced with the danger of planetary diffusion of a **new pathology**, the World Health Organisation contacted various Research Institutions with the aim of sounding out the possibility of financing one or more groups of researchers for the execution of a project destined for obtaining a medicine capable of curing or limiting the effects of the cited illness.

Once the appropriate negotiations were over, it decided to pay special attention to the **sets of researchers** E_1 and E_2 whose centres were found in the U.S.A. and Japan. The elements of each of these sets are:

$$E_1 = \{P_1, P_2, P_3, P_4\}$$
$$E_2 = \{Q_1, Q_2, Q_3, Q_4\}$$

Each one of these researchers may be described by means of the following attributes:

A = Research experience.

B = Results achieved in previous projects.

C = Scientific articles and conferences on the subject.

D = Work capability and tenacity in the persecution of objectives.

$E =$ Creative imagination.

$F =$ Connections with researchers from other countries.

The information available allowed the establishment of descriptors by means of the following fuzzy subsets:

	A	B	C	D	E	F
$\underset{\sim}{P}_1 =$	.4	.8	.9	.6	.5	.1
$\underset{\sim}{P}_2 =$	.8	.8	.6	.4	.9	.6
$\underset{\sim}{P}_3 =$	.9	.8	.7	.9	.3	.8
$\underset{\sim}{P}_4 =$	.6	.9	.8	.5	.8	.5

	A	B	C	D	E	F
$\underset{\sim}{Q}_1 =$	1	.5	.9	.6	.8	.9
$\underset{\sim}{Q}_2 =$	.9	.8	.5	.4	.8	.9
$\underset{\sim}{Q}_3 =$	.7	.9	.4	.8	.7	.9
$\underset{\sim}{Q}_4 =$	.8	.4	.9	.9	.4	.4

Experts in the corresponding field of biology considered the threshold $\propto \geq 0.8$ as sufficient to determine that the researchers had the desired level **in all** of the qualities. This gave place to that the previous fuzzy subsets could be transformed into the following boolean subsets:

	A	B	C	D	E	F
$P_1^{(0,8)} =$		1	1			1
$P_2^{(0,8)} =$	1	1			1	
$P_3^{(0,8)} =$	1	1		1		1
$P_4^{(0,8)} =$		1	1		1	

	A	B	C	D	E	F
$Q_1^{(0,8)} =$	1		1		1	1
$Q_2^{(0,8)} =$	1	1			1	1
$Q_3^{(0,8)} =$		1		1		1
$Q_4^{(0,8)} =$	1		1	1		

With this information the specialists in Operative Techniques of Management began their work deciding to **use the theory of affinities**.

In the first instance they homogeneously grouped the researchers of set E_1 among themselves and those of group E_2 also among themselves. For this they obtained the boolean matrices which relate researchers with attributes. These are the following:

	A	B	C	D	E	F
$P_1^{(0,8)} =$		1	1			1
$P_2^{(0,8)} =$	1	1			1	
$P_3^{(0,8)} =$	1	1		1		1
$P_4^{(0,8)} =$		1	1		1	

	A	B	C	D	E	F
$Q_1^{(0,8)} =$	1		1		1	1
$Q_2^{(0,8)} =$	1	1			1	1
$Q_3^{(0,8)} =$		1		1		1
$Q_4^{(0,8)} =$	1		1	1		

From these matrices they followed the stages which the **algorithm of maximum inverse correspondence** marks.

a) Of the researchers of set E_1.

- The "power set" of E_1 is obtained:

$$\{\varnothing, \{P_1\}, \{P_2\}, \{P_3\}, \{P_4\}, \{P_1, P_2\}, \{P_1, P_3\}, \{P_1, P_4\}, \{P_2, P_3\}, \{P_2, P_4\}, \{P_3, P_4\}, \{P_1, P_2, P_3\}, \{P_1, P_2, P_4\}, \{P_1, P_3, P_4\}, \{P_2, P_3, P_4\}, E_1\}$$

- The relationship between each one of these elements of the "power set" with the representatives of the attributes are established, and for each distinct group of attributes the greatest grouping of investigators is obtained.

$$\begin{aligned} \varnothing &\Rightarrow E_1' \\ \{P_1\} &\Rightarrow \{B, C, F\} \\ \{P_2\} &\Rightarrow \{A, B, E\} \end{aligned}$$

$$\{P_3\} \Rightarrow \{A, B, D, F\}$$
$$\{P_4\} \Rightarrow \{B, C, E\}$$
$$\{P_1, P_2\} \Rightarrow \{B\}$$
$$\{P_1, P_3\} \Rightarrow \{B, F\}$$
$$\{P_1, P_4\} \Rightarrow \{B, C\}$$
$$\{P_2, P_3\} \Rightarrow \{A, B\}$$
$$\{P_2, P_4\} \Rightarrow \{B, E\}$$
$$\{P_3, P_4\} \Rightarrow \{B\}$$
$$\{P_1, P_2, P_3\} \Rightarrow \{B\}$$
$$\{P_1, P_2, P_4\} \Rightarrow \{B\}$$
$$\{P_1, P_3, P_4\} \Rightarrow \{B\}$$
$$\{P_2, P_3, P_4\} \Rightarrow \{B\}$$
$$E_1 \Rightarrow \{B\}$$

$$\{B\} \Rightarrow E_1$$
$$\{A, B\} \Rightarrow \{P_2, P_3\}$$
$$\{B, C\} \Rightarrow \{P_1, P_4\}$$
$$\{B, E\} \Rightarrow \{P_2, P_4\}$$
$$\{B, F\} \Rightarrow \{P_1, P_3\}$$
$$\{A, B, E\} \Rightarrow \{P_2\}$$
$$\{B, C, E\} \Rightarrow \{P_4\}$$
$$\{B, C, F\} \Rightarrow \{P_1\}$$
$$\{A, B, D, F\} \Rightarrow \{P_3\}$$
$$E_1' \Rightarrow \varnothing$$

- The corresponding Galois lattice is constructed:

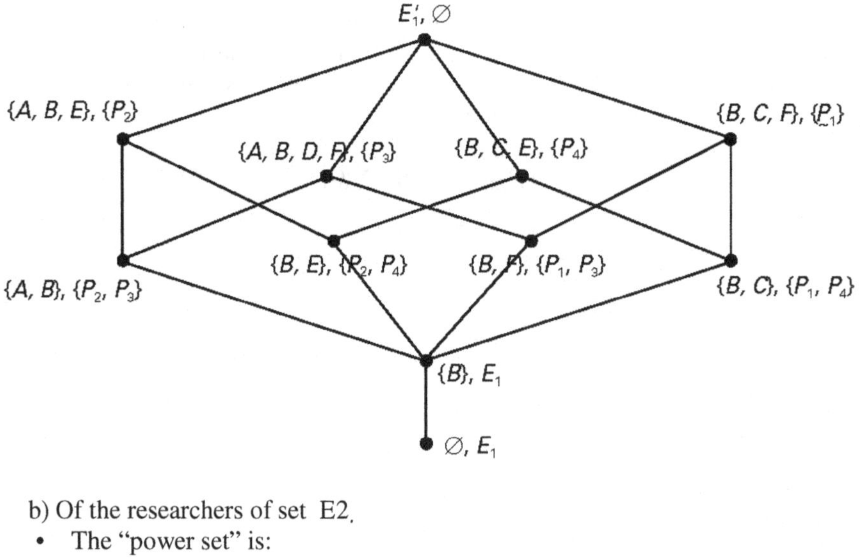

b) Of the researchers of set E2.

- The "power set" is:

$$\{\varnothing, \{Q_1\}, \{Q_2\}, \{Q_3\}, \{Q_4\}, \{Q_1, Q_2\}, \{Q_1, Q_3\}, \{Q_1, Q_4\}, \{Q_2, Q_3\}, \{Q_2, Q_4\}, \{Q_3, Q_4\}, \{Q_1, Q_2, Q_3\}, \{Q_1, Q_2, Q_4\}, \{Q_1, Q_3, Q_4\}, \{Q_2, Q_3, Q_4\}, E_2\}$$

- The correspondence of each of these elements of the "power set" is established, and the intersection of the elements of the set of qualities, as in the maximum inverse correspondence.

$$\begin{aligned}
\varnothing &\Rightarrow E_2' \\
\{Q_1\} &\Rightarrow \{A, C, E, F\} \\
\{Q_2\} &\Rightarrow \{A, B, E, F\} \\
\{Q_3\} &\Rightarrow \{B, D, F\} \\
\{Q_4\} &\Rightarrow \{A, C, D\} \\
\{Q_1, Q_2\} &\Rightarrow \{A, E, F\} \\
\{Q_1, Q_3\} &\Rightarrow \{F\} \\
\{Q_1, Q_4\} &\Rightarrow \{A, C\} \\
\{Q_2, Q_3\} &\Rightarrow \{B, F\} \\
\{Q_2, Q_4\} &\Rightarrow \{A\} \\
\{Q_3, Q_4\} &\Rightarrow \{D\} \\
\{Q_1, Q_2, Q_3\} &\Rightarrow \{F\} \\
\{Q_1, Q_2, Q_4\} &\Rightarrow \{A\} \\
\{Q_1, Q_3, Q_4\} &\Rightarrow \varnothing \\
\{Q_2, Q_3, Q_4\} &\Rightarrow \varnothing \\
E_2 &\Rightarrow \varnothing
\end{aligned}$$

$$\begin{aligned}
\varnothing &\Rightarrow E_2 \\
\{A\} &\Rightarrow \{Q_1, Q_2, Q_4\} \\
\{D\} &\Rightarrow \{Q_3, Q_4\} \\
\{F\} &\Rightarrow \{Q_1, Q_2, Q_3\} \\
\{A, C\} &\Rightarrow \{Q_1, Q_4\} \\
\{B, F\} &\Rightarrow \{Q_2, Q_3\} \\
\{A, C, D\} &\Rightarrow \{Q_4\} \\
\{A, E, F\} &\Rightarrow \{Q_1, Q_2\} \\
\{B, D, F\} &\Rightarrow \{Q_3\} \\
\{A, B, E, F\} &\Rightarrow \{Q_2\} \\
\{A, C, E, F\} &\Rightarrow \{Q_1\} \\
E_2' &\Rightarrow \varnothing
\end{aligned}$$

- The corresponding Galois lattice is:

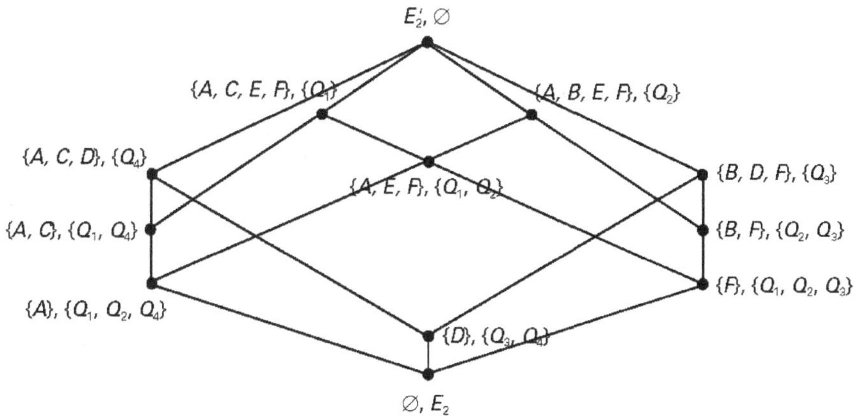

Up to this point with this they achieved the homogenous grouping, depending on their qualities, of the researchers of each one of the Institutes with each other, in a way with one separated from the other. The question which we now ask ourselves is directed to seeing if it is possible to establish homogenous relationships between the scientists of one and the other group together. To use an algorithm of affinities the first thing which should be demanded is the availability of **a matrix** which relates the scientists of group E_1 with those of group E_2. The way to obtain this matrix is one of the aspects which may provoke controversy.

We will propose a way which initially uses the **concept of distance**, in the understanding that many more alternatives exist which may be equally valid.

In the first instance we begin with the obtaining of the distances of each fuzzy subset which describe the investigators of E_1 and those of E_2.

$\delta\{\underset{\sim}{P}_1, \underset{\sim}{Q}_1\} = 1{,}3/6$ $\delta\{\underset{\sim}{P}_1, \underset{\sim}{Q}_2\} = 1{,}7/6$ $\delta\{\underset{\sim}{P}_1, \underset{\sim}{Q}_3\} = 1{,}4/6$ $\delta\{\underset{\sim}{P}_1, \underset{\sim}{Q}_4\} = 1{,}8/6$

$\delta\{\underset{\sim}{P}_2, \underset{\sim}{Q}_1\} = 1{,}5/6$ $\delta\{\underset{\sim}{P}_2, \underset{\sim}{Q}_2\} = 0{,}4/6$ $\delta\{\underset{\sim}{P}_2, \underset{\sim}{Q}_3\} = 1{,}3/6$ $\delta\{\underset{\sim}{P}_2, \underset{\sim}{Q}_4\} = 1{,}9/6$

$\delta\{\underset{\sim}{P}_3, \underset{\sim}{Q}_1\} = 1{,}5/6$ $\delta\{\underset{\sim}{P}_3, \underset{\sim}{Q}_2\} = 1{,}4/6$ $\delta\{\underset{\sim}{P}_3, \underset{\sim}{Q}_3\} = 1{,}2/6$ $\delta\{\underset{\sim}{P}_3, \underset{\sim}{Q}_4\} = 1{,}2/6$

$\delta\{\underset{\sim}{P}_4, \underset{\sim}{Q}_1\} = 1{,}4/6$ $\delta\{\underset{\sim}{P}_4, \underset{\sim}{Q}_2\} = 1{,}0/6$ $\delta\{\underset{\sim}{P}_4, \underset{\sim}{Q}_3\} = 1{,}3/6$ $\delta\{\underset{\sim}{P}_4, \underset{\sim}{Q}_4\} = 1{,}7/6$

We join these distances in the following matrix:

	$\underset{\sim}{Q}_1$	$\underset{\sim}{Q}_2$	$\underset{\sim}{Q}_3$	$\underset{\sim}{Q}_4$
$\underset{\sim}{P}_1$	.217	.283	.233	.300
$\underset{\sim}{P}_2$	.250	.067	.217	.317
$\underset{\sim}{P}_3$	.250	.233	.200	.200
$\underset{\sim}{P}_4$	.233	.167	.217	.283

If a threshold such as $\propto \leq 0.233$ is established the following boolean matrix is found:

	$\underset{\sim}{Q}_1$	$\underset{\sim}{Q}_2$	$\underset{\sim}{Q}_3$	$\underset{\sim}{Q}_4$
$\underset{\sim}{P}_1$	1		1	
$\underset{\sim}{P}_2$		1	1	
$\underset{\sim}{P}_3$		1	1	1
$\underset{\sim}{P}_4$	1	1	1	

From this matrix the now known algorithm is begun, with the consideration of the "power set" of one of the sets E_1, E_2. We arbitrarily choose that of E_1.

$$\{\emptyset, \{P_1\}, \{P_2\}, \{P_3\}, \{P_4\}, \{P_1, P_2\}, \{P_1, P_3\}, \{P_1, P_4\}, \{P_2, P_3\}, \{P_2, P_4\},$$
$$\{P_3, P_4\}, \{P_1, P_2, P_3\}, \{P_1, P_2, P_4\}, \{P_1, P_3, P_4\}, \{P_2, P_3, P_4\}, E_1\}$$

The correspondence between the elements of this "power set" and those of the "power set" of E_2 is the following:

$$
\begin{aligned}
\varnothing &\Rightarrow E_2\\
\{\underset{\sim}{P}_1\} &\Rightarrow \{\underset{\sim}{Q}_1, \underset{\sim}{Q}_3\}\\
\{\underset{\sim}{P}_2\} &\Rightarrow \{\underset{\sim}{Q}_2, \underset{\sim}{Q}_3\}\\
\{\underset{\sim}{P}_3\} &\Rightarrow \{\underset{\sim}{Q}_2, \underset{\sim}{Q}_3, \underset{\sim}{Q}_4\}\\
\{\underset{\sim}{P}_4\} &\Rightarrow \{\underset{\sim}{Q}_1, \underset{\sim}{Q}_2, \underset{\sim}{Q}_3\}\\
\{\underset{\sim}{P}_1, \underset{\sim}{P}_2\} &\Rightarrow \{\underset{\sim}{Q}_3\}\\
\{\underset{\sim}{P}_1, \underset{\sim}{P}_3\} &\Rightarrow \{\underset{\sim}{Q}_3\}\\
\{\underset{\sim}{P}_1, \underset{\sim}{P}_4\} &\Rightarrow \{\underset{\sim}{Q}_1, \underset{\sim}{Q}_3\}\\
\{\underset{\sim}{P}_2, \underset{\sim}{P}_3\} &\Rightarrow \{\underset{\sim}{Q}_2, \underset{\sim}{Q}_3\}\\
\{\underset{\sim}{P}_2, \underset{\sim}{P}_4\} &\Rightarrow \{\underset{\sim}{Q}_2, \underset{\sim}{Q}_3\}\\
\{\underset{\sim}{P}_3, \underset{\sim}{P}_4\} &\Rightarrow \{\underset{\sim}{Q}_2, \underset{\sim}{Q}_3\}\\
\{\underset{\sim}{P}_1, \underset{\sim}{P}_2, \underset{\sim}{P}_3\} &\Rightarrow \{\underset{\sim}{Q}_3\}\\
\{\underset{\sim}{P}_1, \underset{\sim}{P}_2, \underset{\sim}{P}_4\} &\Rightarrow \{\underset{\sim}{Q}_3\}\\
\{\underset{\sim}{P}_1, \underset{\sim}{P}_3, \underset{\sim}{P}_4\} &\Rightarrow \{\underset{\sim}{Q}_3\}\\
\{\underset{\sim}{P}_2, \underset{\sim}{P}_3, \underset{\sim}{P}_4\} &\Rightarrow \{\underset{\sim}{Q}_2, \underset{\sim}{Q}_3\}\\
\{\underset{\sim}{P}_1, \underset{\sim}{P}_2, \underset{\sim}{P}_3, \underset{\sim}{P}_4\} &\Rightarrow \{\underset{\sim}{Q}_3\}
\end{aligned}
$$

$$
\begin{aligned}
\varnothing &\Rightarrow E_1\\
\{\underset{\sim}{Q}_3\} &\Rightarrow \{\underset{\sim}{P}_1, \underset{\sim}{P}_2, \underset{\sim}{P}_4\}\\
\{\underset{\sim}{Q}_1, \underset{\sim}{Q}_3\} &\Rightarrow \{\underset{\sim}{P}_1, \underset{\sim}{P}_4\}\\
\{\underset{\sim}{Q}_2, \underset{\sim}{Q}_3\} &\Rightarrow \{\underset{\sim}{P}_2, \underset{\sim}{P}_3, \underset{\sim}{P}_4\}\\
\{\underset{\sim}{Q}_1, \underset{\sim}{Q}_2, \underset{\sim}{Q}_3\} &\Rightarrow \{\underset{\sim}{P}_4\}\\
\{\underset{\sim}{Q}_2, \underset{\sim}{Q}_3, \underset{\sim}{Q}_4\} &\Rightarrow \{\underset{\sim}{P}_3\}\\
E_1 &\Rightarrow \varnothing
\end{aligned}
$$

The corresponding Galois lattice is then constructed.

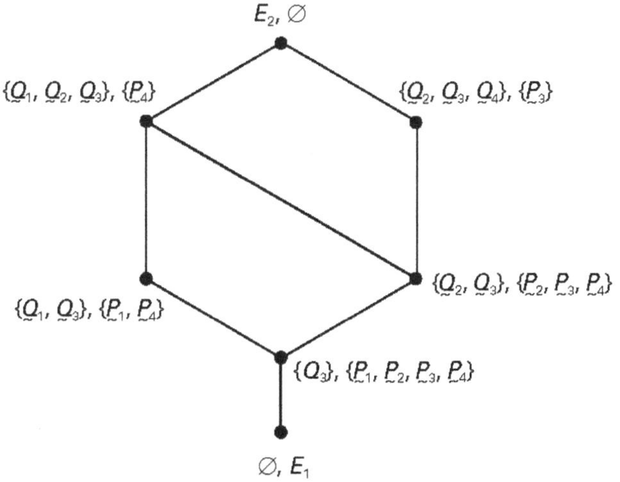

This lattice visually and clearly shows the **homogeneous teams** which may be formed when counting on researchers from either of the groups E_1 and E_2, in such a way that as we move upwards the components of E_2 increase at the cost of those of E_1 which decrease.

In effect, it is palpable that the following homogenous groupings at the level $1 - \propto = 0.767$ may be established, depending on necessity or conveniences:

a) Researcher Q_3 with all of the researchers of E_1.

b) Researchers Q_1, Q_3 with P_1, P_4.

c) Researchers Q_2, Q_3 with P_2, P_3, P_4.

d) Researchers Q_1, Q_2, Q_3 with the researcher P_4.

e) Researchers Q_2, Q_3, Q_4 with the researcher P_3.

We do not believe it necessary to insist on the fact already indicated of the strictly formal validity of the vertices $(\emptyset, E_1)$ and $(E_2, \emptyset)$ of the Galois lattice.

Finally, we wish to state that we have decided to include in this exposition the summary of an application which at its time merited the approval of the bodies responsible for carrying out the global project which had the scheme presented as one of the first sections. It is evidently clear that that which has been developed here may be easily transposed to the fields of economics and management of businesses and institutions and in a very special way into the area of economics and management in which we find ourselves integrated since the beginning of our educational and researching activities.

Some Final Considerations

As a summary we will extract some sufficiently significant elements from our development for a greater understanding of the complex realities which inundate the present economic climate, an indispensable condition for a more adequate explanation of existing phenomenology and as a solid base for the treatment of the problems which arise in a context marked by a growing level of uncertainty.

We are fully aware that it has not been possible to reach all of the objectives which we had set in such an ambitious way. Our target was so high that the harsh reality has placed it in a modest position. Honest work will always place us in that position where ambition is converted into humility, pride into understanding, arrogance into servitude. We once again feel like servants of the daily work of those scientists who, with the sacrifice of each day, each hour and each minute, try to aid others. It is from this perspective that we have dared to firstly tackle and then to work on this intricate corner of knowledge, always looking to reach new goals which are constantly further away when they seemed to be only a hair's breadth away.

However, it is also true that the sacrifice and the illusion of this work has its compensation in the form of achievements, partially if wanted, but achievements all the same which have revived our hope of achieving in the future that which has not been possible to reach up to now. The quantity of that which we are able to offer will be judged by those who have it in their hands to do so. But, ultimately, only the evolution of knowledge will dictate the definitive and unappealable resistance, that which permits, or not, that our names make an indelible mark in a small corner of one of the infinite pages of history.

The studies of pretopology and topology from a combinatorial perspective are new. What is more, their development in the last decades has been magnificent and has merited the admiration and applause of scholars, eager for new formal structures which have a high level of generality. Our work has been placed in precisely these contributions, which, being of unquestionable value, do not satisfy the necessities of those waiting for, in our case, an arsenal of instruments capable of treating new economic and management structures immersed in a world full of uncertainties.

It is true that since several lustrums we have been taking from some theoretical and technical elements, selected from here and there which, in some way are found to be connected with some of the axioms of pretopological and topological spaces. Proof of this is that in the work which we are now presenting we have rediscovered self aspects of the later denominated theory of clans and theory of affinities. Without doubt we were at least unaware of the existence of an "architecture" wide and solid enough to hold all of the bonds between the existing knowledge in this field and, therefore, the work of creation of the new instrument desired results as difficult when not unapproachable. This was our initial work: collect the existing snippets to form a single body with them, trying to homogenise distinct denominations which described the same concept and procuring the establishment of a single nomenclature from the diverse ones employed by the researchers of the distinct schools.

We have been able to confirm that throughout the second half of the 20th century the notion of topology has evolved from an idea based on intuition with the substrate of almost ancestral problems (the Möbius band, the bridges of Köningsberg, the four colour map) to a formulation through axioms with abstract rationalism. We have made a brief reference to the first of these focuses to bind ourselves in a definite way to the second, which we have followed uninterruptedly throughout the rest of our work. With this we believe that we have achieved the presentation of the following proposals:

I) From the distinct axiomatics, already known in traditional studies of pretopology, they have been generalised to give them validity in contexts of uncertainty. For this work we had an interesting precedent available to us due to Arnold Kaufmann who, in one of his "work notes" repeatedly quoted in our work, proposed a way to fuzzify the determinist atoms, consisting of considering some fuzzy subsets as descriptors of physical or mental objects and establishing as a referential set the elements of the referential of the cited fuzzy subsets. We develop this aspect by amplifying, where possible, the proposal of our maestro and at the same time we indicate the difficulties and limitations which this approach brings. We verify with satisfaction the validity of the models and algorithms which arose from the adopting of the hypothesis of Kaufmann for the solution of economic and management problems.

We have risked beginning a second way to move towards a context of uncertainty. This consists of considering as a referential the own fuzzy subsets descriptors of physical or mental objects. This way of conceiving the process of moving to fuzziness allowed the establishment of the "power set", set of all of the parts, with elements which are fuzzy subsets or groupings of fuzzy subsets represented in this way as much by physical or mental objects as by sets of physical or mental objects. It is not possible to hide the interest of this way of knowledge when the "objects" may be economics products or instruments, operations, structures or phenomena of the economic-financial field.

In the work presented a parallelism has been followed between one focus and the other, as the axioms were identified or new concepts were incorporated. It is also in this way that this part of our proposal contains a comparative analysis susceptible to being a source of inspiration and a guide for the moment in which it is necessary to choose the most adequate tool for the treatment of the problems posed by a complex economic reality.

Having taken into account the high level of abstraction we have considered it opportune to accompany the concepts and other technical elements with examples, directed to ourselves and to those whose task it is to judge our work. To those of us who have created this work it has been useful in order to consider our own ideas and, at times, to correct inaccuracies which have occurred during the course of our work. We also hope that it has aided and speeded the reading of others.

II) Topological spaces have centred our attention in the part of the work dedicated to conceptual aspects and formal elements which configure the idea of space. Taking the same path adopted in the case of pretopologies we have followed, step by step, the two aspects of fuzzification, although they have acquired their own nature as a consequence of the fact that while pretopologies are defined through an application, topological spaces are defined through a subset of elements, which is to say by means of a space of positions. We repeat that the methodology has followed parallel paths, which has greatly aided our work and has allowed us to focus our efforts on the search for useful concepts to generate topologies on the one hand and also to study the most ideal processes in order to relate or connect uncertain topological spaces.

As for the generation of uncertain topologies the definition of filter-base, with some very simple axioms, has facilitated the obtaining of topological spaces apt for the treatment of economics and management problems in conditions of uncertainty upon passing to the notion of filter engendered by a base. On the other hand, the figure of neighbourhood has been introduced into our work. We consider that upon inserting these two elements into the study of uncertain topological spaces we have collaborated in moving forwards, even if by only a small step, in the path of a greater understanding of the complex structures which dominate the economic panorama.

In that which refers to the comparison of uncertain topological spaces, our task has taken us to dive into the field of functional applications to get from this those theoretical and technical elements which could, with adaptation, be integrated into the work performed. Step by step we have unravelled the concepts which brought us to consider a homeomorphism susceptible to being contemplated in the relationship between uncertain topological spaces. With total humility we dare to say that upon reaching this point we felt a great relief when verifying the presence of a theoretical conglomeration of which we had dreamt for many years, when we wished to have models or algorithms available to be used in economics and management problems in which the uncertainty was so much that it was honestly not even possible to use the most elemental of the uncertain numbers. The then faltering non-numerical mathematics of uncertainty came to our aid with provisional arguments. Today we have a solid structure of knowledge at our disposition, which we hope will be completed in future works.

III) As a continuation of that which we have considered more formal studies we have directed our efforts to the creation of an instrument which could be used directly in the treatment of economic-financial proposals. With this we moved from fundamental mathematics to applied mathematics.

A first comparison between the axioms of topological spaces and some partial studies of what had become known as the theory of clans has induced the search for those common bindings and separating aspects with the aim of establishing the conditions under which a unitarian vision would be possible. We believe that we have achieved this in such a way that we have presented the process through which by means of beginning from a notion of family, a topological space may be found which is at the same time a clan.

The possibilities of this instrument of management, known as clan, are wide. We have tested these on various occasions with interesting results in the social, economics and management spheres. But the notion of clan has not ended here. The success achieved by the exploitation of files and searches for people in large populations from previously specified determined characteristics is well-known in specialised circles. We believe that our contribution has allowed the establishment of a connection between clan and topology, which specifically means the discovery of the axiomatic which is common and therefore generalises the idea of clan.

IV) In the European Congress on Operational Research celebrated in Aachen in 1991, professors Kaufmann and Gil Aluja presented an address in which they announced the birth of a theory of affinities, which was mainly based on two elements: Moore's closing and Galois' lattice. The proposal was suggestive but, in the words of its own authors, it was found to be lacking in links which would authenticate it as belonging to a wider block of knowledge. It was alive, that much is true, as is the fact of the result of almost habitual use by scholars of fuzziness in non-numerical fields, above all in the processes of segmentation, grouping and ordering of groups. We have always shown a deep respect towards our maestros, which has not impeded the strength necessary to try to advance more along the paths already begun by them. It is from this perspective that the last results presented in this work have been inserted. We have aimed, and we hope that we have partially achieved, to inscribe this theory as an isotone pretoplogy which should fulfil one complementary condition: idempotency. From here we have directed our efforts to establishing the ways which may be taken to find the desired pretopological space. The Moore family, on the one hand, and the fuzzy relationship existing between the elements of the referential on the other have been the starting elements from which it has been possible to move into that which we have named a Moore fuzzy pretopology.

Once this base was set, we have turned to the properties stated in the second part of our work in order to find the open subsets and the closed subsets, which allowed the establishment of the corresponding lattices and later the functional application between these which, as it could be in no other way, resulted as bijective, inverse and antitone and, as a consequence of this, the lattices were isomorphic and dual. We have finally achieved to state that upon performing this transition from uncertainty to determinism by means of decomposition by levels, these properties are maintained. Generalisation when an initial element of a rectangular fuzzy relationship is adopted instead of a square fuzzy matrix has not presented any problems whatsoever. On the other hand, its formulation is full of great possibilities for future development of which the application within our work is just a small sample.

Please allow us some final words to express our conviction that, in spite of the fact of that which we have done with a lesser or greater degree of fortune, there is a long way ahead until the desired goal is reached: an architecture of knowledge capable of harbouring pretopological spaces and uncertain topologies, with some axioms which span from the most general of the spaces to the most diverse singularities and the deduction from this architecture of the technical elements which configure models and algorithms capable of being used in the widest sphere possible of the economics and management situations of today, but above all of the future, a future which we can sense as mutable, complex and uncertain. Our strongest intention is to continue working in the lines which are traced by those who leave their marks in the path which leads towards Knowledge.

Bibliography

Bibliography

Adamek, J., Herrlich, H., Strecker, G.E.: Abstract and Concrete Categories. Wiley, New York (1990)

Adnadjevic, D., Sostak, A.P.: On inductive dimensions for fuzzy topological spaces. Fuzzy Sets and Systems 31 (2002)

Aristóteles: Obras. Lógica. De la Expresión e Interpretación. Aguilar, Barcelona (1977)

Beer, G.: Topologies on Closed and Closed Convex Sets. Kluwer Academic Publishers, Dordrecht (1993)

Bourbaki, N.: Eléments de Mathématique. Livre III Topologie Générale. Hermann & Cie. Editeurs, Paris (1940)

Brümmer, G.C.L.: Topological categories. Topol. Appl. 18 (1984)

Courtillot, M.: Structure canonique des fichiers. AIER-AFCET 7 (1973)

Chang, C.L.: Fuzzy topological spaces. J. Math. Anal. Appl. 24 (1968)

Qiu, D.: Fuzzifying topological linear spaces. Fuzzy Sets and Systems 147 (2004)

Das, P.: A fuzzy topology associated with a fuzzy finite state machine. Fuzzy Sets and Systems 105 (1999)

Dubreil, P.: Leçons d'algèbre moderne. Dunod, Paris (1964)

Erceg, M.A.: Metric spaces and fuzzy set theory. J. Math. Anal. Appl. 69 (1979)

Ertürk, R.: Some results on fuzzy compact spaces. Fuzzy Sets and Systems 70 (1995)

Franz, W.: Topología general y algebraica. Selecciones Científicas, Madrid (1968)

García Maynez, A.: Introducción a la topología de conjuntos. Trillas, Mexico DF (1971)

George, A., Veeramani, P.: Some theorems in fuzzy metric spaces. J. Fuzzy Math. 3 (1995)

Gil Aluja, J.: MAPCLAN, model for assembling products by means of Clans. In: Proceedings of the Third International Conference on Modelling and Simulation Ms 1997, Melbourne (1997)

Gil Aluja, J.: Interactive management of human resources in uncertainty. Kluwer Academic Publishers, London (1998)

Gil Aluja, J.: Elements for a Theory of Decision in Uncertainty. Kluwer Academic Publishers, London (1999)

Gil Aluja, J.: Investment in uncertainty. Kluwer Academic Publishers, Boston (1999)

Gil Aluja, J.: La pretopología en la gestión de la incertidumbre. Speech for the purpose of entering as Doctor Honoris Causa of the University of León. Publ. Universidad de León (2001)

Gil Aluja, J.: Elementos pretopológicos básicos para la modelización de la incertidumbre. Work included in the posthumous homage book for doctor Pifarré. RACEF, Barcelona (2003)

Gil Lafuente, A.M.: Nuevas estrategias para el análisis financiero en la empresa. Ariel, Barcelona (2001)

Gil Lafuente, A.M.: Fuzzy logic in financial analysis, vol. 175. Springer, Heidelberg (2005)

Gil Lafuente, J.: Model for the homogeneous grouping of the sales force. In: Proceedings of Congress MS 2001, Changsha (Hunan, R. P. China) (2001)

Gil Lafuente, J.: Algoritmos para la excelencia. Claves para el éxito en la gestión deportiva. Milladorio, Vigo (2002)

Giles, J.R.: Introduction to the analysis of metric, Cambridge (1987)

Goguen, J.A.: The fuzzy Tychonoff theorem. J. Math. Anal. Appl. 43 (1973)
Harris, C.J.: The stability of input-output dynamical. Academic Press, New York (1983)
Höhle, U.: Many Valued Topology and its applications. Kluwer Academic Publisher, Dordrecht (2001)
Höhle, U., Rodabaugh, S.E., Sostak, A.: Special Issue on Fuzzy Topology. Fuzzy and Sets Systems 73 (1995)
Höhle, U., Sostak, A.: Axiomatic foundations of fixed-basis fuzzy topology. In: Höhle, U., Rodabaugh, S.E. (eds.) Mathematics of Fuzzy Sets: Logic, Topology, and Measure Theory. Handbook of Fuzzy Sets Series, vol. 3. Kluwer Academic Publishers, Dordrecht (1999)
Höhle, U., Rodabaugh, S.E. (eds.): Mathematics of Fuzzy Sets: Logic, Topology and Measure Theory. The Handbook of Fuzzy Sets Series, vol. 3. Kluwer Academic Publishers, Dordrecht (1999)
Huttenlocher, D.P., Klanderman, G.A, Rucklidge, W.J.: Comparing images using the Hausdorff distance. IEEE Trans. Pattern Anal. Mach. Intelligence 15 (1993)
Hutton, B.: Normality in fuzzy topological spaces. J. Math. Anal. Appl. 50 (1950)
Jameson, G.J.O.: Topology and Normed Spaces. Chapman and Hall, London (1974)
Xiao, J.-Z., Zhu, X.-H.: Topological degree theory and fixed point theorems in fuzzy normed space. Fuzzy Sets and Systems 147 (2004)
Johnstone, P.T.: Stone Spaces. Cambridge University Press, Cambridge (1982)
Kaufmann, A.: Introduction à la théorie des sous-ensambles flous à l'usage des ingénieurs. Compléments et nouvelles applications. Masson, París (1977)
Kaufmann, A.: Prétopologie ordinaire et prétopologie floue. Note de Travail 115 (1983)
Kaufmann, A., Gil Aluja, J.: Técnicas operativas de gestión para el tratamiento de la incertidumbre. Hispano Europea, Barcelona (1987)
Kaufmann, A., Gil Aluja, J.: Selection of affinities by genes of fuzzy relations and Galois latices. In: Proceedings of the XI Euro O. R. Congress, Aachen (1991)
Kaufmann, A., Gil-Aluja, J.: Técnicas de gestión de empresas. Previsiones, decisiones y estrategias. Pirámide, Madrid (1992)
Kaufmann, A., Gil Aluja, J.: Técnicas especiales para la gestión de expertos. Milladoiro, Santiago de Compostela (1993)
Khedr, E.H., Zeyada, F.M., Sayed, O.R.: On separation axioms in fuzzifying topology. Fuzzy Sets and Systems 119 (2001)
Kisielewicz, M.: Differential inclusions and optimal control. Kluwer Academic Publishers, Dordrecht (1991)
Klein, E.: Mathematical methods in theoretical econ. Academic Press, New York (1973)
Kuratowski, K.: Introducción a la teoría de conjuntos y a la topología. Vicens Vives, Barcelona (1972)
Lehning, H.: Analyse en dimension finie avec exercices. Masson, París (1986)
Lipschutz, S.: Topología general. McGraw-Hill, México (1970)
Liu, Y.M., Luo, M.K.: Fuzzy Topology. World Scientific, Singapore (1998)
Lowen, R.: A comparison of different compactness notions in fuzzy topological spaces». J. Math. Anal. Appl. 64 (1978)
Malik, D.S., Mordeson, J.M.: Fuzzy Discrete Structures. Physica-Verlag, New York (2000)
Martin, H.W.: Weakly induced fuzzy topological spaces. J. Math. Anal. Appl. 78 (1980)
Mascaró, F., Monterde, J., Nuño, J.J., Sivera, R.: Introducció a la topología. Universitat de València, València (1997)
Menger, K.: Statistical metrics. Proc. Nat. Acad. Sci. 28 (1942)
Munroe, M.E.: Introduction to measure and integration. Addison-Wesley, Reading (1953)

Ponsard, C.: Un modèle topologique d'equilibre économique. Dunod, París (1969)
Rodabaugh, S.E.: The Hausdorff separation axiom for fuzzy topological spaces. Topology Appl. 11 (1980)
Rodabaugh, S.E.: Categorical foundations of variable-basis fuzzy topology. In: Höhle, U., Rodabaugh, S.E. (eds.) Mathematics of Fuzzy Sets: Logic, Topology, and Measure Theory. Handbook of Fuzzy Sets Series, vol. 3. Kluwer Academic Publishers, Dordrecht (1999)
Sapena, A.: A contribution to the study of fuzzy metric spaces. Appl. Gen. Topology 2 (2001)
Schaefer, H.H.: Espacios vectoriales topológicos. Teide, Barcelona (1974)
Shinbrot, M.: Teoremas del punto fijo. Matemáticas en el mundo moderno (various authors). Blume, Madrid (1974)
Simmons, G.F.: Introduction to topology and modern analysis. McGraw-Hill, New York (1963)
Steen, L.A., Seebach, J.: Counterexamples in Topology, 2nd edn. Springer, New York (1978)
Sugeno, M.: Fuzzy measures and fuzzy integrals, a survey. In: Gupta, M.M., Saridis, G.N., Gaines, B.R. (eds.) Fuzzy Automata and Processes. North-Holland, Amsterdam (1977)
Sutherland, W.A.: Introduction to metric and topological spaces. Oxford Science Publications, Oxford (1975)
Tjur, T.: Probability based on random measures. Wiley & Sons, New York (1980)
Veeramani, P.: Best approximation in fuzzy metric spaces. J. Fuzzy Math. 9 (2001)
Wang, E.Y., Chang, T.C.: Canonical structures in attribute based on file organization. CACM (1981)
Wang, G.J.: Theory of topological molecular lattices. Fuzzy Sets and Systems 47 (1992)
Xiao, J.Z., Zhu, Z.H.: On linearly topological structure and property of fuzzy normed linear space. Fuzzy Sets and Systems 125(2) (2002)
Ying, M.S.: A new approach to fuzzy topology (I). Fuzzy Sets and Systems 39(3) (1991)
Zadeh, L.: Fuzzy Sets. Information and Control 8 (1965)
Zadeh, L.A.: A Fuzzy-algorithmic approach to the definition of complex or imprecise concepts. Int. Jour. Man-Machine Studies 8 (1976)
Zadeh, L.A.: A theory of approximate reasoning. In: Hayes, J., Michie, D., Mikulich, L.I. (eds.) Machine Intelligence, vol. 9. Halstead Press, New York (1979)
Zadeh, L.A.: The Role of Fuzzy Logic in Modeling, Identification and Control. Modeling Identification and Control 15(3) (1994)
Zadeh, L.A.: Outline of Computational Theory of Perceptions Based on Computing with Words. In: Sinha, N.K., Gupta, M.M., Zadeh, L.A. (eds.) Soft Computing & Intelligent Systems, pp. 3–22. Academic Press, New York (2000)
Zimmermann, H.J.: Results of empirical studies in fuzzy set theory. In: Klir, G.J. (ed.) Applied General Systems Research. Plenum Press, New York (1978)